ANCIENT DIVINE LIGHT

Master Plan of the Universe - Crystallographic Geometry

and

the "Theory of Everything: RABIA (Quartet)

Tamer Kutdoğa

ANCIENT DIVINE LIGHT

Master Plan of the Universe - Crystallographic Geometry

and

the "Theory of Everything: RABIA (Quartet)

Tamer Kutdoğa

Author & Publisher   : Tamer Kutdoğa
Translator   : Abidin Büyükköse
ISBN for e-book   : 978 605 65318-6-6
ISBN for print book : 978 605 65318-7-3

About the Author:

TAMER KUTDOĞA [1938-]

He was born in 1938 in Istanbul. He graduated the elementary school in Istanbul in 1950, the secondary and high school in Adana, a city in the south of Anatolia, in 1957.

After completing his military service in 1963, he decided to find an institute researching for a more livable world and work there for life.

In 1968, he graduated the Academy of Economic and Commercial Sciences in Ankara and went to Germany. During two to three years he spent in Germany, he contacted some of the 300 addresses in Belgium, Germany and Switzerland, which he found in address books, and concluded that they were ostensible humanistic institutions founded purely for political purposes.

He achieved his goal, which he set in 1963 to some extent exactly half a century later, in 2013, when the previous version of this book was published in America as an e-book as a result of his efforts after his return from Germany in 1978.

In 2018, he felt the need to finalize his book by changing the title, the front cover, and the content of the book where RABIA was emphasized.

The author states that he wrote in this book what the people of the world would like know but do not know; what they wanted to say but could not do so; however he believes that the main reaching the sufistic stage of "upheavel with Allah" (*kıyam billah*) and heading towards "perpetuity with Allah" (*Bekabillah*) and thus initiating the golden age, which is the ideal of mankind could only be possible if the academicians, who could be defined as philosophers in the cosmology departments in the universities of today acknowledge that **RABIA** is truly a the **"Theory of Everything"** and continuosly update the issues in this book in the light of the improvements and thus he believes to have given the Sumerian, Babylonian and Egyptian priests credit for rendering unto Caesar the things which are Caesar's.

PREFACE

If we would analyse the evolution of knowledge of the humans from the ancient times up to today, it would be enlightening to determine that; evolution of knowledge has followed the route of **Central Asia, Sumerian Empire, Egypt, Greece, India, Islam and Europe**. The western civilizations of today stick to their habit of starting everything from the ancients Greeks due to the impact of the struggles between the Crusaders and Islam in the Middle Age. **There are still so many things in the accumulation of global knowledge, collected during and before Al-Andalus and still not valued even today, though traveled via al-Andalus route to Europe and started the Renaissance. One of the most important points is that hundreds of Islamic scholars (namely philosophers) who lived after the VIIIth Century, engaged in all sciences of their age and revealed a synthesis of their knowledge in their works. Ibn-i Sina (Avicenna) is one of the most important examples. Both these scholars, and the persons who read their manuscripts and who were in a position to enlighten the public (namely *wise old men*) were followed. An example is the Seljuk Sultan who was in tears because of waiting too long for talking to Mawlana (Rumi) in Konya in the XIIIth century. Who can regret the saying of Ataturk "Our true mentor in life is science." But today, in the 21st century, how many "mentor professors" are there, who learned the other sciences of their ages besides their own fields so much as to be called a Philosopher? Who explained the synthesis they obtained in their works as to give a message to the ordinary people? We need to ask that.**

The holy books cite that, before creating the universe, Allah wrote everything on the "Levh-i Mahfuz" (*Protected Plate*), which is referred to as "Nur-u Kadim" (*Ancient Divine Light*) in Sufism.

In this book, we tried to gather the knowledge to reach a modern, universal, easy understandable synthesis for universality and finding the secrets of universe as a guidance to the humans, devoted to the purpose of a better world and humanism. Thus, it was tried to draw conclusions from important studies of 21st century like: **Islam Mysticism of XIIIth Century** in other words '**Kesrette Vahdet'** (Unity in Plurality); the thesis of Einstein: "Four forces in atom have become as one force in Big Bang" which he tried to prove for 30 years, but could not manage to do; and the theories of "**Quantum Mass Gravity and Combined Field**" and "Mass Gravitational Electromagnetic Field" (Annex Ia and in the middle of front cover). Finally through the **THEORY OF EVERYTHING: RABIA,** given in Part 4, the purpose has been passed over and as a result of a study with the aim to enlighten the things yet unknown, it has been concluded that the essence and keyword of our universe is **RABIA end DOUBLE RABİA**.

Now let us give the meanings of some scientific descriptions:

SCIENCE: Its purpose is finding the laws of universe. These laws are searched in two ways.

1- THEORY: Intangible **knowledge**, in other words in the field of **thought**.

2- APPLICATION (Practice): Tangible knowledge, in other words in the field of **matter and motion**.

If there is theory, there is 'Application', too. If there is no theory, then there is no 'Application', either. They cannot be separated.

Matter, kinetics, time; all of them are different states of **energy** and they are integral.

IRREGULAR KNOWLEDGE AND THOUGHT: It is non-connected knowledge.

SCIENTIFIC KNOWLEDGE AND THOUGHT: It is semi-connected knowledge.

PHILOSOPHICAL KNOWLEDGE AND THOUGHT: In a sound expression, **"universal knowledge"** or **"cosmology"** is the full-connected knowledge. In other words, knowledge is a lacking and incomplete process, which keeps going towards the accurateness. At the end of the fourth chapter of this book, it was revealed that the completed Knowledge was Rabias.

PHILOSOPHER: He is a person who connects the knowledge of ages within the mental principles and who establishes a system.

However, today how many universities in the world have Cosmology lecture? I do not know. It is up to the philosophers. However, they are in the fight of "Are minds, minds; are ideas, ideas?" However, philosophy should be made particularly through mathematics by adding modern knowledge to the old ideas and knowledge. As far as we know, **"Humans"**, the most perfect creatures that Allah has created, have to know first the Universe and next himself as a whole, in order to remove the problems on the world, to build a happy social order suitable to both the perfectness and symphony of the

Universe, and his own perfectness. All evil and negations are caused by lack of knowledge or misinformation. Who can find fault with the mistaken, unless the Universal facts are told to the humans?

They gathered the blinds, they have brought an elephant, each of the blinds touched a different part of the elephant and when they were asked to describe the elephant, each of them had a different description of an elephant. In today's world, humans use initiative only according to the ear or tail of the elephant. If we cannot understand what the Universe that we live inside and which we are a part of it, is; and what the humans are; the chaotic situation will go on in the World. **Cosmology Science** should have explained the whole Universe for the **humans** to be able to use initiative, in the correct way. Only then, we can expect the authorities to build a happy social order based on **justice** (Balance, rhythm, harmony, regularity, order, equation, symmetry) and **knowledge** (Cosmology, Rabias, the core knowledge of the universe) in our beautiful world, ideal for the perfectness of Universe and humans. **(P.10)** Otherwise, no one can blame the ones who make the world unlivable, with wrong initiatives.

Today, the answer of question: "True? For whom? For what?" which will eliminate the chaos in the world of today, might be **Mathematics**. Today, there is anyone, who does not accept that mathematics is superior to all sciences. However, the ancient Sumerian, Babylonian and Egyptian priests concluded after the studies of thousands of years that the essence of Mathematics is numerically the numbers 1, 2, 3, 4 and geometrically the **Triangular Pyramid**, that is **RABIA** in their words.

In the first part of the book, contemporary knowledge about **universe**, and in the second part, about **humans** is given. In the third part, as per the contemporary knowledge in the previous two parts, it is concluded that **the synthesis of universe and humans is the "triangular pyramid", i.e. RABIA, which represents the human (i. e. micro cosmos) and which is also the geometry of Mawlawi Dervish**. As a result of a more comprehensive study, it was again both the "Triangular pyramid" (Theory of Everything) and **"Double triangular pyramid" (DOUBLE RABIA),** which is based on additional tables at the end of the book and is not available in contemporary literature and is the salvation algorithm of the Earth as presented in the fourth part of the book.

In general, in response to the scientific community that does not take geometry seriously; The geometry of the RABİA is that it is based on the geometry of concrete element crystals. Like the particles that make up the Rabia on the front cover, the universe is already composed of three-dimensional spheres and particles as far as we can perceive from the astral scale to the subatomic scale.

Thus, it was concluded that the essence of the **Ancient Divine Light**, the **Master Plan of the Universe**, the **Theory of Everything**, and the **Order of the Universe**, is the **RABİA**. I wonder whether Sumerian, Babylonian and Egyptian priests want to tell ud that the path leading to **Allah** was **RABİA**, since the word "RABB" refers to ALLAH both in Arabic and Hebrew languages? According to Pythagoras, the facts that the ratio of the frequencies of the tuneful voices in music are 1, 2, 3, 4 and their multiples; the measurements of 1, 2 and 3 in the mathematics of the golden ratios, as well as the golden ratios of the thuluth script (1/3) in Arabic letters and 3 main colors, namely blue, yellow, red, which are the source of thousands of colors, cannot be considered separately from the number 4 of **Rabia**. The tables, **"Our Universe between Two Big Bangs"** given in **Annex-Ia, b, c, d** put the **Calculus Arc and Equation** which is the unreached peak of High Mathematics up to day, by making it

THE CRYSTAL OF PEACE IN THE UNIVERSE AND IN THE WORLD

able to calculate the basic functions of Universes. Although today it seems to be rather unprofessional, I believe that the **"Calculus Equation of Universes"** will retrieve the geometries presented in this book from being a fantasy, and it will have the last word in the future. Even though the things written in the book appeal to everybody, since they also appeal to the scientists in order to prove the assumptions in part 4, readers may omit the details, which they have difficulty to understand. The bottom-line is the conclusion that **RABIA** can explain everything in the Universe and in our world as a simple geometry that everyone can understand. Making accurate translations and revealing all of these works one by one, may allow the Humanity to find the "**Universal Truth**", and thus the purpose of Mawlawi Dervish for reaching Allah through **mystical ways**, may be achieved through knowledge and so Science and Religion combine; atheism and anarchism can be eliminated. According to Einstain, the best solution is the simplest end According to Newton, the truth is hidden in simplicity. While trying to collect everything about the universe and human in the world literature for 56 years, I was always wondering what will happen in the end. Let's list some of the Rabias we have given as of page 174 below:

1) THE CRYSTAL OF PEACE IN THE UNIVERSE AND IN THE WORLD:
Rabia in the upper half of the double triangular pyramid (P-42) is the
"Paradise for Society" algorithm and **"Paradise for the individual"** in the lower half.This two-dimensional star of David may also have a three-dimensional and "double triangular pyramid" interpretation.

2) SOCIAL GEOMETRY OF PEACE IN THE UNIVERSE AND THE
 GOLDEN AGE IN THE WORLD (P.175)
According to this geometry seen on the side, Double semicircular pyramid is seen in this semiotic (P-96) sphere, which is a common geometry of the individual, society and the State for the establish-ment of a **"UNITED STATES OF WORLD"** being "for the community" upward and "for the individual" downward are the algorithm of Paradise both on Earth and in the Hereafter. These two Rabia are the main syntheses of all the information in universities and libraries around the world and in this book. If we can combine the correct words in the world with the correct geometries like atoms, we can take the peace in the Universe as an example and start the golden age peace in the world.

3) THE NUMERICAL RABIA IN THE WORD "SILM"
  IN ISLAM::
There are 4 main types of peace in Islam: (P.179)

1. **PEACE AMONG ALL CREATIONS IN THE UNIVERSE;**
2. **PEACE AMONG THE WORLD NATIONS AND STATES.**
3. **PEACE AMONG THE FAMILY MEMBERS**
4. **PEACE BETWEEN THE SELFISHNESS (NEEDS) AND THE MIND OF THE INDIVID-UAL.**

4) RABIA FOR SURVIVAL OF HUMANITY, UNITY IN PLURALITY AND PERPETUITY WITH ALLAH:
On the head of the Mawlawi Dervish on the front cover of the book;

1) **ONE GOD**
2) **ONE UNIVERSE**
3) **ONE WORLD**
4) **ONE HUMANITY**

On the front cover of this book; the $E=MxC^2$ formula of Einstein under his feet seems to contain all information in the libraries of the whole world, that is, the key to the creation of the universe is hidden in it. Through the repetition of the triangles between the head and the hands of the Mawlawi Dervish, all two-dimensional shapes, including the Circle and through the repetition of the triangular pyramids, all tridimensional shapes, including the Globe can be achieved. If such a simple summary of the Universe, which everyone could easily understand, is generally accepted by the academic au-thorities, it may be the key to a paradise we can live on Earth. **"Qiyambillah"** (*standing up for Allah*) in Islamic Sufism means to stand up and turn to the level of **"BekaBillah";** after transcending all the Spiritual, that is, the ETHİC ranks of humanity, Allah c.c.. Thus, it was aimed to explain that the survival of humanity depends on spiritual values.

Now, please allow me to invite everyone to say "For the survivability of humanity on Earth, for starting peace and golden age long live **"RABİAS, THE KNOWLEDGE OF THE PEACE IN THE UNIVERSE"**.

Tamer Kutdoğa

PART ONE "UNIVERSE"

A) ASTRAL UNIVERSE:

I)   HISTORICAL EVOLUTION OF OUR UNIVERSE:

13, 7 Billion years ago	Universe started to be formed by means of Big Bang (big collision)
13, 2 Billion years ago	Enlightened (it was dark before)
7 Billion years ago	Universe is as gas clusters made of the first atom: hydrogen (Nebula)
4, 5 Billion years ago	Formation of Sun and Planets
3 Billion years ago	The first living plant was formed from a single celled algae coal.
565 Million years ago	The first moving marine animal
300 Million years ago	The first plants
225-75 Million years ago	Dinosaurs
260-180 Million years ago	Europe and America separate from each other
100 Million years ago	The first insects
55-30 Million years ago	"Mayohipis" (cleft foot): Ancestors of ancestors
7 Million years ago	"Rif cavity" from the central South of Africa to Syria and Hatay. Tanganyika lake (4 thousand meters deep) was formed.
B.C. 3 – 200	Mountains started to appear
B.C. 1 million –600000	The oldest ice age (Nile River Valley is covered with lake)
B.C. 600000 – 540000	I. Ice age (Nile River Valley dried up)
B.C. 540000 - 480000	480 thousand hot age (the first wild horse)
B.C. 480000 – 430000	II. Ice age
B.C. 400000- 240000	II. Middle Ice age
B.C. 240000- 180000	III. Ice age
B.C. 180000 – 120000	Hot age
B.C. 60000- 10000	**Last middle ice age. Northern hemisphere is covered with ice from Himalaya Mountains to the Alps, to Istanbul and Central America. Only Central Asia has the livable climate since it is far from the effects of oceans. There is a lake in the Central Asia, which is wider than Caspian Sea.**
B.C. 10000	**Ice melts in the northern hemisphere. World seas are rising 80 meters. Today (A.D. 2003) the oxygen rate in the atmosphere is 21%.**

As indicated above, when the universe was started to be formed by means of Bing-Bang **13, 7 billion years ago** (10^{10} years ago), the temperature was 5 billion °C. It was enlightened **13. 2 billion years ago**. The other particles were started to develop from the **Photon** particles constituting the light. **300-700 thousand years after Big Bang, the** first atom: **Hydrogen** appeared. Today, still **80%** of the Universe is formed by **Hydrogen**. After 5-10 Billion years, when Hydrogen which the Sun burns inside comes to an end, its diameter will be 100 times enlarged, it will be a "**Red Giant**", its brightness will increase 1000 times, the near planets (Mercury and Venus) will melt because of its heat and the oceans on Earth will be evaporated. Later, the Sun will finish its fuel completely, its outer cover will disappear and it will be a "**White Dwarf**". Today, one-tenth of stars (Suns) in our galaxy Milky Way are in this condition. Their specific weights are a teaspoon, one tone, in low brightness and blue-white in color. Later density will go on increasing. **Electrons** and **Protons** will combine to constitute **Neutrons** and **Anti-Neutrinos**. Neutrinos will run away to the stars and become distant. The remaining neutrons will go on densifying and the Sun will become a "**Neutron Star**". Still, 200 of 100-200 billion stars in Milky Way are Neutron Stars. They will also go on densifying and shrinking. When they reach to three times the mass of the Sun, they will become "**Black Holes**" and then they will start to vacuum everything inside. The ones being vacuumed by a Black Hole will either disappear, or they will become "**White Hole**," and a new Universe will start to be formed in the space by means of "**Big Bang**".

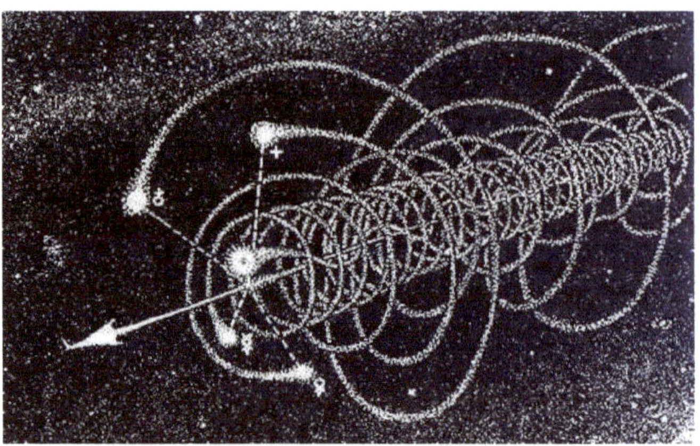

Motion of Sun and Planets in the Milky Way

		DIAMETER (KM)	DISTANCE FROM THE SUN (MILLION KM)	ROTATION AROUND THE SUN	**ROTATION AROUND ITSELF**
SUN		1. 391. 000			**24 DAYS**
MERCURY		5. 140	57	88 DAYS	88 DAYS
VENUS		12. 190	108	225 DAYS	225 DAYS
EARTH		12. 756	149	365 DAYS	**24 HOURS**
MARS		6. 860	227	687 DAYS	245 HOURS
ASTEROIDS					
JUPITER		192. 700	778	12 YEARS	9 HR 53 MIN
SATURN		120. 800	1427	29, 5 YEARS	10 HR 14 MIN
URANUS		49. 700	2871	84 YEARS	10 HR 42 MIN
NEPTUNE		53. 000	4499	165 YEARS	15 HR +48 MIN
PLUTO		6. 600	5910	250 YEARS	?

II) SOLAR SYSTEM:

Planets rotate around the Sun on a plane orbit. This orbit is ellipse shaped. Sun is on one of the focal points of this ellipse.

Three different motions of Sun:

1. Sun completes the left to right motion around itself in **24 days**.

2. It completes its motion around the Milky Way together with its planets, making a spiral arc motion from right to left in **230 million years**.

3. Its common motion with the other surrounding stars.

Temperature in the center of Sun is **14 million °C**. In case the heat was 5 billion °C, **Neutrino and Anti-Neutrinos** would run away to the space and the remaining **Neutrons** would have chain explosions. While the solar system travels in the Milky Way, it enters in a dark cloud once in every 10 million years. The world comes to an Ice age when the solar system enters in this dark cloud. While the solar system rotates in spiral motions in the opposite direction of Milky Way, from right to left, its speed is 216 km/seconds. Solar system is one in **160 billion** of Milky Way.

a) Moon: Being 384 thousand km far from the Earth, it rotates around the Earth from left to right. **50 million years later,** Moon will move away from Earth and the location of continents will change on the Earth.

b) Venus: The surface temperature is 480 °C. It is the only planet, rotating in the opposite direction of Earth, around the Sun. All of the other planets rotate from right to left like the Earth.

c) Comets: Being 15^{17} **meters far from us,** they are the cloudbanks rotating around the Solar System. That is, the Sun rotates not only together with its planets, but also together with its comets around the Milky Way.

When the comets come closer to the Sun, they get tails behind them. When they move away, they seem like round planets again. There are 42 comets rotating around the Sun in periods between three years and thousands of years.

d) Earth: It rotates around the Sun on an ellipse orbit, which is 300 million km long. It rotates around itself from right to left with a speed of **465 m/sec.** and around the Sun with a speed of **29760 m/sec.** In approximately **50 thousand to 1 million years,** the poles of the Earth will translocate. The nearest star is 40^{17} **m away** from the Earth and its lights reaches us in **4.5 years.** Lunar and solar eclipse occurs once in **18 years and 11 days** in the Earth.

STARTING DATES OF SEASONS IN THE WORLD:		
SEASONS	**NORTH HEMISPHERE**	**SOUTH HEMISPHERE**
SPRING	MARCH 23rd	DECEMBER 22nd
SUMMER	JUNE 22nd	MARCH 21st
FALL	SEPTEMBER 23rd	JUNE 22nd
WINTER	DECEMBER 22nd	SEPTEMBER 23rd

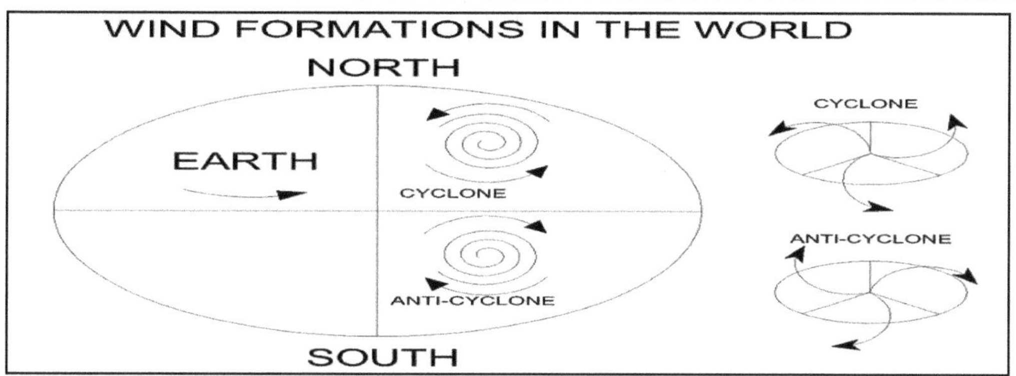

III) OUR GALAXY MILKY WAY:

Its diameter is **110.000 light-years,** (1 light year= 9450 billion km) and thickness is **20.000 light-years.** There are nearly 200 billion stars similar to the Sun in the Milky Way. Milky Way completes its rotation around itself from left to right in 250 million years.

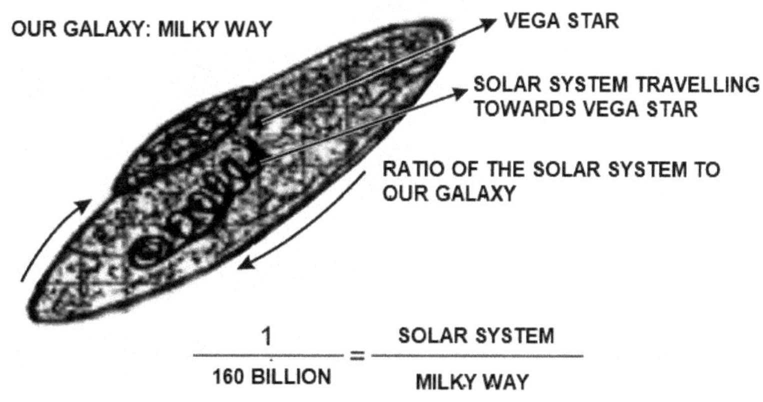

IV) OTHER GALAXIES:

The closest one, "**Andromeda**" is spiral shaped and 750 thousand light years far from the Earth. The farthest is more than 1 billion light years away. It means that today we see the galaxies as they were 1 billion years ago. **6 billion years later,** our galaxy Milky Way will collide with the closest galaxy Andromeda and they will interpenetrate. But their stars will not collide, their surroundings will be overheated. Galaxies are in groups, which have their own gravity areas. Our galaxy Milky Way is also in such a galaxy group. Among them, we know that there are 500 galaxies in the "**Virgo**" group. It can be determined in 3D by special telescopes that billions of galaxy clusters and super galaxy clusters in the universe are interconnected with Filaments like the nerves in the human body. Filaments are beams formed by Photons, which are the elementary particles of our universe, which enlarges because of the electromagnetic force. Also, the galaxies have their **Satellite Galaxies**. Every single galaxy in the Universe moves away from each other with a speed of 61. 000 km/sec. It means that the **Universe is expanding increasingly faster**. The speed of the orbs increases as they come closer to the center of gravity.

V) QUASARS:

They radiate more energy than 100 big galaxies in the space. In fact, they are space objects, which are only two-three times bigger than the Sun. Quasars are considered as the farthest visible objects of the Universe. The quasars were formed after Big Bang and some of them were a hundred trillion times brighter than the Sun. All of them disappeared later. But since the rays are still on their ways, they are visible even today. **Some of the 350 Quasar**s **visible today,** are 10 billion light years away from us. "**Blue Star Quasar**" is 4 billion light years away from the Earth. Since such a big energy as Quasars generate could not be a nuclear reaction, it is supposed that they are constituted from the big images as a result of "**Material-Anti-Material collision**". (This shows the accuracy of our thesis that **the universe before ours was the antimatter universe starting with a Bing-Bang** as given in Part Four). According to the theoreticians, the quasars were the black holes in the center of the galaxies before, and the kinetic energy that they generated after imploding everything around them and surpassing the speed of light, becoming sole energy and exploding, is still seen as light today. (again we can see the accuracy of the thesis which we explained in the Part Four that: the previous universe before Big Bang concentrated due to the potential energy and became a black hole and then exploded and expanded due to kinetic energy, is seen here as well). The last Quasar in the Universe died about 1 billion years ago.

VI) BLACK HOLES:

When the mass of black holes become smaller, they become warmer. When they become warmer, their radiation increases and they lose mass. They become warmer and they radiate further. At last, when they come to saturation, they explode terribly. A Black hole as big as the Sun explodes **13.7 billion x 10^{60}** years after it first starts to radiate. But a black hole whose mass is as big as a small mountain would explode in a period as long as the age of **Universe (13. 7 billion)**. The exploded black holes radiate **Gamma Rays** (γ). (Here, again let us go to the "Theory of Everything" (**P. 167**) in the Part Four. All of the scientists agree that Big Bang did not start in one location, but everywhere in the Universe at the same time. It means that at the last phase of the **previous antimatter universe,** the entire Universe became a black hole. This black hole pulled all of the Anti-Materials at the same time and demolished them. No mass was left and also no mass gravity power was left. As a result of the **big kinetic energy of antimatter-matter interaction,** first, the quasars, which were formed by subatomic particles, appeared and then **galaxies and Star, which were formed by the first atoms,** appeared. I think it's worth to investigate whether the BBS (Big Bang Star), which has been discussed in the annexes at the end of the book, constitutes the geometry of the core of quasars and our universe.

ADDITIONAL FIGURES and DRAWINGS TO ASTRAL UNIVERSE:

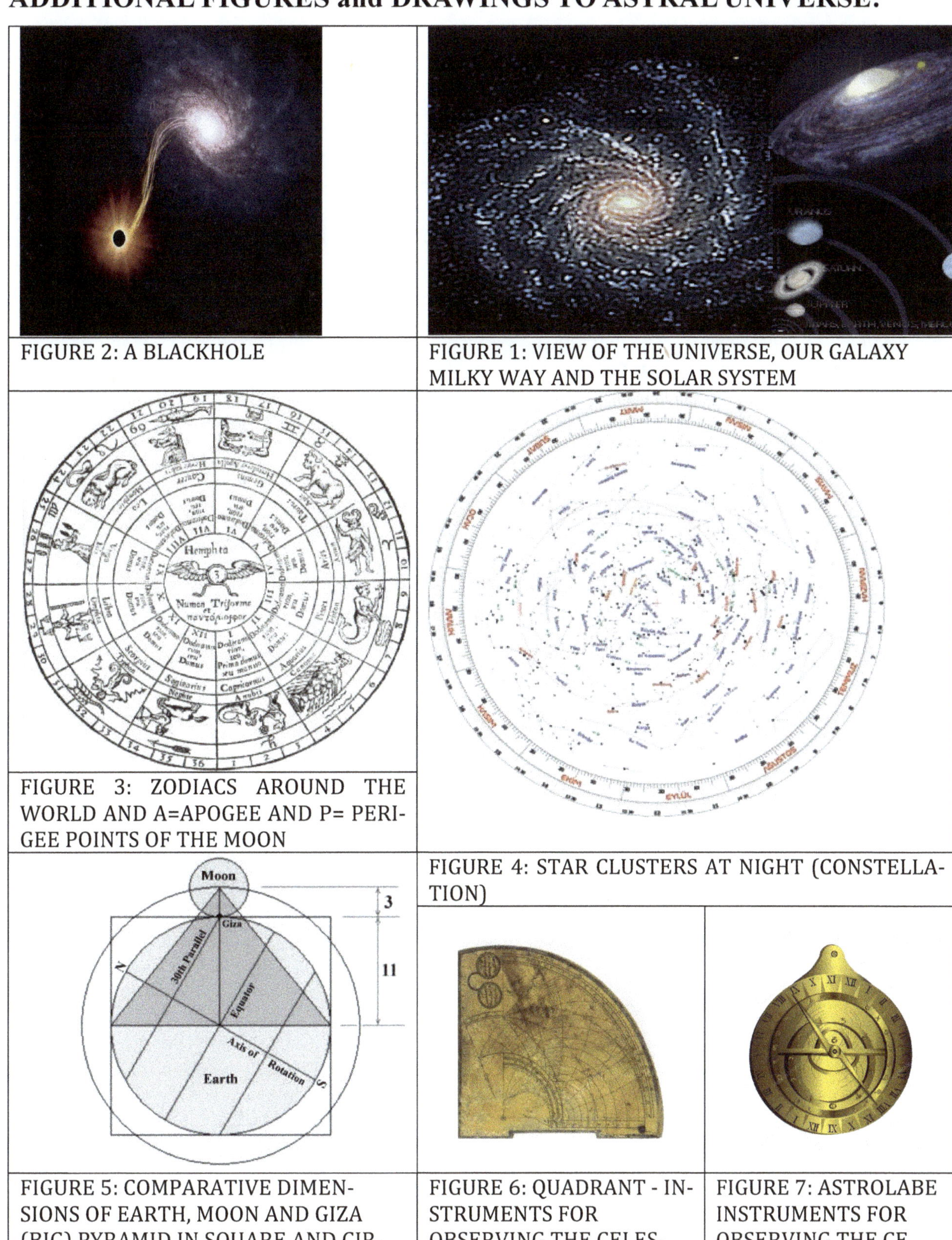

FIGURE 2: A BLACKHOLE	FIGURE 1: VIEW OF THE UNIVERSE, OUR GALAXY MILKY WAY AND THE SOLAR SYSTEM
FIGURE 3: ZODIACS AROUND THE WORLD AND A=APOGEE AND P= PERIGEE POINTS OF THE MOON	FIGURE 4: STAR CLUSTERS AT NIGHT (CONSTELLATION)
FIGURE 5: COMPARATIVE DIMENSIONS OF EARTH, MOON AND GIZA (BIG) PYRAMID IN SQUARE AND CIRCLE	FIGURE 6: QUADRANT - INSTRUMENTS FOR OBSERVING THE CELESTIAL BODIES / FIGURE 7: ASTROLABE INSTRUMENTS FOR OBSERVING THE CELESTIAL BODIES

B) ATOMIC UNIVERSE:

I) SUBATOMIC PARTICLES:

In the year 1932, "**Cyclotron**" was the only device to spin and accelerate the triple particles of atom (See **Annex-IIi** at the end of the book) to a very high speed and throw on the target, using the Electric and Magnetic fields. After the Second World War, **Accelerators** like "**Betatron**", later "**Synchrotron**" and "**Cosmotron**" were developed. Thanks to such devices, which can accelerate the elementary particles of atom up to nearly the speed of light and collide one with the other, new particles as a result of these collisions were examined and knowledge about Atom was improved.

For example, two quark particles (named "**meson**") appeared because of the collision of proton with another proton. Since these were unstable, they were breaking up after 1/1 million seconds and they were transforming to **electron** (Θ) and **photons** (γ). Some of dozens of types of these ephemeral mesons are shown in the "dual particles of atom" table in **Annex-IIh**. When a proton was collided with another proton in a faster and more severe way, ephemeral triple particles named **baryon** (Annex-IIi) which were heavier than proton and neutron, appeared by means of resulting bigger energy and they transformed to protons and photons in a short time. **Mesons** and **baryons** are together called as **hadron**, and strong nucleus force controls the interaction between them.

Most of the particles indicated in the table of elementary particles of atom (**Annex-IIe, h, i**) have been introduced by means of particle collision tests in the accelerators. Also today, **particle** and **anti-particle** pairs can be produced in the laboratories using energy.

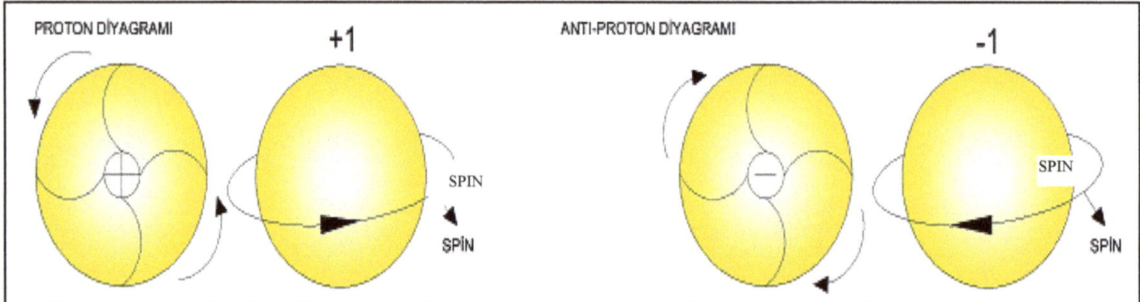

With the rotation of the planets around the sun in the universe and rotations electrons around the nucleus of atom, **astral universe** is reflected over the **atomic universe**. Like the planets rotating both around themselves, and around the sun at the same time; electrons also rotate around themselves while they are rotating around nucleus at the same time, and this is defined with the term "spin" in the table of elementary particles (See **Annex-IIe, f, g, h, i**). As it can be seen in the tables, subatomic particles are not only the **electron**, rotations of other all **particles** and **anti-particles** around themselves have their own properties magnetically and angularly, according to the rules of quantum mechanics.

As indicated in the above drawing of **proton and anti-proton diagrams**, the rotation of material particles around themselves is from right to left; the rotation of anti-material particles is from left to right. Rest mass, spins and average life of a particle and an anti-particle are the same; their **electrical, nucleus charges and magnetic moments** are the same in value, but their polarities are opposite. In case **matter and antimatter atoms collide, both disappear in 1/ 10 million seconds.** Energy comes out or mesons (dual particles) are formed and proton and anti -proton are formed from mesons. The particles known and accepted today are shown in blue color and defined "**real**" in the table of "Single Particles of Atom" in **Annex-IIe** at the end of the book**.** The others are the yet unknown ones, which I foresee to be found in the future. This foresight comes from the assumption of that the first 10 neutrinos are composed of **10 elementary particles (7 bosons (+) 3 leptons)** and later 20 quarks should have been formed by 20 particles. Now, let us give some information about the elementary particles one by one:

1) GRAVITONS and ANTI GRAVITONS: They are the particles of the "**Mass Gravity Force**" which is one of 4 forces of atom.

2) PHOTONS (γ): They are the particles of the "**Electromagnetic Force**" which is one of 4 forces of atom. **Electromagnetic interaction** is provided by the photon (γ) exchange between the positive electric charged protons and negative electric charged electrons (Θ). Photons, which are light particles that travel from the sun to the earth in 8 minutes (light speed).

3) NEUTRINOS: While energy is generated during the fusion of hydrogen nuclei in the center of sun (fusion), a big amount of "**neutrino**" spreads. The difference between the neutrinos and photons is that neutrinos have more penetration power than the electromagnetic radiation. Similar to the photons, neutrinos also reach from sun to the earth in 8 minutes. They flow unimpeded through the Earth. They travel in the space freely. 100 billion neutrinos pass from every cm^2 on Earth in one second. Neutrinos are the particles of the "**Weak Nucleus Force**" which is one of 4 forces of atom and they interact by means of this force. They are inviolable against "**strong nucleus force**". It has been proved by experiments that neutrino whose charge and mass are zero, is composed of two components; and that the second of these interacts with one proton to form one neutron and one positron. One of these two components is **neutrino,** the other is **anti-neutrino**.

$$\frac{MUON \qquad = \mu^+ \rightarrow e^+ + \overline{\vartheta_\mu} + \vartheta_e}{ANTI\ MUON = \mu^- \rightarrow e^- + \overline{\vartheta_e} + \vartheta_\mu}$$

4) MUONS: They are the particles whose mass is between the mass of **electron** (Θ) and mass of **nucleon** (207 Θ). They can be either positive or negative. When muons disintegrate, they are also called as **heavy electrons**.

5) ELECTRONS (Θ): While they push each other, an "**energy quantum**" exchange happens between them. Since the quantum of electromagnetic force is **photon**, (light is also called as electromagnetic radiation), the "**Electromagnetic Force**" between charged particles means a basic "**photon exchange**".

6) QUARKS (Q): They are the particles of strong nucleus force and this force controls them. Strong nucleus force is being carried by the particles named "**gluon**" (g). In other words, gluons connect the quarks together. **The power between quarks does not decrease with the distance between them; on the contrary, it increases.** Every quark has their **red, blue** and **green** colors (not in the meaning of colors that we know). Also there are quarks with **anti-red** (purple), **anti-blue** (yellow), **anti-green** (orange) colors. Gluons related with the color of quarks can be found in **Annex-IIe**, under the Table of Single Particles of Atom. The charges of quarks are equal to 1/3 or 2/3 of the charges of proton or electron.

Just like the materials are composed of **elements**, the elementary particles in the atoms are assumed to be formed by **quarks**. It is believed that the **strong nucleus force** keeping proton and neutrons together is generated by the **gluons, which** are exchanged between quarks, composing the proton and neutrons. This is similar to the exchange of mesons between the hadrons or the situation of neutral atoms in the molecules. The permanent and continuous interaction between matters is provided by means of such a particle exchange in the universe. The interaction of electric charged particles occurs by means of **gluon** exchange. The three quarks of **baryons** are always in different colors. **Hadrons** (baryons and mesons) are colorless. Also **leptons, photons and W^+, W^-, Z^0 bosons** have no color charges. That's why they are not interactions of strong nucleus force.

7) MESONS: They are the particles with two quarks, which are given in **Annex-IIh**. One or more of the meson clouds surround the protons and neutrons in the nucleus of atom. The energy exchange in the nucleus of atom occurs by means of "**Mesons**" which are the exchange particles of **strong nucleus force**. The mesons around protons and neutrons are neutral. But (+) and (−) mesons are exchanged between them. For example, a **neutron** gives up a negative meson and becomes a **proton**. Picking up the negative meson, proton becomes a neutron. In case **pi-meson** is left alone, (**Annex-IIh**) a **muon** becomes a **neutrino** particle in 1/40 million seconds. If a **positron** collides with an **electron** (Θ), it becomes **pi-meson (ud)**. Pi-meson is used for the exchange of **strong nucleus force**. in the nucleus, a meson travelling from one nucleon to the other, causes the energy of the nucleon it has left to decrease and the energy of nucleon which picked it up to increase. The effective distances of nucleus force and mesons are the same. (10^{-15} m) (**Annex-IIh**)

Examples of disintegration of some particles and transform to other particles:

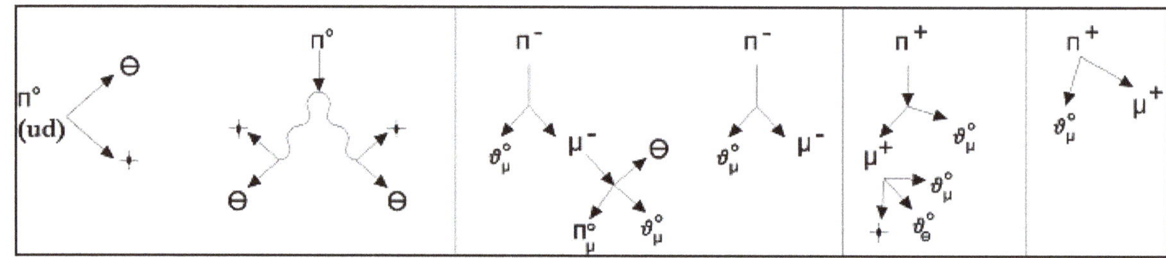

Pi-Mesons:

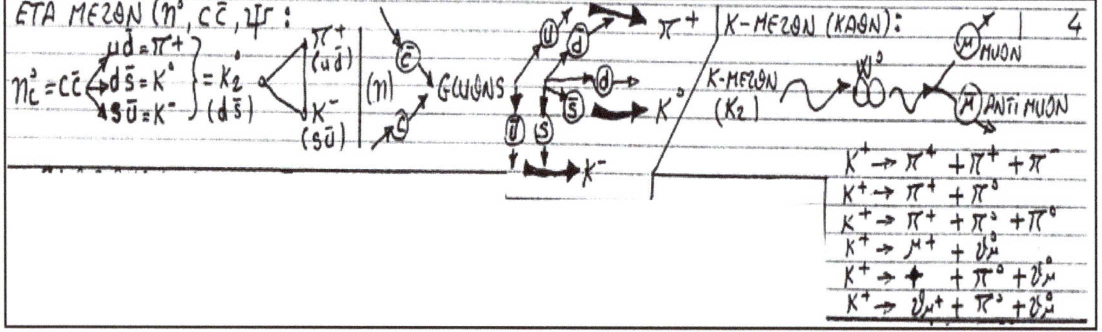

8) BARYONS: They are the triple particles as given in **Annex-IIi**.

The particles, which are listed in the table of "Single Particles of Atom" in Annex-IIe at the end of the book, had been reproduced from **graviton** and **photon** particles, which were present in the **Big Bang** collision, which is the beginning of the universe. The tables have been started with **bosons**, which are the single particles after the gravitons known today in the first place; and then the **fermions** (leptons and quarks) which are again single particles have been given. The **mesons** are formed from the binary combinations of **quarks and anti-quarks;** and **baryons and nucleons** are formed from the triple combinations. As known, atoms are formed of nucleon combinations, **elements** are formed of atom combinations, **molecules** are formed of combinations of atoms of various **elements**, **compounds** in the universe are formed of the **combinations** of molecules and **creatures** are formed of the combinations of compounds.

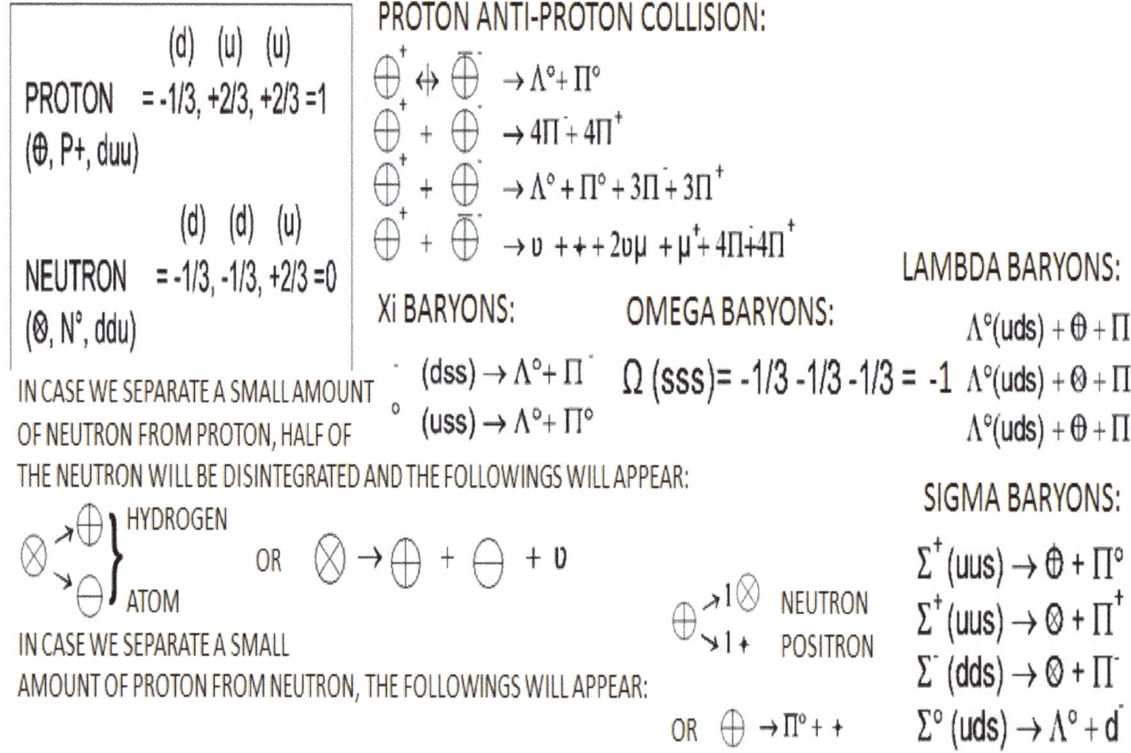

II)   VIRTUAL GEOMETRIES, WHICH ARE FORMED BY SUBATOMIC PARTICLES:

DIMENSIONS FROM SUB-ATOM PARTICLES TO VIRUSES AND CELLS

ELECTRONS OF ATOM			VIRUS / CELL	MOLECULE	ATOM	NUCLEUS	DOUBLE-TRIPLE PARTICLES	SINGLE FUNDAMENTAL PARTICLES
ORBITS	ELEKTRON SAYILARI	ORBIT RADIUS						
1. ORBIT K	$1^2 \times 2 = 2$	$0{,}5 \times$ $A°$	10^7 MR				MESONS	BOSONS
2. ORBIT L	$2^2 \times 2 = 8$	$0{,}5 \times 4 = 2°$ A	CELL					$O°, H°, g°,$.
3. ORBIT M	$3^2 \times 2 = 18$	$0{,}5 \times 6 = 3°$ A						$Z°, W^-, W^+$
4. ORBIT N	$4^2 \times 2 = 32$	$0{,}5 \times 16 = 8°$ A						LEPTONS
5. ORBIT O	$5^2 \times 2 = 50$	$0{,}5 \times 25 = 12{,}5°$ A					BARIONS	$e, \mu, T,$
6. ORBIT P	$6^2 \times 2 = 72$	$0{,}5 \times 36 = 18°$ A	10^{3-6} MR					$\vartheta_e°\ \vartheta_\mu°\ \vartheta_T°$
7. ORBIT Q	$7^2 \times 2 = 98$	$0., \times 49 = 29{,}5°$ A						QUARKS
								$-Q, -Qd, -Qs$
								$+Qt, +Qu, +Qc$
1 MICRON	$= 1\backslash 1\,000$ mm							
1 MICRON	$= 1\backslash 1\,000\,000$ MR		10^{3-3} MR	10^6 MR	10^{10} MR	10^{11} MR	10^{15} MR	10^{18} MR
1 MICRON	$= 1\backslash 1\,000\,000\,000$ KM		10^{8} mm = 1 ANGSTROM (A°) = $1\backslash 10\,000$ MICRON					

These virtual geometries are given in Part Four, Section H (P. 167) under the heading **"The explanation of the Theory of Everything: Rabia, based on the numbers from 1 to 4, and Tetrax"**.

III)   RADIATIONS:

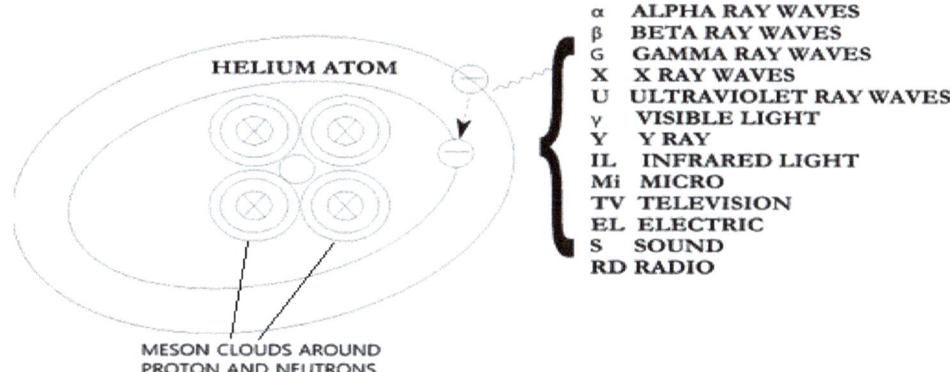

You can review the tables in **Annex-Ie, f, h, i** about this subject. Below are some particles of atom, various radiations and the areas they are generated, have been shown on a "**helium atom**" model with two protons, two neutrons and two electrons.

The particles of weak nucleus **W⁺, W⁻, Z°** and **neutrino** forces, which are effective only in the nucleus, provide the interaction in the nucleus.

If there are (+) and (–) **electric charges** on a **static** object, then an **"electric field"** forms around this object. If there are (+) and (–) **electric charges** on a **dynamic** object, then a **"magnetic field"** forms around this object. The effect of a magnetic field is only on the travelling electric charges.

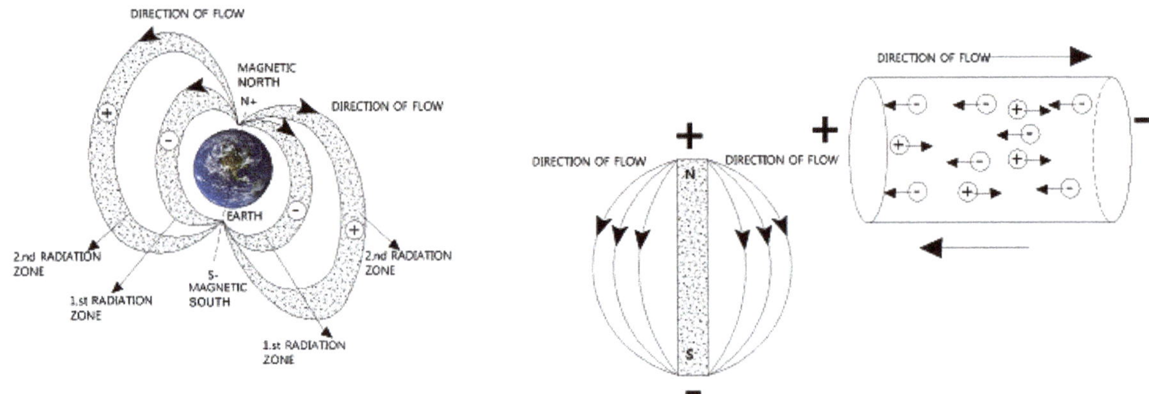

The **two radiation belts** (Van Allen belts) **developing around the Earth** because of the interaction of **cosmic rays** (particles) from the space with the **magnetic field** of the Earth are shown above. They prevent the high energetic and harmful cosmic rays from reaching to the Earth.

1)   ELECTRO/MAGNETIC WAVES:

A change in the **magnetic field** on any points in the space immediately causes a change in the **electric field** on the same point. In case the changes in the electric and magnetic fields happen periodically on a point, it causes "**electromagnetic wave**" propagation in the space. This wave can be seen in the following drawing. Electric fields (**E**) and magnetic fields (**M**) affect each other by both pushing and pulling.

The electromagnetic waves; between 1mm and 10cm are "**micro waves**", between 10cm and 10m "**television waves**"; between 10m and 100m "**short waves**", and between 100 m and 1000 m "**medium waves**". These waves are used to transmit the radio broadcast to very distant locations on Earth by reflecting from the **Ionosphere layer of Atmosphere**. The radio waves with a wavelength of 1 km to 100 km are the "**long wave**" broadcasts.

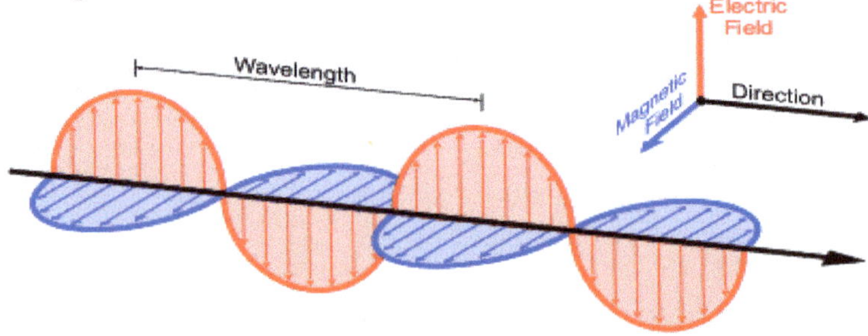

2)   PHOTON and GRAVITON:

PHOTON: (γ) The electrons rotating on specific orbits do not propagate radiation energy. As in the helium atom in the previous page, **radiation** occurs only as a photon propagation to the outside equal to the energy difference between two levels during the electrons rotating on the high energetic upper orbit pass to the lower energetic lower orbit.

PHOTON:

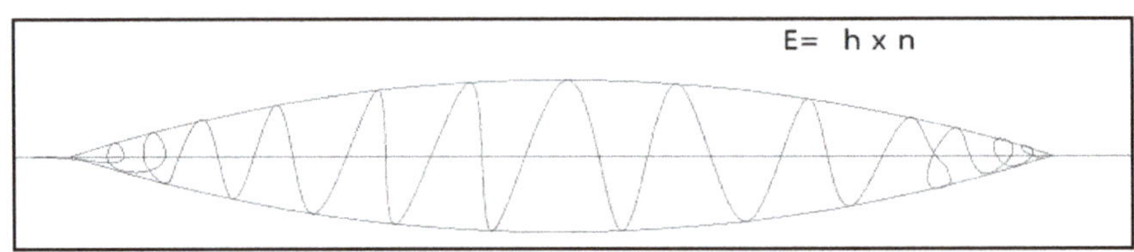

Moreover, every object with a temperature over the absolute zero (absolute zero= 273°C) propagates **radiation** (radiation, electromagnetic wave). The energy within the atom comes out as photon particles. Some of the radiations from the **radio waves**, having **the longest electro-magnetic waves,**

such as **radio waves, television waves, radar waves, micro waves, infrared light, y-ray waves, light visible by the humans** (wavelength between 4000 A° and 7500 A°), **ultraviolet light, x-ray** and **gamma ray,** (wavelengths>0. 1 A^0) are photons with different energy levels (**Annex-If**).

PHOTONS; As can be seen above, they consist of **electromagnetic wave packages** (light quantum). As the energies of these wave packages increase, their **frequencies** (number of vibrations in one second) also increase, whereas their **wavelengths** get shorter and rather show "**particle**" properties. The more their energy decreases, the more their frequency decreases, their wavelengths becomes longer and they rather show "**wave**" properties. While the "**light**" seems like travelling on a straight line, in the micro level, this line is in quantum (discontinuous) structure, in packages or pieces. Every quanta includes energy as much as **E= h x n** where **n**= Frequency of light, **h**= Quanta constant= 6. 62×10^{27} ERG/seconds (ERG/sec= energy unit).

GRAVITON: They are the particles of the "**mass gravity force**" which is one of 4 forces of atom. Gravitation areas affect each other only on the **gravitation** direction. In the micro objects, this gravitation is in trace quantity; in the macro objects the bigger the mass, the stronger the gravitation force.

3) BLACK BODY:

The rays entering from the hole on the wall of an object whose inner walls are painted in black (as shown above), are absorbed completely by the walls as a result of reflections on the inner walls of the object and they cannot go out. Black body is the strongest radiation source at the same time. The more the heat of Black body increases, the shorter the wavelength of the light it radiates, and the higher its energy. The energy of **radiation energy is maximum for 4700 A° wavelength** and **6000-6500 K° temperature**. Black body radiates or absorbs in all possible frequencies. All of the light particles (photons) in the black body have the same speed (speed of light). But as can be seen above in the black body, these light particles have different **energy, frequency and wavelengths**. They do not collide with one another. **Black body absorbs energy as much as it radiates.** The number of photons in the object does not remain constant. Every time new photons can be formed, as well as some of them disappears. In other words, they are absorbed.

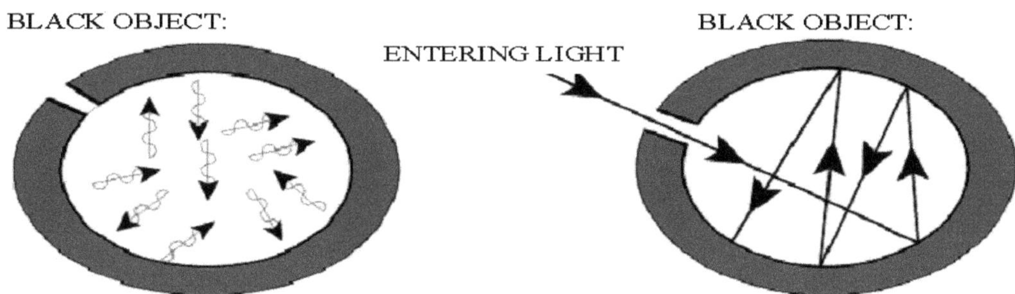

4) VISIBLE LIGHT:

It is the radiation in the form of light. "**Visible light**" which is the visible part of **electromagnetic waves**, is a combination of seven colors. Below are the wavelengths of colors of the light:

VISIBLE LIGHT

COLOR	WAVELENGTHS
PURPLE	4000-4500 A°
BLUE	4500-5000 A°
GREEN	5000-5700 A°
YELLOW	5700-5900 A°
ORANGE	5900-6100 A°
RED	6100-7500 A°

The combination of these colors seems us like the normal light.

5) ULTRAVIOLET LIGHT:

These are the outermost orbits of atom, the source of light. In case an electron is given up, the remaining electrons connect to the nucleus stronger. To break off another electron from the atom, if we did not give enough energy to break off the electron from the atom, electron gives up radiation as **ultraviolet ray** while it goes back to the upper orbit. In case it returns to the lower orbit which the previous electron has emptied, electron emits radiation as **visible light**.

6) INFRARED RAYS:

They consist of the vibrations between atoms and molecules. It is the radiation coming from Sun. We sense it as heat. Infrared light is obtained by ceramic plates made of special soil and which are heated by conducting electricity from the resistances inside them. Heat is energy similar to sound. Heat appears from all motions of atoms and molecules.

7) RADIOACTIVITY (Radioactive Disintegration):

There are 118 elements in the table of elements starting from **hydrogen** atom which includes one proton and one electron (**Annex-IIIa**); taking hydrogen as 1, as the proton numbers increase (Z=proton number) (atomic number) different elements are obtained. Proton number of hydrogen (in other words its atomic number) is 1; it is 2 for helium and 3 for lithium.

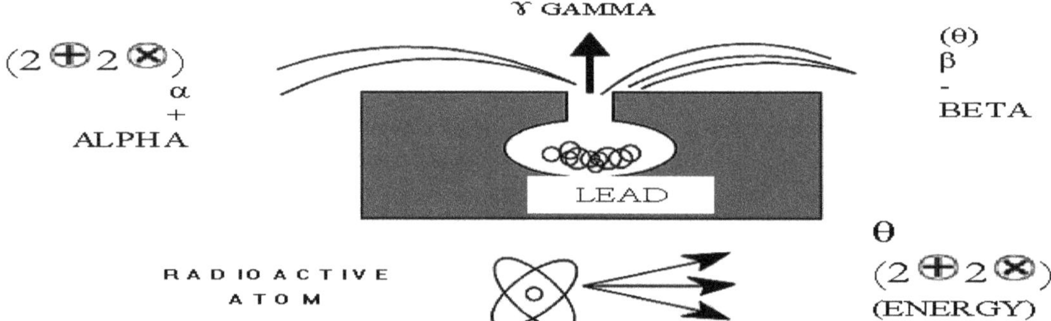

However, **polonium** with the atomic number thus the number of proton 84 and the subsequent elements (**Annex-IIIb**) are instable due to the high energy loads in the same rate with the multiple Protons and Neutrons they carry in their nuclei and occasionally, like in the **radium** (atomic number 88) atom shown above, **the rightwards negatively charged beta particles**, i.e. electrons (Θ), **leftwards positively charged "alpha particles"** (a = 2 proton (+) 2 neutron = helium nuclei) and **straight upwards uncharged**, i. e. zero charged gamma particle are accompanied by 1 neutrino. All of the nuclei of radioactive elements tend to be transformed into a more stable element by occasionally exposing one, several or all of the above particles, called **radioactive decay** and **fragmentation**.

ATOMIC NUMBER: Proton number (means the atomic number) in the nucleus of an element is **Z**. The neutron number in its nucleus is **N**. (**Annex-IIIb**) **Z+N=A** is the **mass number** of the element. So:

a) When a nucleus gives up "**alpha particle**" and disintegrates, a new nucleus appears whose "**atomic number**" is **2 less,** and "**mass number**" is **4** less.

b) When a nucleus gives up "**beta particle**" and disintegrates, it transforms to a new nucleus whose "**atomic number**" is equal to the atomic number of the previous element.

c) When the nucleus causes **gamma radiation,** atom and mass numbers do not change, but there happens a decrease in the energy level of nucleus.

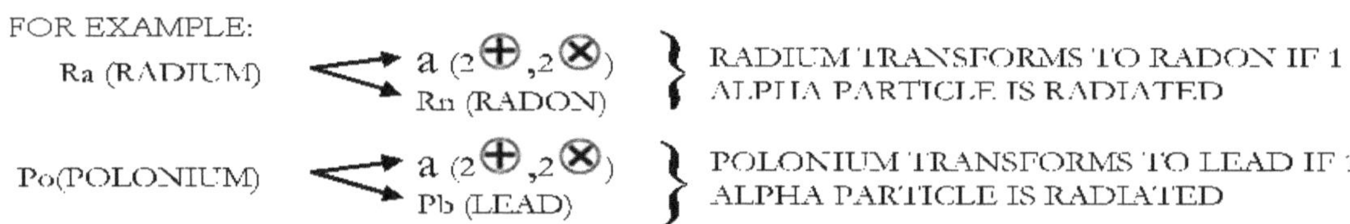

8) GAMMA RAYS:

Their wavelengths are >0, 1A0 (Angstrom) or shorter. When a radioactive nucleus **alpha particle** spreads and disintegrates, the resulting "**maiden nucleus**" becomes activated and spreads "**gamma photons**" while it comes to the lower energy level or basic form. This condition is similar to the promulgation of a "**light photon**" or an "**x-ray photon**" from the electron orbits, while returning of an activated atom to the basic form. The most important difference is that the energy level intervals in the nucleus are much bigger than the electron level steps, the energies and frequencies of gamma photons caused by the nucleus are very high, and wavelengths are very short, accordingly.

Although the **electron level steps are approximately 1eV** (electro Volt), the energy level intervals are **approximately 0. 1Mev** (micro electro Volt) in the **nucleus**. In other words, it is 100 thousand times bigger.

Beta rays: They are the electrons whose energy level is approximately 10Mev.

Alpha rays: They are the helium nuclei (2 protons, 2 neutrons) whose energy level is approximately 20Mev (Micro Electro Volt).

Gamma rays are photons whose energy level is minimum 1. 02 Mev (Micro Electro Volt).

When a high-energy "**gamma photon**" enters into the radius of a heavy nucleus, it disintegrates by giving up 1 electron and 1 positron. When **a positron combines with an electron, both of them disappear** and **2 gamma ray photons come out. "Alpha particles"** which are spread by the radioactive elements have approximately the same speed and energy. This fact reminds us the possibility that the particles are collected in the nucleus on specific levels, as the electrons do around the nucleus. In other words, the possibility of that the particles can be rotating on some of the orbits one around another. Of course, there is a dense collection in the nucleus. It can be compared with the center of the galaxies. The most stable ones in all nuclei are the ones whose proton number is equal to the electron number (**N=Z**) and the atom nuclei with the lowest energy.

9) X-RAYS:

X-Rays are composed of photons. When high energetic electrons collide in a high speed with a metal surface, they transfer all or some of their energy to the electrons of the metal they have collided. Picking up the energy, electrons of the metal leap out. The energy, which is lost during the electron on the upper orbit returns to the emptied lower orbit, comes out as **X-Ray**. X-rays are electromagnetic waves, which have **a small wavelength, very high frequency and high energy**. [Wavelengths are between 0. 1 and 100 A^0 (Angstrom)] They can penetrate through extra thick and liquid media since they are neutral; have a very short wavelength and high energy.

10) RADIO WAVES and MICROWAVES

(Annex-If): **Radio Waves** are the electromagnetic waves, which have the longest wavelength (between 1 mm and 100 km). Among the radio waves; **short waves** have a wavelength between 100m and 10m, **medium waves** have a wavelength between 100m and 1000m, **long waves** have a wavelength between 1 km and 100 km.

Microwaves have a wavelength between 1mm and 1m. They are used for radar and radio communication. Radio waves are used to transmit radio broadcast to very long distances around the world by reflecting the waves from the ionosphere layer of the atmosphere. These waves are not always as a result of an energy yield between atom and molecules. They are formed as a result of an electric current passing on a conductor whose strength and direction change relative to time; and thus, as a result of changing of the electric and magnetic fields periodically. At the same time, being influenced by a periodical force, the molecules in the solid, liquid and gas environments tend to diverge because of the field alterations and condition of the environment. They are formed by the motions of charged particles of atom. Also, some of the electrons leap out because of very fast rotation of **neutron stars**. But they return with the gravity. The energy they lose spreads to the space as **microwaves**. In case the frequency of waves sent from a microwave transmitter is equal to the vibration or rotation frequency of the molecules constituting the environment, absorption reaches to its maximum value.

11) TELEVISION WAVES:

They appear by transforming the energy of electron motions in the transmitter systems into waves. Television Waves have a wavelength between **10 cm and 10m**.

12) COSMIC RADIATIONS (UNIVERSE RADIATIONS):

"**Sub-atom particles**" in other words "**cosmic rays**" coming from the space in very high speed and energy collide with the world atmosphere. 90% of them are **protons, %9 are helium nuclei** (alpha particles = 2 protons (+) 2 neutrons) and 1% are the nuclei of heavier elements. New particles like electron, proton, photon, meson and neutron and gamma rays are formed by the collision of these rays on the upper layers of atmosphere with **nitrogen and oxygen nuclei** and air **molecules**. Energy of cosmic rays is as much as **1012-1020 eV** (Electro volt). Although today the maximum energy we can obtain in the strongest particle accelerators in nuclear physics laboratories is approximately **108 eV-400 GeV** (Giga electro volt), this value is very smaller than the energy of cosmic rays.

Cosmic rays interact with the magnetic field of the world and they form **radiation belts** around the world (Van Allen belts). A part of the cosmic rays is composed of charged particles from Sun, and a bigger part is composed of charged particles appearing from the explosions of greater stars in the space (Page 20).

13) LASER RAYS:

The light amplifiers emitting activated electromagnetic radiation obtain them. In the amplifier, there are two half-silvered mirrors located vertically in the direction of the path of light rays which are obtained by means of an "electric discharge". The distance between mirrors varies between several centimeters to several meters. Thus, the energy of the light waves accumulates between 2 mirrors and it reaches to a very high energy density. Their accumulated energies are either collected in a transparent solid object (chrome atoms in the ruby crystals) or in a suitable gas mixture (like CO_2). The electric field caused by the **laser rays are so strong that parallel rays coming out of laser get closer to each other and get focused, they are collected as filaments**. (Similar to the filaments between the galaxies) These combined Red Ray beams create ultraviolet waves whose frequencies are equal to two times of their own frequencies. (Harmonics of Laser Rays)

IV)   ATOMS

If the nucleus of an atom was as big as the mass of Sun, the mass of one of the electrons rotating around it would be as big as the mass of the world. But the distance of electron from nucleus would be 10 times far away from the most distant planet of the Sun. Another example: In case the nucleus diameter of an atom is increased by 1 mm, electrons as big as 1/1836 of the mass of nucleus would rotate 10-100 m far from the nucleus on the first orbit; 490 and 4900m (approximately 5 km) far from the nucleus on the seventh and last orbit, around the nucleus. %99. 4 of the mass of **Hydrogen atom** is in the nucleus. The weight of nucleus, which is as big as a thimble, is 400 million tons. (48 tons) Atom nuclei do not have an effect on the interactions between atoms. Because the surrounding electrons prevent them from coming closer. Thus the chemical properties of atom are determined only by the number of electrons, in other words by the "**Z**" charge number. (**Z**=Atomic number **N**= Neutron number, **A**= Mass number = Proton (+) Neutron total number).

Universe was composed of an energy dense Black hole before the Big Collision (Big Bang). All of the energy of Universe was only collected in the **Mass Gravity Force** and its particle **gravitons**.

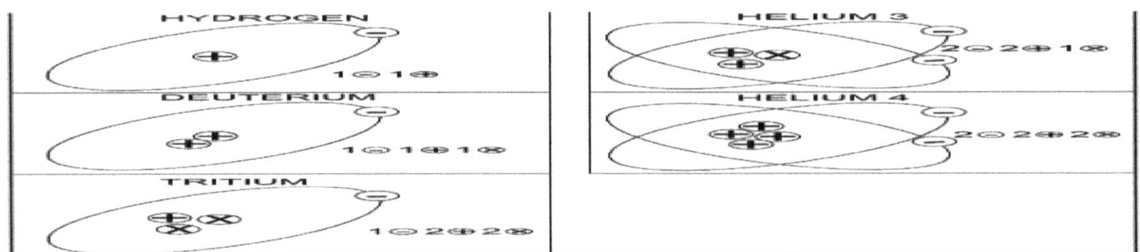

After the big collision, as can be seen in the previous Astral Universe section, with the step-by-step creation of the three other forces of atom and subatomic particles; protons from baryons which were the top phase of particles, combined with **electrons** to form the first element, **hydrogen**. As can be seen above, later, isotopes of **hydrogen** and **helium were formed**. These formations continued as can be seen in the table "**Periodical Elements and Isotopes**" (**Annex-IIIb**) and all elements in the Periodic Table of Elements (**Annex-IIIa**) were formed up to day. Today, still hydrogen nuclei in high temperatures are combined to form helium nuclei in the stars and Sun. All elements in the Periodic Table of Elements (**Annex-IIIa**) are lined up according to their atomic numbers. They are divided in 7 periods (7 orbits) from up to down. While these elements rotate around the nuclei of electrons, it means that they follow orbits at different distances from nucleus. In the periods, elements of similar properties are written one under the other, all elements from left to right are collected in 18 groups. The rightmost group no 18 from up to bottom; **He, Ne, Ar, Kr, Xe and Rn** elements with atomic numbers **2, 10, 18, 36, 54 and 86**, all have 8 electrons on their outer orbits except helium. As can be seen below, 8 electrons on the outer orbit is the most stable condition of an atom. That's why they are chemically ineffective.

In the Periodic Table of Elements (**Annex-IIIa**), the number of electrons is the same on the outer orbits of element groups, which are lined up from up to, down:

INERT GASES:

HELIUM	2
NEON	2, 8
ARGON	2, 8, 8
KRYPTON	2, 8, 18, 8
XENON	2, 8, 18, 18, 8
RADON	2, 8, 18, 32, 18, 8

For example, all elements in the 18th group except **inert gases** tend to become stable like the inert gases, in other words they tend to make the number of electrons on their outer orbits as 8. In the Periodic Table of Elements, "**alkali metals**" in the 1st column on the left give up 1 electron on their outer orbits; "**alkaline-earth metals**" in the 2nd column on the left give up 2 electrons on their outer orbits to make compounds with the "**a-metals**" in the 13th and 17th groups. Because the a-metals tend to receive 3-7 electrons to have 8 electrons in total. In an atom, the number of lost electrons gives the "**positive valence**"; and the number of received electrons gives the "**negative valence**".

Metallic property of (+) valence metals on the left side of Periodic Table of Elements increases from right to left and from up to down. (**Annex-IIIa**) The most real metal is **francium** (proton number is 87) on the lower left corner. However, since it is a very fast disintegrating radioactive metal, **cesium on it** is the most real (the most positive) metal. For the **a-metals** in the upper right section of the table, a-metal property increases from left to right and from down to up. The most real (the most negative) a-metal is **fluorine** on the upper right. **Fluorine** is the most active element in receiving electrons, in other words negative valence; **cesium** is the most active element in giving up electrons, in other words positive valence. A-metals make ion compounds by receiving electron from the metals. Metals cannot make compounds between each other. But a-metals make compounds with "**covalent bonds**" by using their electrons commonly without giving up or receiving electrons. The number of positive valence and negative valence sum of an element is equal on the table. This number is the **atomic number** of the element.

The status of electrons rotating around the nucleus of an atom is determined by **4 quantum numbers**:

n = is related with the distance of its orbit to the nucleus. It is one of the 7 orbits, which are shown with the letters **K, L, M, N, O, P and Q** in the table of elements (**Annex-IIIa**) on the left side and right side from up to down.

L = is related with its orbit. Also it is one of the sub orbits lined up near every orbit with the letters **s, p, d, f, g, h and j** on the table.

Ml =is the quantum related with the magnetic field which electron creates while rotating around the nucleus and it is called as **spin** (Annex-Ic). If it rotates from right to left its spin value is **+1/2**, if it rotates from left to right, its spin value is **-1/2.** Even the electrons, which are rotating in the same direction, rotate with a speed of 50 thousand kilometers in a second, they wrap around the nucleus like a ball of string and they do not collide to each other. Because in an atom, there are not two identical electrons, in other words, there are not two electrons whose 4 quantum numbers are the same. The electrons, which rotate on every single orbit, have different levels of energies. The energy levels of electrons, which rotate on the upper orbits, are higher. But the rays with the highest energy (like X-rays) spread from the electrons on the orbits (**K, L, M**) which are closest to the nucleus with the lowest level. It means that, it is possible to say that, the rays with the highest are hiding in the electrons, which are the closest to the nucleus whose energy level is low. Gravitation is maximum on the lowest orbit where the energy of the system is minimum. May be that's why, the rays with the highest energy appear in the electrons in this area.

While the **electrons** rotate around the nucleus, they radiate electromagnetic waves (rays). Thus, kinetic energy of the electrons decreases gradually. Orbits around the nucleus are expected to get closer to the nucleus and they are expected to combine at the end. But, some kind of balance appears between the gravitation force which positive charged nucleus applies on the negative charged electron and centrifugal force, which the electron gains by means of gravitation force. And the continuity of rotation is achieved.

NUCLEAR STABILITY GRAPH:

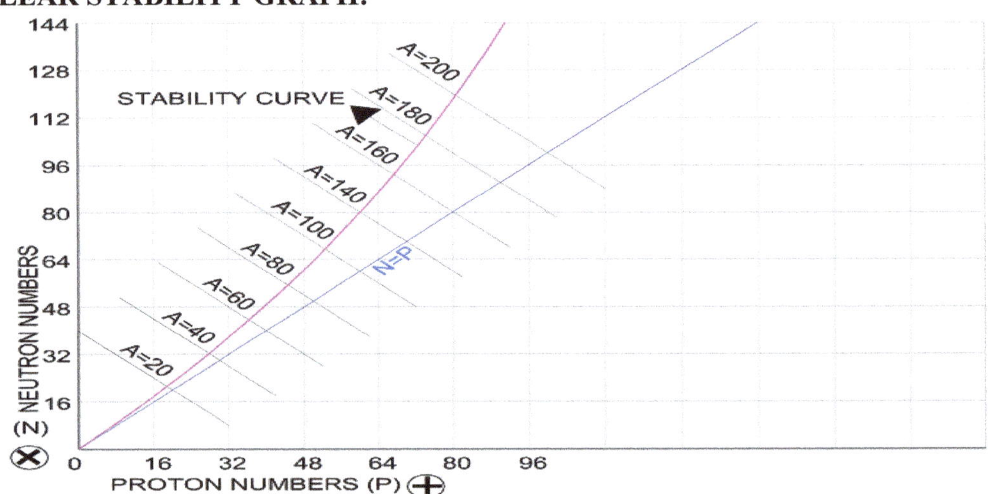

The ratio of proton and neutrons in the atom nuclei of elements develops on the "**Stability Curve**" which is shown above. There are no elements whose number of Protons is equal to the number of neutrons after calcium whose atomic number is 20 (**Annex-IIIb**). If the **atomic numbers** (p) and **mass** (A) of the elements grow, then we need Neutrons more than the number of protons in order to obtain the balance between electrons and protons in the nucleus. There are conditions, which do not suit with the ratios in the "**Stability Curve**" above.

PROTON DISINTEGRATION:

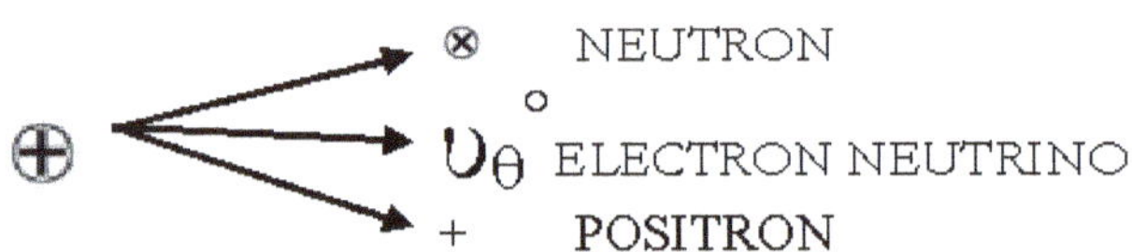

Radioactivity; In case one neutron is missing for the nucleus of an atom to pass from unstable condition to a stable one, in other words to gain the ratio on the Stability Curve, this is named "**proton disintegration**". In other words **1 positron and 1 electron neutrino** are given up from 1 proton out of the nucleus and proton becomes a neutron. Since one proton is given up in the nucleus, it transforms to a nucleus whose atomic number is one less than itself. If there are more Neutrons, "**neutron disintegration**" happens. In other words, **1 electron** (beta particle) and **1 anti–electron neutrino** are given up from 1 neutron out of the nucleus, as can be seen below, and neutron becomes a proton. Since one proton is received, it transforms to a nucleus whose atomic number is one more than itself.

NEUTRON DISINTEGRATION:

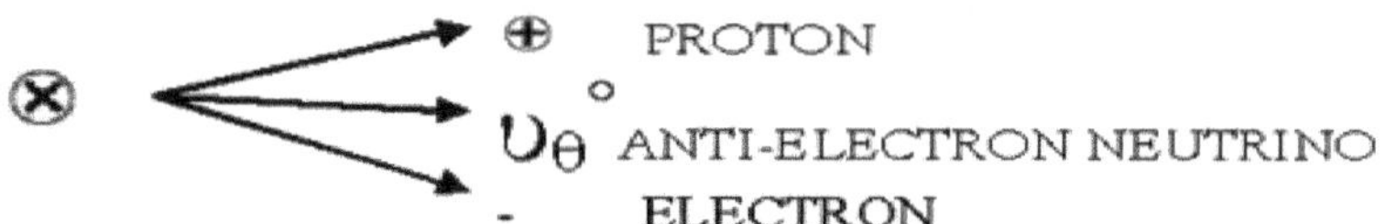

Sometimes 1 alpha particle (helium nucleus which is composed of 2 protons (+) 2 neutrons) is given up from the radioactive nucleus, nucleus atomic number transforms to two small nuclei and gamma rays spread. As radium (atomic number 88) transforms to radon (atomic number 86) on Page **22,** when a nucleus spreads alpha particle and disintegrates, the new formed nucleus is in activated condition and **gamma photons** spread while it is transferred to the lower energy level. This is similar to light photons and x-rays spreading of the electrons of an activated atom. The reason of the difference is the high level of gamma photon energies, since energy level ranges in the nucleus are higher than the energy levels between the electron orbits. While the electron energy levels are approximately 1 eV, it is assumed that there are energy levels and orbits of nucleons in the nucleus, like the electrons of atom have energy levels and orbits. Accordingly, since the nuclei including proton and neutron are very stable, it is concluded that there are 7 orbits in the nucleus on which 2, 8, 20, 28, 50, 82 proton and neutron and 126 neutron (**Annex-III a, b**) rotate. As in the energy levels of electrons, **stable nucleus** structures appear as a result of saturation of energy levels of nucleons. Also as +1/2 and -1/2 **spin** electrons (spins are in the opposite direction) around the nucleus are located on the orbits in pairs, it is concluded that Nucleons are also rotating on their orbits in the nucleus as +1/2 and -1/2 **Spin** pairs in the opposite directions to each other.

Nearly 90 of 118 elements in the periodical table of elements (Annex-IIIa) **are present in nature. The others are obtained in the laboratories.** Among the 90 elements in nature; **2 of them are liquid (bromine and mercury); 11 of them are gas** [(Inert gases **He, Ne, Ar, Kr, Xe, Rn Uuq)** **and chlor, fluorine, oxygen, nitrogen and hydrogen**]; and the remaining of them are solid.

V)   4 FORCES WITHIN AN ATOM:

The table in **Annex-IIe, f, h and i** can be reviewed for the elementary particles of atom which we will discuss in this section. The forces, which arise from the mutual interactions of elementary particles of atom and as a result, which regulate the behaviors of elementary particles, are divided in 4 groups:

1) MASS GRAVITY FORCE (Magnetic force): This force arises from the "**graviton**" particle exchange between the materials. Being minimum in the nuclei of atoms and maximum in the center of black holes, high gravity in the center of stars and planets (Earth) decreases the distance between nucleus of atom and electrons. In the center of stars (Sun), electrons travel freely between the nuclei. Gravity arises as a result of "**Graviton**" exchange between the World and pulled materials. As in its name, gravity force has only a gravitation affect. Although it is the weakest force in the atom, it conveys its gravity force affect to the materials of universe, which are on an infinite distance by means of "**graviton**" particle exchange. It is its own anti-material. Gravitons and photons interact by means of being exchanged between material and anti-material. The energy which graviton carries compared with photon is as indicated here in the 2nd column. Graviton is assumed to be massless and its impact distance is assumed to be infinite in the universe.	PARTICLES GRAVITON (Θ) 10-39 MeV MASSLESS $\dfrac{Photon}{Graviton} = \dfrac{1}{10^{-39}}$

2) WEAK NUCLEUS FORCE: It cannot overflow out of nucleus. An electron would not exist without this. It manages the interactions between **muons** and changes, transformations in the nucleus. It is effective in the **beta** (electron) **disintegrations, pion, muon disintegrations, neutrino dispersions and radioactive disintegration of many sub-atom particles and nucleus**. Also it is the fundamental force causing the Sun and stars come into existence. This force appears by the exchange of **W⁺, W⁻, Z⁰** Particles. **W** particle has two neutral carriers; one of them is **"photon"** which is the massless energy package which can also carry electromagnetic force; and the other is the **"Z⁰"** particle which is neutral, but whose mass is approximately 95 times of proton mass.

W^+, W^-, Z° and NEUTRINOS (υ) 10^{-11} MeV

Neutrinos: They only enter into the formation of this force. Some of the properties of Neutrons are:

They give up 1 electron and 1 neutrino, transform into a proton, and gain weak nucleus force. Since neutrinos are not charged, they do not interact electromagnetically. Besides they have immunity against the strong nucleus force. Material is transparent with neutrinos to a great extent. Neutrinos interact with the materials only by weak nucleus force. Apart from neutrinos, other leptons (like **muon, tau, electron**) are the particles, which interact by this force. The impact duration of this force is 10^{-10}–10^{-8} seconds; its maximum impact distance is 10^{-18} m.

IMPACT DURATION: 10^{-10}-10^{-8} SEC MAXIMUM IMPACT DISTANCE: 10^{-18} meters

3) STRONG NUCLEUS FORCE: It cannot overflow out of nucleus. It holds nucleus together. This force arises by the exchange of **meson particles**, which are assumed to be located around the **nucleons** (proton and neutrons) as a cloud. In other words, energy exchange in the nucleus is by means of the exchange of mesons, which are the particles of this force. Nucleus interactions, which cause the inter-nucleon interactions and meson creations, are obtained by this force. Pions of nucleons are neutral. But (+) and (−) charged **pions** are exchanged between them. Electromagnetic forces out of nucleus have an effect on the forces independent from charges in nucleus; and it causes a mass difference in meson particles. For example the rest mass of charged pi-mesons (**Annex-IIh**), is bigger than the rest mass of neutral Pi-Mesons. The effect from out of the nucleus, in other words from electrons can be (+) or −. giving up − pion, a neutron becomes a proton. receiving − pion, a proton becomes a neutron. As a result of last developments, it is being assumed that this force appears because of "**gluons**" (color particles, glue particles) which are exchanged between **quarks**, which create proton, and neutrons. The effects of this force appear **in approximately 10^{-22} -10^{-23} seconds.** But maximum impact distance is very short like 10^{-15} **m.**	MESON ITS ENERGY: 10^3 MeV GLUONS IMPACT DU-RATION: 10^{-22}-10^{-23} SEC IMPACT DIS-TANCE: 10^{-15} meters
4) ELECTROMAGNETIC FORCE: It is effective between the electric charged particles. It creates (+) and (-) affects between **proton** and **neutrons**. It is neutral. The electromagnetic interactions, which are caused by this force, are obtained by the "**photon**" exchange **between the** charged particles like **electrons**. In other words, this force appears by means of Photon exchange. **It pulls the electrons and makes atoms and molecules come together.** It is the force, which creates all visible rays and waves of Electromagnetic spectrum. The effects of this force appear **in 10^{-20}-10^{-18} seconds.** Impact distance is infinite in the space and radiating speed is approximately **300 thousand km/seconds** (speed of light). "**Photon**", the particle of this force has been told in the chapter 'Electromagnetic Waves'.	PHOTON (γ) ITS ENERGY 1 MeV IMPACT DU-RATION: 10^{-20}-10^{-18} SEC IMPACT DIS-TANCE: INFINITE

The interaction of 4 forces of the atom given above happens by the exchange of their own particles (**Quantum**). According to the Quantum Mechanics, all energy units are 'quantized', in other words they are composed of small tangible particles named "**Quanta**". Scientists who tried to prove that the 4 forces above were the images of the same main force could not succeed up to date in this project named "**Combined Field Theory**". Albert Einstein dedicated 30 years of his life for this study.

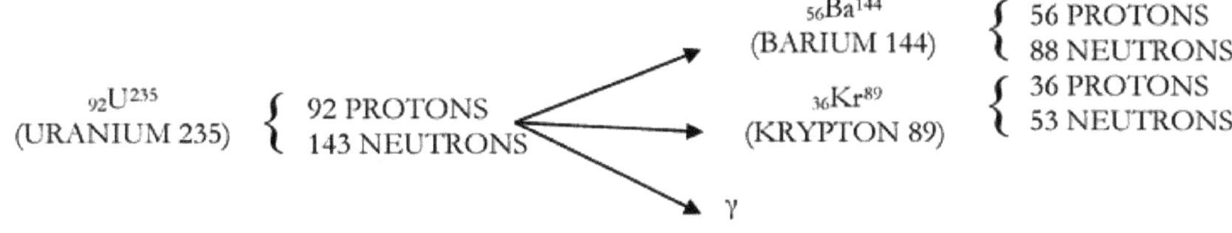

VI) FISSION:

It is the fission of Atom nucleus as can be seen in the previous page:

1) CONTROLLABLE FISSION:

(Atomic energy plants and radioactivity) in 1937, "**uranium nuclei**" was bombarded with neutrons and it was found out to be fissioned into the elements like **neptunium** ($93\ NP^{239}$), **plutonium** ($94\ PU^{239}$), and **americium** ($95\ AM^{239}$). In 1939, **uranium 235** was bombarded with neutrons and it was fissioned to **barium 144** ($56BA^{144}$) and **krypton 89** ($36KR^{89}$) nuclei. Since it looked like the division of cells in biology, it was called as "**fission**". Later, it was found out that fission was not only in uranium, all elements with the atomic number between 95 and 139 (in other words the ones whose number of neutrons is bigger than the number of protons) were themselves (in other words without bombarding with neutrons) radiating beta rays (β) in the unstable atom nuclei in short periods; and a neutron, proton balance was being established and they were transforming to stable nuclei. Examples of radioactive materials are:

$$\text{(ANTIMONY 133)}\ _{51}Sb^{133} \xrightarrow[50]{\beta} \text{(TELLURIUM 133)}\ _{52}Te^{133} \xrightarrow[60]{\beta} \text{(IODINE 133)}\ _{53}I^{133} \xrightarrow[22\ \text{DAYS}]{\beta} \text{(XENON 133)}\ _{54}Xe^{133} \xrightarrow[5\ \text{DAYS}]{\beta} \text{(CESIUM 133)}\ _{55}Cs^{133}$$

In addition, it was found out that **uranium 238** is being fissioned itself in 10^{16}-10^{17} years of half-life without being bombarded to give fission products. The reaction starting with fissioning a single uranium nucleus with a single neutron in the atomic energy plant; regarding the continuously increasing number of neutrons as a result of fissioning of the uranium nuclei on the neighborhood by the released neutrons; also increasing the number of nuclei in the Fission, a self-perpetuating "**chain reaction**" appears and a very big amount of energy comes out. The working principle of **fission reactors** (nucleus reactors) obtained from atomic energy is attributed to the "**controllable chain reactions**". **Uranium 235**, **uranium 238, plutonium 239** and **thorium 232** are used as Fission materials. Today in many countries of the World, there are several "**nucleus reactors**" of various types. A big amount of them are used to produce electricity, some other part are used to move the nuclear ships and to train the personnel on this field.

2) UNCONTROLLABLE FISSION (Atomic bomb)

Any one of **uranium 235, plutonium 239** or **uranium 233** is used as Fission material in the atomic bomb. When an atomic bomb explodes, temperature rises to approximately **10 million °C**, and the destructive power of pressure waves reaches to unbelievable levels. After the atomic bombs exploded in **Hiroshima** and **Nagasaki on** 1945, extraordinary fatalities happened with the death of 100 thousands of people instantly, and the disability and death of 100 thousands of humans with the radiation affect. The detrimental effects of the resulting radioactive materials are going on even today. In 1960s, it was mentioned that an atomic bomb would cause a 300-500 meters of grave at the dropping point and after 6 hours, on a 500 km distance, the ashes raining on the humans would have radioactive effects. It is not difficult to imagine that today's atomic bombs will be more horrible.

VII) FUSION (combining atomic nuclei):

Thermo-Nuclear Reactions: Like energy comes out when nucleus breaks down, approximately 4 times more energy comes out when the **nuclei combine (Fusion).** Fusion depends on the temperature, which should be as high as the temperature of Sun and stars (**approximately 10^7 K°**). That's why fusion is also called as "**thermo nuclear reaction**". The process at the beginning of the Universe starting with (**Big Bang**), and first of all formation of **elementary particles of atom**, later formation of **helium nuclei** by combination of these particles, is still going on today in the Sun and stars. Still, 80% of the Sun is composed of **hydrogen**, 19. 1% is **helium,** and the remaining part; a small amount of heavier nuclei like **carbon**, **nitrogen** and **Oxygen**.

Starting with Big Bang in the universe, fusion, in other words formation by combination, is still going on by first of all formation of "elementary particles" and later formation of "**element atoms, molecules, compounds and creatures**". In the Sun, a big amount of energy comes out from the Fusion reactions, which start with Proton-Proton cycle or Carbon-Nitrogen cycle and form **helium 4 nucleus** (Alpha Particle) as a result. "**Helium 4 nucleus**" is one of the most stable nuclei in the nature. While 93% of the energy from **Sun radiates** as heat and **light**, and 7% of radiates as energy, they are spread to the space as **neutrino particles**.

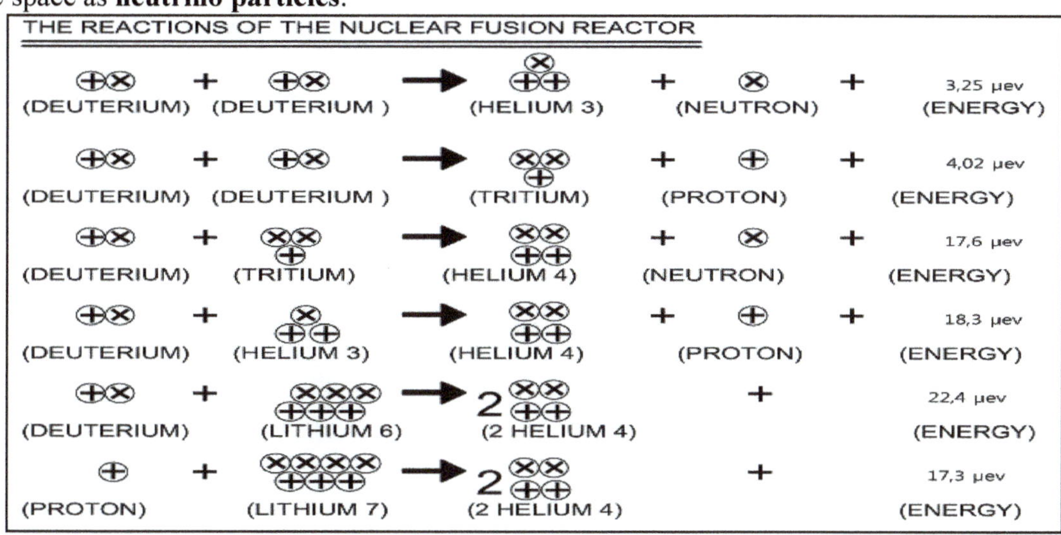

I) CONTROLLABLE FUSION:

Studies related with satisfying the energy needs of humans by accomplishing the fusion in the world like it occurs in the Sun and stars, is not succeeded yet. If it can be accomplished, our energy need might be satisfied up to the end of time. For the controllable fusion to sustain itself and set off a "**thermonuclear reaction**", for example the temperature inside "deuterium plasma (solution)" should be increased up to several hundred million K° (Kelvin degrees). In this temperature, nuclei would leave the electrons and get closer to each other as to achieve the fusion reaction. In the last 20-30 years, there are several studies on "nuclear fusion reactors" for combining the hydrogen nuclei in America, Japan, England, Russia and Germany.

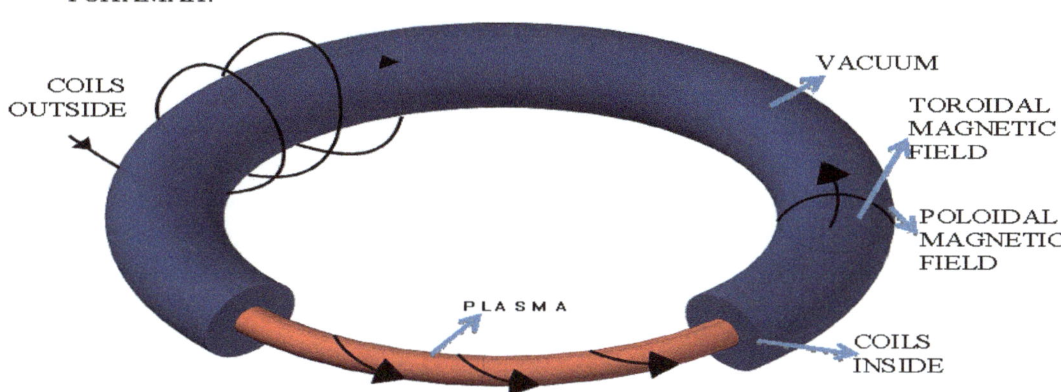

One of the tried techniques "TOKAMAK" (It means empty ring-shaped magnetic room, in Russian Language) (Torus shaped) can be seen above. **Poloidal magnetic fields, which are produced by the coils out of the cabinet while applying vacuum on Tokamak and Plasma is,** kept on the central line of Torus, away from the walls, by means of the compound effect of **Toroidal Magnetic field** parallel to walls surrounding the vacuum area, which is formed by the coils while heating the plasma. Otherwise, the walls would be evaporated instantly.

Second technique- "Fusion by Laser": In this technique, rays of laser or atom particles are sent to Deuterium particles, which are smaller than a millimeter within a **lithium-coated room**. Under this bombard, particles are jammed very fast; and a temperature arises which is enough for fusion. By taking the particles into the room quickly and one after the other, an uninterruptible power production is obtained. Melting Lithium is used to obtain the vapor to keep turbines working.

FUSION REACTIONS OF HYDROGEN BOMB

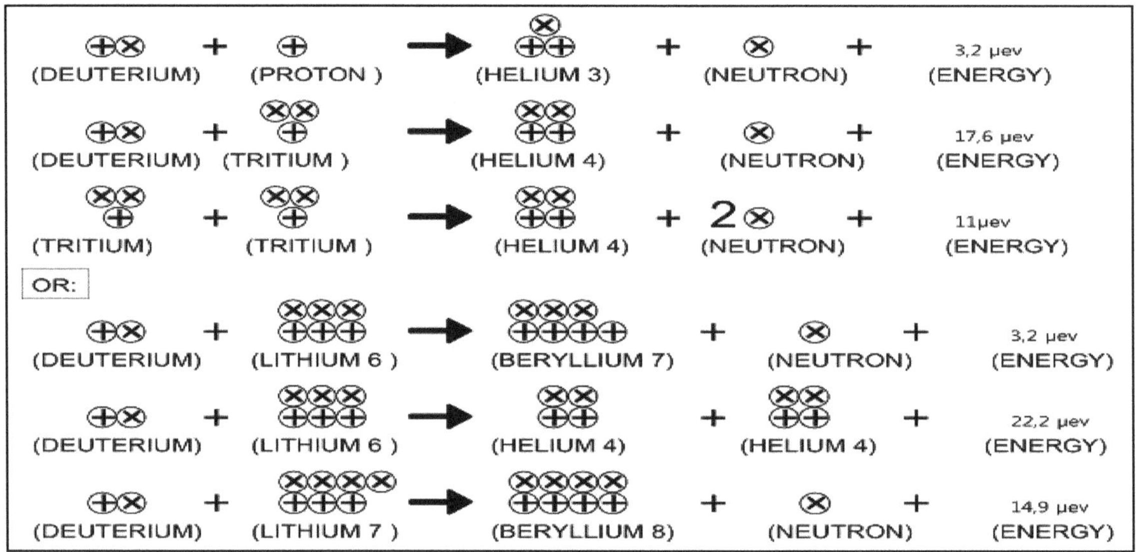

2) UNCONTROLLABLE FUSION (Hydrogen Bomb):

Although they have been going on naturally in the stars for millions of years, the first human made artificial **fusion reaction** was realized when the **Hydrogen Bomb** was exploded in the year 1955. The main temperature necessary to detonate the bomb can be supplied by the **atomic bomb,** in other words its fission (breaking into pieces). After 0.1 microseconds, the reaction of fusion materials (**deuterium** and **tritium)** starts. Although the destructive power of the atomic bomb is limited, there are no limits for the destructive power of the **hydrogen bombs**, unfortunately. **It means that it can also demolish the World.** So, the tragical age of humans can be seen in the world of battles of the 21st century.

VIII) MOLECULES:

Atom groups, which are formed as a result of same or different types of atoms coming together in various forms, are called as "**molecules**". **Molecules** come together to form "**COMPOUNDS**". Molecule is the smallest part of a compound, which has its own properties. The atom number of molecules is available from 1 to 1 billion. (Like the stars in a galaxy) In the Periodical Table of Elements (**Annex-IIIa**), Inert gases on the right hand**,** and metal molecules on the left hand have 1 atom; on the upper right corner, a-metals on the left of inert gases have variable number of atoms, which is between 2 and 8. Molecules can be formed by the atoms of a single element or by the atoms of different elements. While gases are composed of two atoms of the same element, **protein molecule**, which is an organic material, can have 100 thousand-200 thousand atoms.

Molecule size is 10^{-8}-10^{-7} **cm**, and it is assumed to be a sphere like an **atom**.
Diameter of an oxygen molecule is 2. 9 A° (Angstrom) 2 atoms
Diameter of a nitrogen molecule is 3. 1 A° (Angstrom)
Diameter of a hydrogen molecule is 2. 7 A° (Angstrom) 2 atoms
Diameter of a helium molecule is 2. 6 A° (Angstrom)
Besides phosphor molecule has 4 atoms
Sulphur molecule has 8 atoms.
There are spaces between the molecules of all materials (like the spaces between Galaxies). These molecules are moving continuously in a speed increasing with temperature. Like the words in

languages we speak are composed of sonant and consonant letters (Alphabet); all materials in the **Universe** are formed by the combinations of 118 elements in the Periodical Table of Elements, in other words, combination of **positive valence "metals"** on the left hand side of Table of Elements and **negative valence "a-metals"** on the upper right of the table. Inert Gases do not react.

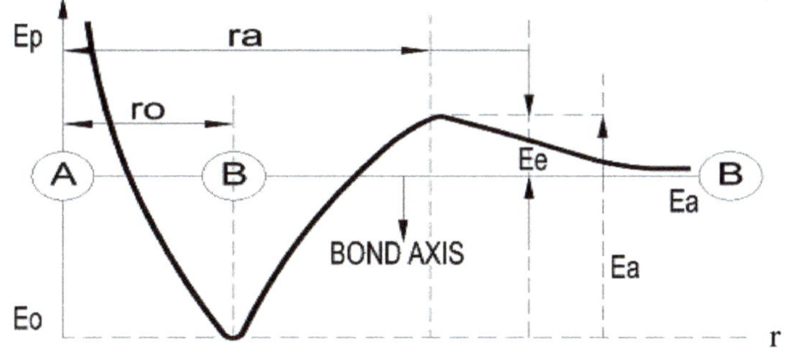

r=	DISTANCE BETWEEN ATOMS
E_p=	POTENTIAL ENERGY
E_a=	DISSOCIATION POTENTIAL ENERGY OF ATOMS
E_e=	INTERACTION POTENTIAL ENERGY OF ATOMS

COMBINATION OF ATOMS TO FORM A MOLECULE:

In the above graphics, there is an example of two atoms making a compound to create a molecule with two atoms. While approaching to atom A, Atom B in the graphics states the change in the potential energy and push-pull forces according to the distance between them. As the atoms get closer to each other, mutual potential energies (**Ep**) increase and hence, atoms push each other. On the point **r = ra, repulsive force becomes maximum**. When the approaching **B** atom exceeds this potential limitation, mutual potential energy starts to decrease and two atoms start to pull each other. When the atom **B** comes to the distance r_0, the energy of the system becomes E_O, for the **Ea** energy, which the inner forces of atom use. In this case, the molecule, which is composed by two atoms, comes to the most stable condition in which the mutual potential energies are minimum. It is described as "**Molecule is in the basic state**".

In case the atom **B** gets too close to the atom **A,** then the potential energy between the atoms increases in a bigger speed for ($r<r_0$ **a**) and a very big thrust is found out to occur between these atoms. Its reason is that the electron layers of two very close atoms start to interpenetrate. Atoms of the elements combine with each other by the electrons on the outer orbits. These electrons can be thought as a hook. Hydrogen atom has 1 electron (in other words 1 hook) on its outer orbit to combine with other atoms; oxygen has 2 electrons, in other words 2 hooks.

THERE ARE TWO TYPES OF HOOKS OR BONDS CONNECTING THE ATOMS TO EACH OTHER TO FORM A MOLECULE:

1) IONIC BONDS: They are the bonds, which appear when some specific atoms give one or more electrons to another atom. Since the atom, receiving **electron becomes** a negative ion, and the atom giving up **electron becomes** a positive ion, a bond is created between them by means of an **electrical gravity force**. **NaCl** (Sodium chloride) and **LiF** (Lithium Fluoride) can be given as examples.

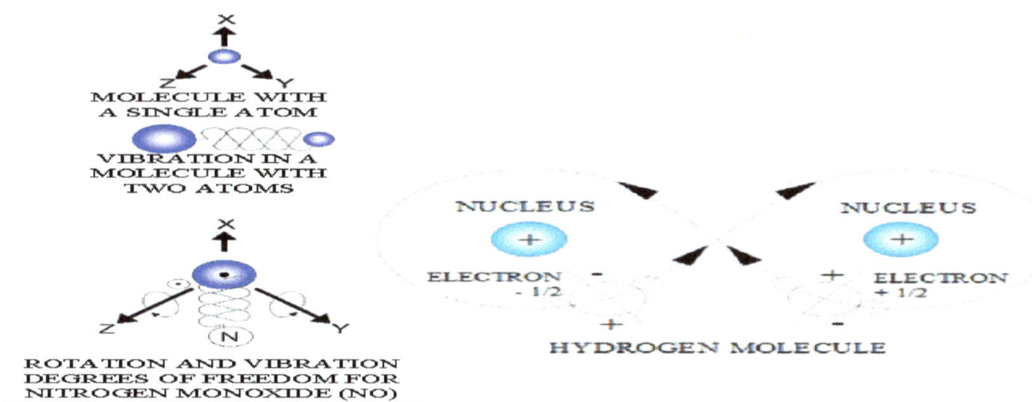

2) COVALENT BONDS: Atoms make "**Covalent bonds**" by means of using the electrons in common without giving up or receiving the electrons, as in the "**Hydrogen molecule**" which can be seen above. H_2 (Hydrogen) CL_2 (Chlor), CH_4 (Metan), CO_2 (Carbon dioxide) and **HCl** (Hydrogen Chloride) can be used as examples. Atoms in the molecules are no longer free atoms. The electrons they have are under the influence of the electron and nuclei of their own and other atoms. As an example, it will be assumed that the orbits of electrons of Hydrogen Molecule, which is composed, of 2 Hydrogen atoms do not change; only these two electrons can be belonging to one of the atoms in specific periods. Electron Spins of two Hydrogen atoms forming the Hydrogen Molecule should be parallel in opposite direction as can be seen above. Otherwise, a gravity force does not appear between the two atoms. Also if the spins (rotating around themselves) of protons forming the nucleus of hydrogen atom, are parallel in the same direction, the molecule they form is called as "**Ortho Hydrogen Molecule**". If the Proton Spins are parallel in the opposite direction, it is called as "**Para Hydrogen Molecule**".

3) DEGREES OF FREEDOM OF THE MOLECULES: There are "Three shifting degrees of freedom" for the **single atom Inert Gas (Hl, Ne, Ar, Kr, Xe and mercury vapor) molecules** on the rightmost side of the Periodical Table of Elements (**Annex-IIIa**) which are able to move freely in the space. Apart from 3 shifting degree of freedom, molecules like Hydrogen (H_2) Nitrogen (N_2), Nitrogen Oxide (**NO**) and Carbon Monoxide (**CO**) with 2 atoms, have the degrees of freedom for vibration and rotation as can be seen above. It is possible to assume these atoms belonging to the molecules as two spheres, which are connected with a flexible but hard string in the direction of centers of mass. These types of molecules can make vibration motions around the center of mass according to the changes like heat and pressure in the environment; and rotating motions around the axes, which pass through the center of mass by means of various interactions between molecules. **Nitrogen Monoxide (NO)** molecule which is a two-atom molecule has "**3 inner degree of freedoms**" like 2 rotations and 1 vibration motion as can be seen on the upper right corner.

IX) COMPOUNDS:

They appear by the **combination of molecules**. Molecule is the smallest part of a compound, which has its own properties. Compounds are materials which are formed by the atoms of "Identical" (similar) molecules and which are composed by the atoms of two or more elements. There are millions of **chemical compounds**, which can be found in the nature or can be obtained in the laboratories. They vary from the basic structures like water molecule (H_2O) which is composed of two Hydrogen atoms connected to one Oxygen atom; to the Nucleic Acids which are composed of thousands of atoms and which have a complex structure. Most of the materials in the nature are mixtures of two or more compounds.

1. **CLASSIFICATION OF COMPOUNDS ACCORDING TO THE ELEMENTS IN THEIR MOLECULES:**

   a) **INORGANIC COMPOUNDS:**

   These are the materials, which are formed by the combination of two or more elements in specific ratios, excluding carbon. They are classified according to their elements. **Chlor, bromides, iodides, fluorides, hydrides** as a compound of hydrogen, **oxides** as a compound of oxygen **and sulphur** are examples of binary compounds, which are classified according to a single element. Two element compounds are called **alloy**.

   b) **ORGANIC COMPOUNDS:**

   In the 19th century, chemists divided the carbon compounds in two categories; they defined the compounds, which they thought to originate only from the living creatures as "**organic**"; and the compounds in the non-living materials like minerals, as "**inorganic**". Although later it was understood that the organic compounds could be prepared in the laboratory from the inorganic compounds, today the section of chemistry related with the organic compounds is still named "**organic chemistry**". **Because of its atom structure,**

carbon element, which is the source of organic compounds, tends to share electrons (covalent bond) instead of giving up or receiving electrons. Since every one of the 4 electron on the outer orbit of **carbon atom** can make a covalence bond, it can also be connected to another carbon atom or other elements and compounds with covalent bonds at the same time. That's why **carbon makes compounds itself more than the other elements do.** The most common compounds of carbon are the ones it makes with **hydrogen, oxygen, nitrogen, chlor and sulphur.** It makes most of the organic compounds like **fats, proteins, carbohydrates, hemoglobin, chlorophyll, enzymes, hormones** and **vitamins** which are the basic compounds of foods. The compounds of carbon with hydrogen are called as **carbohydrates.** Organic compounds are divided in two parts.

b1) ALIPHATIC COMPOUNDS: These are the compounds with a **carbon skeleton** looking like a diamond network. **Aliphatic compounds"** are composed as a result of cooperation of non-cooperated (not covalent bonded) electrons on the outer orbits of carbon atoms, with the electrons of **hydrogen, a-metals** or **various roots** in any carbon skeleton which looks like the diamond network. Below are some examples:

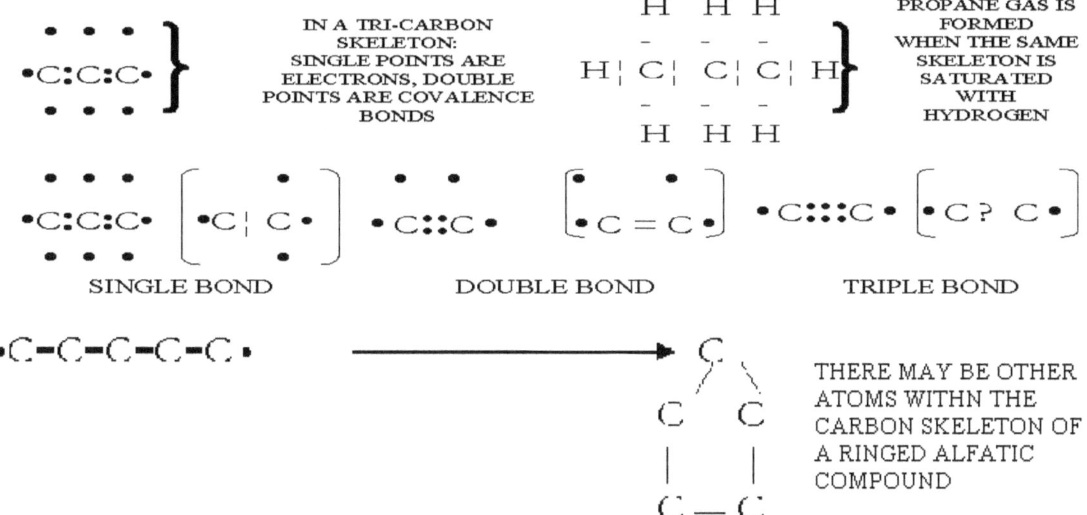

As the carbon atoms can connect with each other through a single covalent bond, they can also connect with two or three covalent bonds in the aliphatic compounds. The compound of carbon atoms in which they are connected with a single bond is called "**saturated**"; the compound of carbon atoms in which they are connected with two or three bonds is called "**unsaturated compounds**". Aliphatic carbon compound are generally open chained as indicated above. But two ends of the chain can combine to form a "**closed chain**".

BENZENE

b2) AROMATIC COMPOUNDS: Since the carbon skeleton looks like a graphite structure in these compounds, they are always in ring structure as can be seen in the above **BENZENE** example. At the same time, one single, one double bond exist in this ring structure consecutively. Since all of these compounds are aromatic, they are called as "**aromatic compounds**". Being the main material of organic compounds, **carbon** is the eleventh most common element on the Earth. The structure of all creatures is composed of carbon to a great extent. It is also a heat and energy source.

c)   ORGANO-METAL COMPOUNDS:

These materials, which metal atoms create with organic compounds, are met frequently in the structure of living creatures. For example; "**hemoglobin**" which is found in the blood of superior animals, and which is a carbon compound combined with ferro atom; and "**chlorophyll**" which is a big organic molecule combined with magnesium in the green plants, are the best-known **organo-metal compounds**.

2.   CLASSIFICATION OF THE COMPOUNDS ACCORDING TO THE TYPE OF CHEMICAL BOND:

a)   ION COMPOUNDS:

They are the compounds which atoms are combined with ion bond. **Salts and Minerals** are in this group. These are generally hard materials with high stable melting point.

b)   COVALENCE COMPOUNDS

(Covalent bonded compounds)**:** Atoms make these compounds by using one or more electrons in common. The most important ones are **carbon compounds**. Since here the gravity force between the molecules is lower, molecules separate from each other with regard to ion compounds. **Polymer** and macromolecules can be formed by means of the combination of thousands or more atom using the covalent bonds. **Plastics, proteins in the living creatures, carbohydrates and nucleic acids** are covalence-combined polymers. Silicium minerals like quartz (SiO_2) and clay (**aluminum silicate**) are also covalence combined inorganic macromolecule compounds. (**silicium**'s being the elementary material of computers, means that it is relative with carbon, since it is the elementary material of structure of living creatures)

c)   COORDINATION COMPOUNDS:

These are the bonds in the compounds, which are composed of one metal atom in the center and surrounding atoms or atom groups. Since only one atom gives up electrons, it is not a covalent bond.

3.   CLASSIFICATION ACCORDING TO THE CHEMICAL REACTION:

a)   ACIDS, BASES, SALTS:

The compounds, which change the color of some identifiers (which redden the blue litmus paper) and which give up hydrogen by reacting with some metals, are classified as "**acid**"; and the compounds, which have an opposite chemical effect on these compounds, are classified as "**base**". Together with the materials they include, the symbols of bases like **hydroxide (KOH)** end with the names given by public like **lime and soda (CaOH, NaOH)**. **Salts** are formed by the reaction of **acids with bases**.

b)   OXIDIZING AND REDUCING AGENTS: With the **removing of electrons** from the atom of an element, that element is oxidized. And it is reduced with the **addition of an electron**. Except this three main groups, compounds are classified as "**gaseous, liquid, solid**" regarding their physical conditions; and "**natural**" and "**artificial**" compounds regarding their sources, during these formations.

X) CRYSTALS:

Crystal means **ice** in Greek language. When the solid materials turn to **liquid** or **gas** state by means of heat or solvent materials, their atoms are in an irregular travelling condition. If cooling happens slowly, the material, which **resolidifies** when cooled, finds time to form the geometric structure of crystal special to that material by the interaction of different forces special to that material in the atoms forming it. An **amorphous** structure appears if the cooling happens very fast. A visible single crystal structure can be obtained under very special conditions. Natural precious stones and a big single Ice crystal can be formed by the freezing of water very slowly. Artificially obtained crystals show an extraordinary order. A crystal solid material has an order and a periodic structure regardless of the irregularity of the material in its gaseous, liquid and glass states. In a crystal, the points where the atoms or molecules are present, is called as **"nodal point"**. Atoms, **ions** (+ or − charged atoms) or **molecules** which form the crystal of every solid material are located on the nodal points as per a geometric sample (**unit cell**).

1)   UNIT CELLS OF CRYSTALS ARE DIVIDED INTO 3 ACCORDING TO THEIR NODAL POINTS:

a)   SIMPLE UNIT CELL: Nodal points are only on the corners and it is represented by the letter "**P**".

b)   VOLUME CENTERED UNIT CELL: Nodal points are both on the corners and on the center of their faces. It is represented by the letter "**I**".

c)   FACE CENTERED UNIT CELL: Nodal points are both on the corners and on the center of their faces. It is represented by the letter "**F**". **Unit cell** is the smallest unit, which carries all of the geometric properties of a crystal. **All of the solid materials** (like Metals, Minerals, Alloys, Compounds, ceramics and some solid organic materials), have different crystal structures which are lined up in equal intervals, composed of millions of duplicates, in the direction of their sides, diagonals and face diagonals of its own unit cells in 3D or in one dimension. The distance between the crystal atoms and **molecules is called as "period"**. This distance can change according to the direction. But it is always the same for the same direction of the same crystal. Shape differences of the crystal are caused by the properties of force, which keeps atoms and molecules together. Some of the metal crystals are visible. This is similar to the **zinc crystals** on the **zinc coated galvanized sheet metal**. Sometimes it is similar to the **iron and steel crystals,** but they are visible under microscope.

APPEARANCES OF CRYSTALS (HABITUS): The factors effecting the formation of a crystal are **temperature, pressure** and **chemical sediment conditions**. Different appearances in the crystals of the same mineral can be caused by different crystallization conditions.

SINGLE CRYSTALS: Geometrical, acicular, capillary, thread shaped, flat blade, etc.

CRYSTAL GROUPS: Their shapes can be; geometrical, rod shaped, **wire shaped, radial**, spherical, **bunch of grapes, kidney shaped, half spherical, branched network shaped, microscopic small pieces, ellipsoid, stalactite, stalagmite, etc.** Three beams, which are called as "Coordinate", are used to identify a point in the space. The intersection point of these three beams in different directions is called as "**origin of coordinates**". For the crystals, these beams are called as "**crystallographic symmetry axis**"; their intersection point is called as "**center of symmetry**".

X, Y, Z-axes shown below:
α (Alpha), β (Beta), γ (Gamma) angles between these axes are indicative for the geometric shape of unit cell. "**Unit cell**" appears in a unit sphere. **The length of Symmetry axes** is the distance between the center of symmetry and the point where the axis pass through the unit sphere.

As in the spherical structure of atoms and molecules, geometric structure of the crystal unit cells, which are composed, of atoms and molecules, generally appears as a result of beams connecting the centers of atoms and molecules composing the unit cell, in 3D structure of Crystal Unit cells. As in the image of eight pieces of ¼ atom located on the corners of cube around an atom which is in the center of a crystal in the cubic structure shown under the image above. Natural image is the atom, molecule spheres in clusters.

2)   3 BASIC RULES OF CRYSTAL UNIT CELLS:

a)   ANGLE CONSTANCY: It is about the outer shape of the unit cell. The two plane angles between the crystal faces of unit cells of the same material are the same and constant.

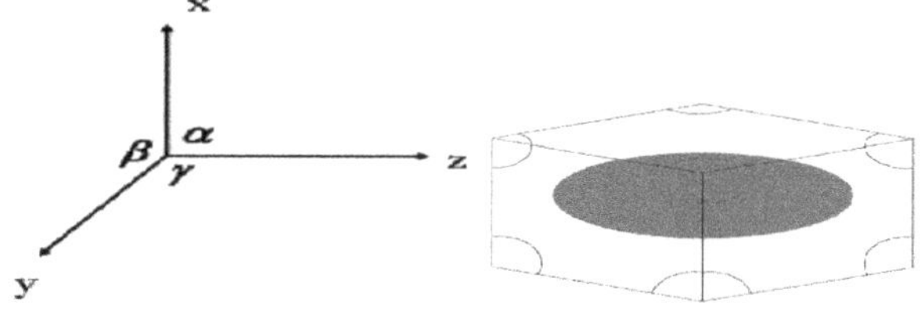

b) PARAMETER CONSTANCY: It is the length of **X, Y, Z** Crystallographic Axes whose drawings were given in the previous chapter. In other words, it is the distance of the points where various faces of unit cells intersect the Crystallographic Axes, to the center of symmetry. Also it determines the shape of the unit cell.

c) SYMMETRY CONSTANCY: The unit cell symmetry of a crystal determined according to three symmetry elements.

c1) CRYSTALOGRAPHIC SYMMETRY AXIS: They are the **X, Y, Z** beams whose drawings were shown in the previous chapter. **n**: Number of axes of a crystal in a plane. In case the **Crystallographic symmetry axis** is rotated around the unit cell with an angle equal to the value $360°/n$ (**n** equals to 2 or a bigger integer), it coincides with its first condition. These axes are assumed to be virtual, to determine the faces of crystal.

c2) PLANE OF SYMMETRY: It is the plane, which divides the unit cell of crystal making one half-similar to the other half's image on a plane mirror.

c3) CENTER OF SYMMETRY: There is always only one center in the unit cells of every crystal and it is the point where crystallographic symmetry axes intersect. **Unit cell** is symmetric according to the point of center of symmetry. It is the point on which all symmetry elements pass through. This point is always fixed while Symmetry works are performed.

c4) SYMMETRY PROCESSES: They are the processes, which make the unit cells of a crystal to match with its initial condition. The important ones are:

1- **Rotation around an axis**,
2- **Reflecting from a plane**,
3- **Inversion** (symmetry with respect to a point)

NOTE: Sometimes several of these processes can be used together.

3) CLASSIFICATION OF CRYSTALS:

a) 230 SPACE GROUP: It is performed according to the "**symmetry elements**" which are the reflection of physical properties (relations between the measurable quantities). According to the following **6 symmetry elements**, crystals are divided into "**230 Spaces Group**".

a1) CENTER OF SYMMETRY: Indicated above.

a2) PLANE OF SYMMETRY: Explained above.

a3) GLIDE PLANE: It is a process to be applied on the unit cells of crystal. It transforms every point of crystal into the location of its image on the mirror. Later, shifts in parallel to the mirror plane.

a4) n TIMES ROTATION AXIS: Provides a rotation of $2\pi/n$. It is the line, which makes the unit cell, intersects with the previous condition.

a5) n TIMES SCREW AXIS: Provides a rotation of $2\pi/n$ around a given axis and then shifts (feed) in parallel to the axis of rotation.

a6) n TIMES INVERTION AXIS: Provides a rotation of $2\pi/n$ and then inversion as per the given point on the axis (finding the point symmetric to that point).

7 CRYSTAL SYSTEM AS PER THE CRYSTALLOGRAPHIC AXES		
IN HEPTAD SYSTEM	IN 32 GROUP	SHAPES AND AXIS
I- TRICLINIC SYSTEM:NO-SYMMETRY SYSTEM: - PARALLEL EDGE-BASE OBLIQUE PRISM - IT HAS 3 AXES IN DIFFERENT LENGTHS INTERSECTING EACH OTHER WITH DIFFERENT ANGLES. BASIC UNIT CELL=(P) - $CuSO_4$, $CaSO_4$ etc.	C1-PEDION Ci-PINACOID 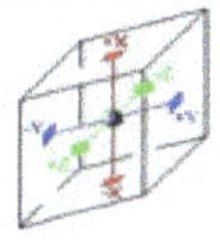 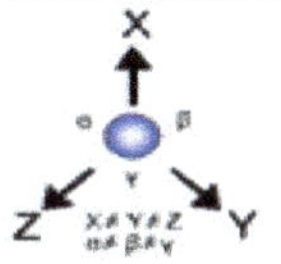	
II- MONOCLINIC (CLINORHOMBIC)(MONO) SYSTEM:: RHOMBOHEDRON-BASE OBLIQUE PRISM - ALPHA (α) ANGLE 90 °, BETA (β) AND GAMMA (γ) ANGLES ARE EQUAL TO EACH OTHER. IT HAS 3 AXES IN DIFFERENT LENGTHS. IT HAS 1 SYMMETRY PLANE. BASIC AND SURFACE CENTERED UNIT CELL =(F,P - OXALIC AND TATARIC ACID, Na_2SO_4 ($\bullet$)	Cs-DOMA C2-SPHENOID C2h-PRISM 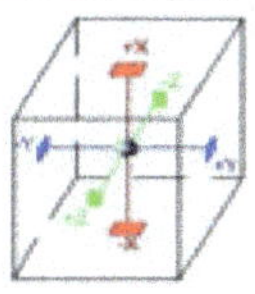 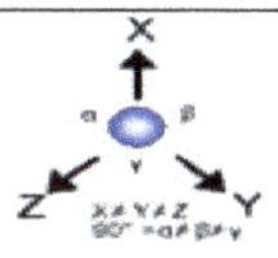	
III- ORTHORHOMBIC (RHOMBIC) SYSTEM: RHOMBOHEDRON-BASE RIGHT PRISM - IT HAS 3 PERPENDICULAR (90°) AXES IN DIFFERENT LENGTHS. IT HAS 3 SYMMETRY PLANES. SURFACE AND VOLUME CENTERED, BASIC UNIT CELL= (F,I,P) - $ZnSO_4$, KNO_3 and S (SULPHUR) etc. ($\bullet$)	V-PRISMATIC TRIANGULAR PYRAMID C2v-PRISMATIC QUADRILATERAL PYRAMID Vh-BIPYRAMIDE 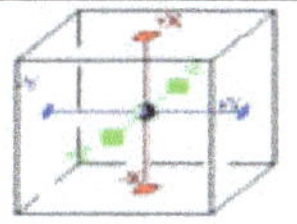 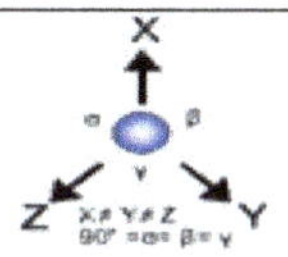	
IV- TRIGONAL (RHOMBOHEDRIC)(TRIPLE) SYSTEM: - EQUILATERAL PARALLELEPIPED - IT HAS 3 SYMMETRY PLANES, 4 AXES (3 OF THEM IN EQUAL LENGTHS INTERSECTING EACH OTHER WITH 60° ANGLE, 1 OF THEM PERPENDICULAR TO THE OTHERS AND IN A DIFFERENT LENGTH - BASIC UNIT CELL ($\blacktriangle$)	C3-TRIANGULAR PYRAMID C3v-DITRIANGULAR PYRAMID C3i-EQUILATERAL PRISM D3-TRIANGULAR TRAPEZOID D3a-TRIANGULAR SCALENOI $x + y = z = z'$ 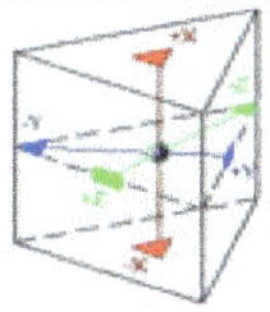 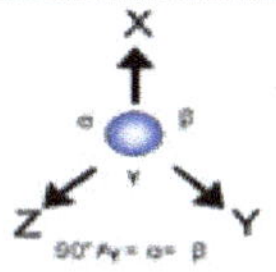	
V- TETRAGONAL (SQUARE) SYSTEM - SQUARE BASED PERPENDECULAR PRISM - IT HAS 3 AXES (2 OF THEM ARE PERPENDICULAR IN EQUAL LENGTHS, ONE OF THEM IN A DIFFERENT LENGTH. IT HAS 5 SYMMETRY PLANES. VOLUME CENTERED AND BASIC UNIT CELL=(I,P) $NiSO_4$, $2\text{-}SiO_4$ (zirconium) etc. (τ)	C4-SQUARE PYRAMID C4v-BISQUARE PYRAMID C4h-TOP AND BOTTOM BISQUARE PYRAMID D4-SQUARE TRAPEZOID S4-BISPHEROID TETRAHEDRON PYRAMID Vd-SCALENOID TETRAHEDRON PYRAMID D4h-TOP AND BOTTOM BITETRAHEDRON PYRAMID 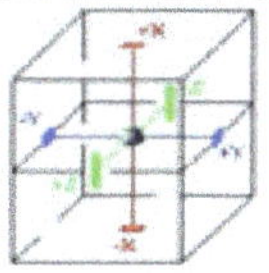 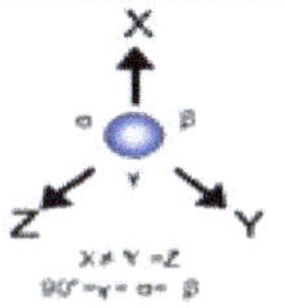	
VI- CUBIC (ISOMETRIC) SYSTEM - CUBE (REGULAR OCTAHEDRON) - IT HAS 9 SYMMETRY PLANES IN EQUAL LENGTHS PERPENDICULAR (90°) IN 3 AXES. - SURFACE VOLUME CENTERED AND BASIC UNIT CELL =(F,I,P) NaCl, Ki (POTASSIUM IODURE) ($\blacktriangle$)	T-REGULAR PENTAHEDRONAL DODECAHEDRON CUBE O-PENTAHEDRONAL TRISOCTAHEDRON CUBE Td-6 PCS TETRAHEDRONAL CUBE Th-12 TRAPEZOID TETRAHEDRONAL CUBE Oh-48 TRIANGULAR TETRAHEDRONAL CUBE 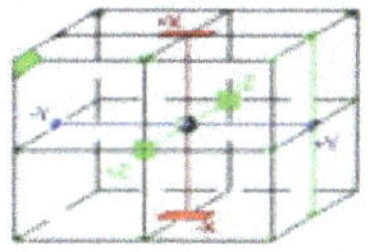 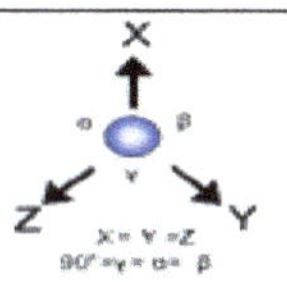	
VII- HEXAGONAL SYSTEM - HEXAGONAL BASED PERPENDECULAR PRISM - IT HAS 4 AXES (3 OF THEM IN EQUAL LENGTHS INTERSECTING EACH OTHER WITH 60° ANGLE, 1 OF THEM PERPENDICULAR TO THE OTHERS AND IN A DIFFERENT LENGTH. IT HAS 6 SYMMETRY PLANES.. - BASIC UNIT CELL = (P) SiO2 (QUARTZ), Bi (BISMUTH) etc. ($\bullet$)	C3h-TOP AND BOTTOM TRIANGULAR PYRAMID D3h-TOP AND BOTTOM BITRIANGULAR PYRAMID C6-HEXAGONAL PYRAMID C6h-TOP AND BOTTOM HEXAGONAL PYRAMID C6v-TOP AND BOTTOM BIHEXAGONAL PYRAMID D6-BIHEXAGONAL TRAPEZOID PYRAMID D6h-TOP AND BOTTOM BIHEXAGONAL PYRAMID 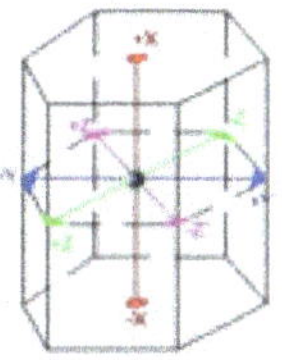 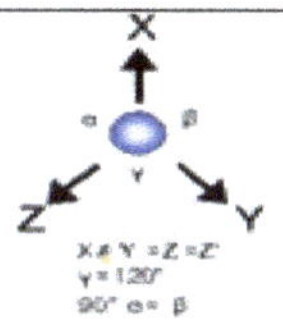	

    b) 32 POINTS GROUP: According to the following **4 symmetry elements,** (Macroscopic Symmetry Elements) crystals are divided into 32-point groups. **Macroscopic 4 symmetry element of the Crystals:**

        b1) Center of Symmetry: explained above.

        b2) Mirror Plane: Explained above.

        b3) 1, 2, 3, 4 or 6 times rotation axes: as indicated above.

        b4) 1, 2, 3, 4 or 6 times inversion axes: As indicated above. On one of the compounds of elements, which are explained above, there is an intersection point of all elements and this point remains constant for all of the symmetry processes. That is why the performed symmetry processes are called as "**Point processes**". Possible combinations of macroscopic symmetry elements are called as "**Point groups**".

c) 7 CRYSTALSSYSTEM: All of thecrystals**230 spaces group** and **32 points group,** all of them gather in "**7 Crystals System**" according to number of Crystallographic Symmetry Axes and their position. On the table above, while 7 crystals system is explained, **32 points group** crystals are listed next to every system accordingly.

32 CLASSES CRYSTAL SYMMETRY			
	STEREOGRAPHIC	INTERNATIONAL	
CENTER OF SYMMETRY	NONE	$\bar{1}$	
MIRROR PLANE	O	m	m= SYMMETRY PLANE
ROTATION SYMMETRY AXES			
1-FOLD ROTATION AXIS	NONE	1	
2-FOLDS ROTATION AXIS		2	
3-FOLDS ROTATION AXIS		3	
4-FOLDS ROTATION AXIS		4	
6-FOLDS ROTATION AXIS		6	
INVERSION SYMMETRY AXES			
1-FOLD ROTATION AXIS ≡ CENTER OF SYMMETRY	NONE	$\bar{1}$	3/m = 3 ROTATION AXES PERPENDICULAR TO THE CENTER OF SYMMETRY
2-FOLDS ROTATION AXIS ≡ MIRROR PLANE PERPENDICULAR TO THE AXIS	O	$\bar{2}$ (≡ m)	
3-FOLDS ROTATION AXIS ≡ 3-FOLDS ROTATION AXIS + CENTER OF SYMMETRY	▲	$\bar{3}$	
4-FOLDS ROTATION AXIS (WITH 1-FOLD ROTATION AXIS INSIDE)	◼	$\bar{4}$	
6-FOLDS ROTATION AXIS ≡ 3-FOLDS ROTATION AXIS + 1 PLANE PERPENDICULAR TO THAT AXIS	⬢	$\bar{6}$ (≡ 3/m)	

d) EXPRESSION OF CRYSTALS WITH INTERNATIONAL SIGNS:

On page 39-40, these crystals are introduced like: their international names in "**Schoenflies code**" including letters and numbers which indicates that crystal on the lower left corner of every crystal cell **in the table of crystals in 32 points group**; 3D geometric shape of the crystal unit cell on the left; information about the symmetry elements of 3D crystal shape, on the circles on the right by means of stereographic signs. Under every circle, there are international signs indicating the crystal. The Unit Sphere with the crystal unit cell in these circles indicates the equator plane. Projection of symmetry points of the unit cell under and on the sphere, are indicated on the sphere with the **Stereographic Signs**.

Crystallographic symmetry axes: They are the virtual beams passing through the unit cells of crystals. Such that when the crystal is rotated around this axis, it repeats itself two times or more in this full rotation. Symmetry axes can be of **1, 2, 3, 4, 6 rotations**. **1** rotation symmetry axis shows that there is no symmetry in the crystal. There are no symmetry axis of **5** rotations or rotations more than **6**.

When the unit cells of 32 points group crystals on the circles in the table of crystals of **32 points group**, are rotated **180°** to indicate their 3D body shapes, if it repeats its initial view, in other words if the plane, edges and angles come to the same places, this axis is a '**two rotations**'. If we take **n= number of rotations,** the first view repeats when a crystal is rotated **2/n** times around a 'n rotation' symmetry axis. As an example, any elements of a crystal with 4 rotation symmetry axes, repeat in every 90^0 when the crystal is rotated around this axis.

Symmetry Planes are indicated by the letter "**m**". The symmetry plane which is perpendicular to the symmetry axis is given with **2/m, 4/m. 2/m** shows the symmetry plane which is perpendicular (90^0) to the symmetry axis on the tables on page42, 43; **4/m** shows the symmetry plane which is perpendicular to the 4 rotations symmetry axis.

Altern Symmetry Axis: This axis is a combined symmetry element. Both **Rotation** and **Inversion** processes exist at the same time. (Rotation Inversion) If an **n** rotation Altern Symmetry Axis exist in a crystal, this crystal repeats itself by rotating **2/n** times around the axis and completing the inversion process as per the center on the axis.

Monoclinic, **Tetragonal**, **Cubic** and **Hexagonal systems:** When expressing with international signs, the first symbol shows "the main symmetry axis". As an example, 4 shows the "main symmetry axis" in the symbol **4mm**.

In the orthorhombic system: International signs give the symmetry elements between **X, Y and Z**-axes. As an example, "**mm2**" symbol indicates that vertical two symmetry planes are passing through **Z** and **Y**-axes; and **X**-axis is a 2-rotation symmetry axis, in the **Rhombic Pyramidal** class. This order is especially important for 230 spaces group crystals.

In the tetragonal system: Second symbol shows the symmetry elements passing through the axes; and third symbol shows the symmetry elements passing through the middle of angles. As an example, there is a 42m symmetry in "**Tetragonal scale no hedron**" class. Here, number **2** indicates the 2-rotation symmetry axis passing through **Z** Crystallographic axis; **m** shows the symmetry planes, which make 45°angle with the axes.

In the cubic system: Second symbol shows 3 rotations symmetry element; third symbol shows the 2 rotations symmetry element. Two rotation symmetry elements can be the Symmetry Axis or Symmetry Plane. We can give "**Hexakis tetrahedron**" in class **42m** (here 'm' indicates the symmetry plane) and "**Pentagon icosi tetraeder**" in class **4, 3, 2** (here two rotation symmetry axis) as an example.

In the hexagonal system: Second symbol shows the symmetry elements passing through the axes; and third symbol shows the symmetry elements passing through the middle of angles. In the **Dihexagonal Pyramidal** class, there are 3 vertical symmetry planes passing through **6mm** axes and there are 3 other vertical symmetry planes making 30°angle with them:

4/m= 1 mirror plane and 4 times rotation axes perpendicular to it.

4/mm= 1 mirror plane and the mirror plane which includes the 4 times rotation axes perpendicular to it.

4/mmm= 1 mirror plane and the mirror plane which includes the 4 times rotation axes perpendicular to it and a third mirror plane making 45^0angle with this mirror plane.

TABLE OF THE CRYSTALS IN 32 POINT GROUP

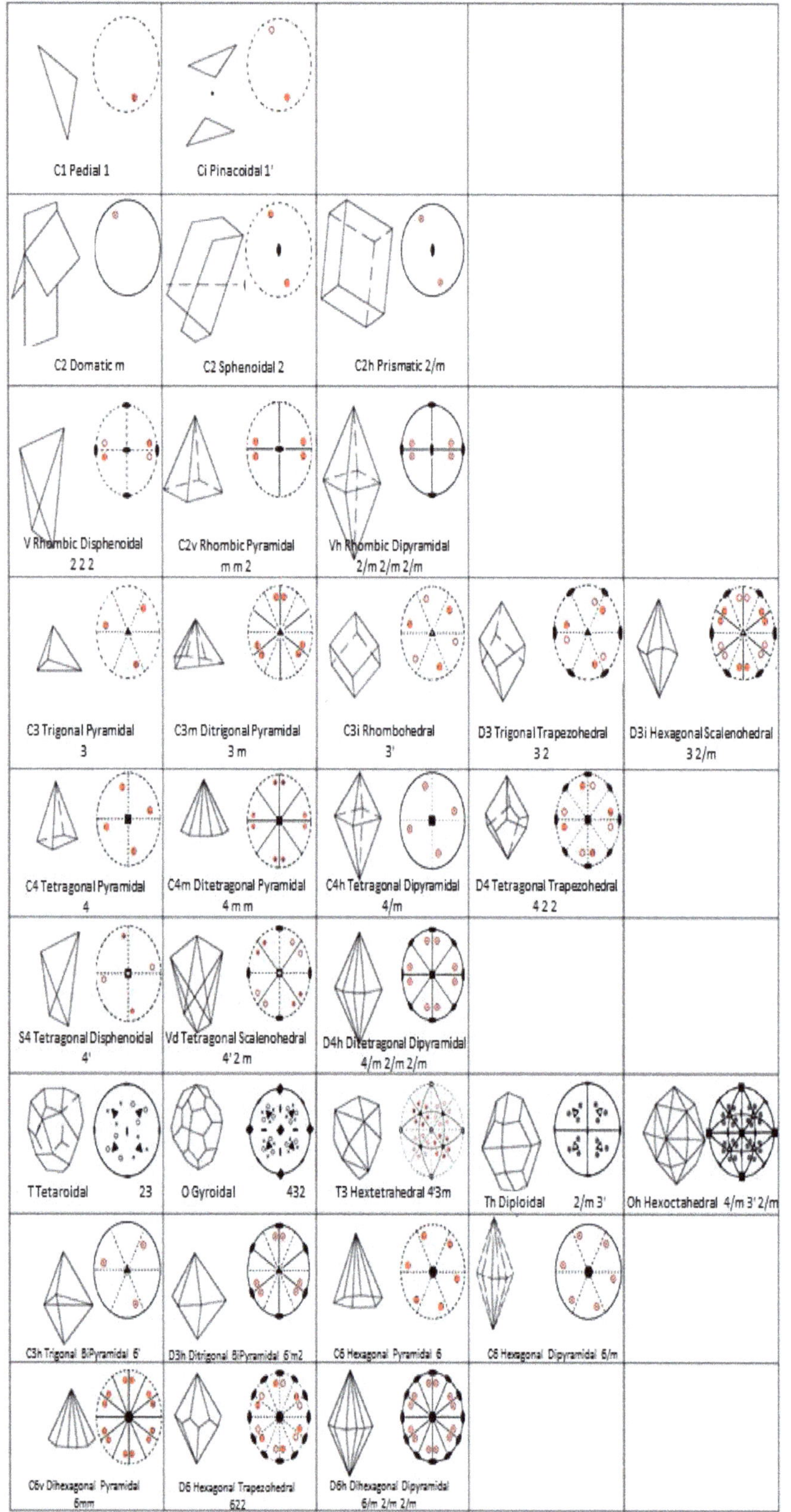

Class symbol		Number of
Intl.	Schoenflies Code	Independent C_{ij} and S_{ij}
1	C_1	21
1'	C_1 (S_2)	21
2	C_2	13
m	C_3 (C_{1h})	13
2/m	C_{2h}	13
222	D_2 (v)	9
mm2	C_{2v}	9
mmm	D_{2h} (v_h)	9
4	C_4	7
4'	S_4	7
4/m	C_{4h}	7
422	D_4	6
4/mm	C_{dv}	6
4'2m	D_{2d} (v_d)	6
4/mmm	D_{4h}	6
Class symbol		Number of
Intl.	Schoenflies Code	Independent C_{ij} and S_{ij}
3	C_3	7
3'	C_{3i}(S_6)	7
32	D_3	6
3m	C_{3v}	6
3'm	D_{3d}	6
6	C_6	5
6'	C_{3h}	5
6/m	C_{6h}	5
622	D_6	5
6/mm	C_{6v}	5
6'm2	D_{3h}	5
6/mmm	D_{6h}	5
23	T	3
m 3	T_h	3
432	O	3
4'3m	T_d	3
m 3 m	O_h	3

ADDITIONAL TABLES FOR THE ATOMICUNIVERSE:

ADDITIONAL FIGURES AND DRAWINGS FOR THE ATOMIC UNIVERSE IN PART ONE

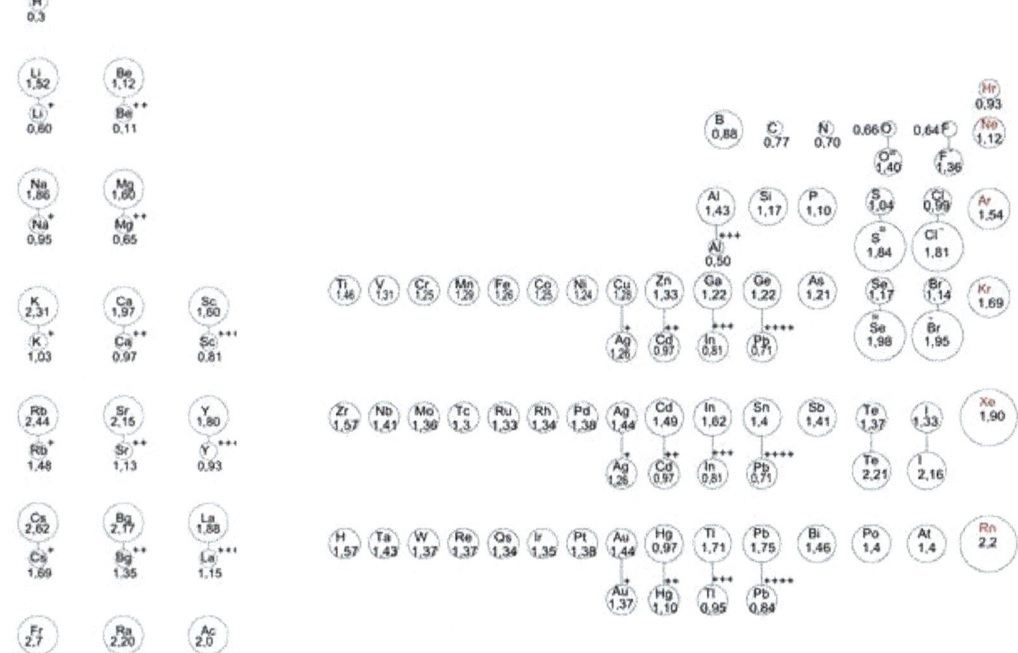

FIGURE 8 – CAMPBELL'S PERIODIC TABLE OF ELEMENTS IN ACCORDANCE WITH ATOMIC AND IONIC AXES [ATOMIC AXES ARE DETERMINED AS PER THE COVALENT BOND (BONDS BETWEEN ELECTRONS) DISTANCES]. PERIODIC TABLE OF ELEMENTS IS PROPORTIONAL WITH ORBIT RADII AS PER BOHR'S LAW.

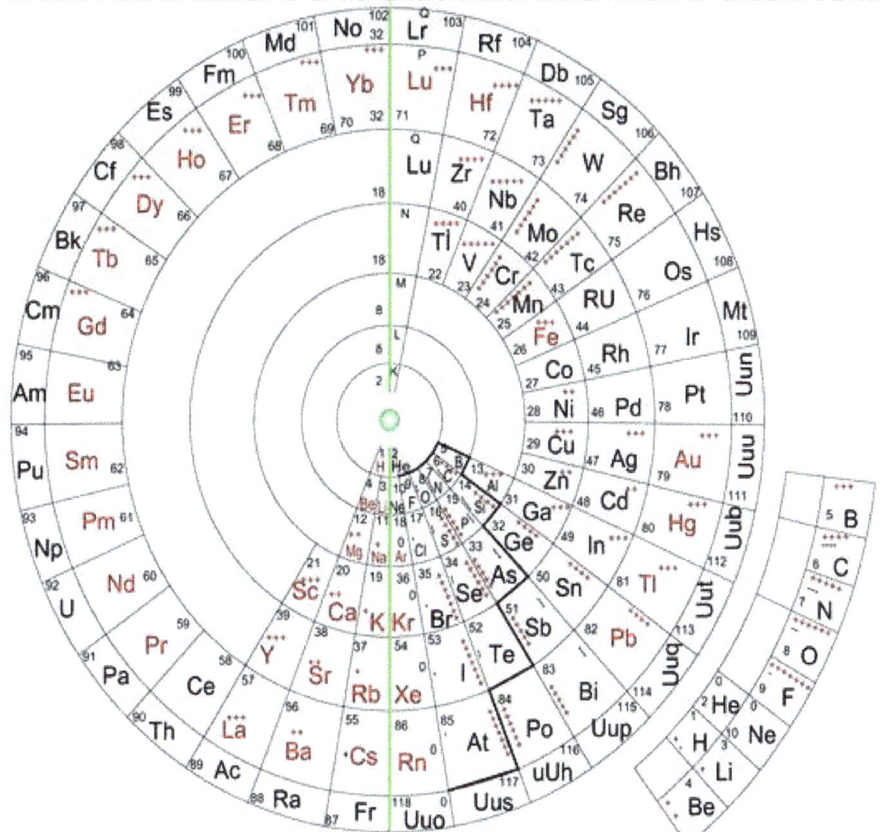

FIGURE 9 – PERIODIC TABLE OF ELEMENTS SHOWN IN NESTED CIRCLES

C) LIVING THINGS:

In the 1/10 nanometers scale (10^{-7} **mm**) structure of non-living or living thing, it is known that there are electrons rotating with a speed of 50000 km/sec. around the nucleus in the middle of atoms of the material. This structure looks like the **planets** rotating around stars in the space. Rotation speed of World around theSunis**29, 76 km/sec.** In other words, it is very slow. Because of the system rotating around each other in the astral and atomic scale, now, gas, liquid and solid materials seem to us as they were stable and fixed on the world. Below are the developments from the formation of Universe by means of Big Bang to the creatures and us, the Humans: (Annex-Ie)

13,7 Billion years ago, the **single, dual** and **triple** particles had been formed respectively from the "**Sub-atomic Particles**" in the universe after the Big Bang. **300-700 thousand years** after Big Bang; since the universe has cooled enough, **Helium nuclei**, which is composed of 1 Proton and 1 Neutron particle, connected with **electrons** to form the first atoms, in other words, "**Hydrogen**" atoms of the universe. Universe has become a gas cluster composed of Hydrogen (**Nebula**).

-**Galaxies1 billion years after the** Big Bang

-**Heavy elements** (atoms) emanated **2 billion years after the** Big Bang in the galaxies.

-**Sun, Planets and World** emanated **9,2 billion years after the** Big Bang in our Milky Way galaxy.

I)   SINGLE CELLED CREATURES:

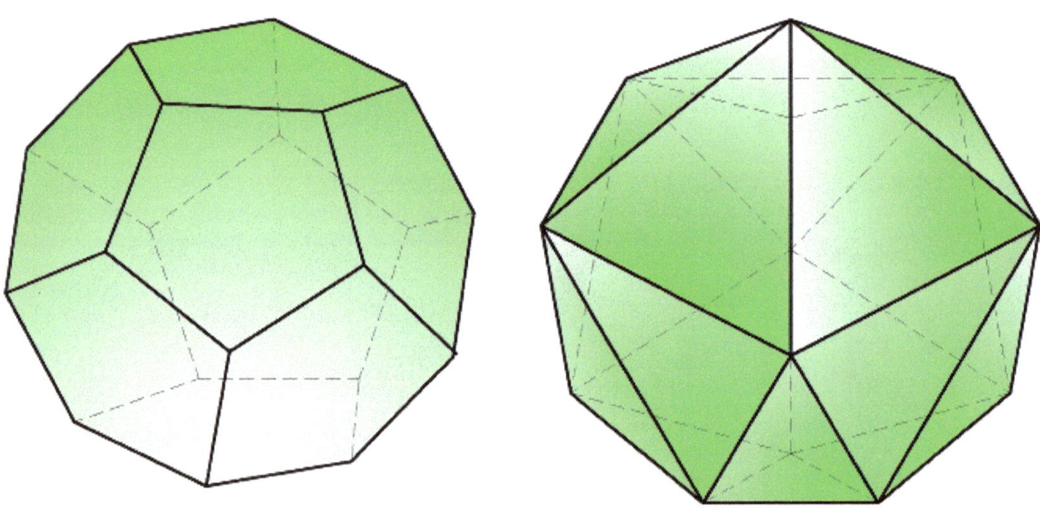

DODECAHEDRON POLIO VIRUS   ICOSAHEDRON POLIO VIRUS

Approximately 12 billion years after Big Bang, World cooled down sufficiently; and on this cooled shell (As thin as the apple skin, in hot lava form), the first living creature (**single celled algae)** emanated in South East Africa. But previously, Inorganic compounds (**water, minerals, acid, base, salt, carbon**) and organic compounds (**carbohydrates, fats, proteins, vitamins, nucleic acids**) which are the main compounds of creatures emanated. Like the sub-atomic particles form the atoms, the first **single celled creatures** emanated on the waters including the main components which were formed previously by means of the electric energy of storms. Today, single celled creatures can be formed in the laboratories by means of giving electricity to the glass container, which includes these compounds. Size of the cell is 10-15 microns. (10^{-3} **mm= 1 micron** ~ 10^{-5-6} **mr**).

II)   MULTI CELLED CREATURES:

STARFISH

FLORAL DIAGRAM

On the **page 19**, it is very interesting that the virus in the **"Dimensions from Subatomic Particles to the Virus and Cell"** table, is in the same structure with the icosahedron of Plato's regular penta hedron in Part 4 of the book (Also the other examples which are given above). In the cell, the above listed organelles, which are composed of inorganic and organic compounds, perform digestion, respiratory, circulation, excretion and reproductive functions of the cell. Thus starting with **single celled algae**, the living life has transformed to multi-celled creatures by the chemical and biological evolution of DNA in the cell nucleus. Multi celled plants and animals emanated first in the sea and later on the earth. At the end, **approximately 60-100 thousand years ago, "Homo Sapiens", the first thinking and speaking** humans, which we are also a member of the group, were created.

Today, in the 21st century, universe provides us for the first time the comprehension of not only the universe we live in, but maybe of the previous and future universes in present-day conditions. But is it only us who have this consciousness in the universe? We do not know that, yet.

Rhythm and **musicality**, which are valid for **sub-atomic particles, atoms, molecules, compounds and astral particles,** are also valid for the **creatures**. **Biorhythm** (cycle) is the period of time between the start and end of a biological incident. These biological rhythms provide the existence of creatures. Coherence of sleeping at night, living in the daytime, heartbeats, etc. **rhythms** (in musical nomination, '**harmony**') indicates a **healthy life; discordance** indicates an illness (Menstruations of women and respiration rate are called as the **biological and physiological clock**). Rhythms depend on the DNA of creatures, in other words, genetic reasons.

As the Sun has "**solar**" and Moon has "**lunar**" rhythms in the astral world; rotation times of the cycles of planets around the Sun with each other, form a "**rhythm**" (cycle).

For the last words, let's make some explanations on different perceptions of various creatures about color as per the Relativity Theory of Einstein, and then we will drop the subject:

Horses see the sky in grey, not blue. Since **bees** are able to see UV rays, they can see many different colors. **Alligators, dogs and field mice** see everything in black and white. Besides, animals are said to perceive the shapes different from humans.

Positive precepts and educations had always been provided by means of religions to the humanity. Today, unfortunately the variety of religions, alterations, and sometimes a cause of conflicts between humans, immature science that cannot explain the whole universe yet; make some of the humans see the world in black and white. If religions and science cannot produce a **"universal science"** or **"Cosmology"** which can be explained by the mysticism understanding (Kesrette vahdet= Unity in Plurality) of humans, nobody can blame humans for behaving like alligators. It will be a pity for the alligators, as well as their victims. Let's put an end to this subject to be continued in the "**Theory of Everything**" at the end of the book.

PART TWO HUMANS

Homo sapiens, in other words, the first thinking and talking humans, which also includes us, emanated in Central Asia approximately 50-100 thousand years ago, in the last ice age. The eyes of today's **humans** are able to sense the waves with the frequencies between **390-780 000 Hertz** (number of waves per second); and our ears can sense the sounds with frequencies between **20-20 000 hertz**. Cats can hear the sounds with a frequency of 40 000 Hertz; and the dogs can hear the sounds with a frequency of 80 000 hertz. Our sense of taste occurs by means of the nerves chemically on our tongue; sense of smell and touch occur also by means of the nerves. As a result, all of these senses are the effects of **electro/magnetic force**, which is one of the 4 fundamental forces of our universe and atoms. In fact, this is the force causing the expansion of our universe, which still expands.

A)   HISTORICAL EVOLUTION OF HUMANS:

50 Million years ago	Paleozoic (Old animal) age.
30 Million years ago	A man-like chimp in the Fayoum Oasis in Cairo.
10 Million years ago	A human skeleton in Thessaloniki.
8 Million years ago	A skull in China.
3 Million years ago	Australopithecus in the Central Africa (Brain volume: 500 cc).
3 Million years ago	"Lucy" - A human skeleton in South East Africa. I. Human type: Ramapithecus. A skull in the Central Asia "Consul"
2 Million years ago	Homo Habilis in the Central South Africa.
1 Million years ago	Homo Erectus. Brain volume 900 cc. II. Human type.
1 Million years ago	The oldest ice age started.
600 Thousand years ago	The oldest ice age ended.
600-540Thousand years ago	I. ice age. The lake dried up in the Nile Valley.
600-150 Thousand years ago	Heidelberger Man in Europe.
540-480Thousand years ago	Hot age.
500Thousand years ago	Pithecanthropus Erectus - Jawa Man is in Jawa Island
480-430Thousand years ago	II. Ice age.
400-240Thousand years ago	Hot age.
370-240Thousand years ago	II. Middle ice age
350 Thousand years ago	Discovery of fire near Beijing.
250 Thousand years ago	Human traces in a cave near Manisa city, Anatolia.
240-180 Thousand years ago	III. Ice age.
180-120Thousand years ago	Hot age
150 Thousand years ago	Neanderthal man is in Asia, Europe and Africa. III. Human type.
120-12Thousand years ago	IV. Ice age.
60- 12Thousand years ago	The last middle ice age.
40Thousand years ago	Today's Homo Sapiens – the first thinking and talking human type in Central Asia. In the last middle ice age, ancestors of a tribe, which still survives today in, the South Africa, came to Australia and Central Asia through Egypt, Arabia, and India and they formed the Homo Sapiens. This truth was proved scientifically by means of DNA tests.
30 Thousand years ago	Migration from Europe to the South because of the cool winds from the North.
16Thousand years ago	Migration from Asia to North America through Bering Strait.
15Thousand years ago	Migration from Asia to the surroundings of Mediterranean.
12Thousand years ago	The last middle ice age ended. In the Northern hemisphere, ice thawed from Himalayas to Alps in Europe, from Central America to the Anatolia. The big lake dried up in the Central Asia. (Noah's ark, after the great flood in Anatolia, near Urfa city, Göbeklitepe and the first temple of the world)
9 Thousand years ago	Alpines are in Europe. "Çayönu Tepesi" village in Ergani town of Diyarbakir in the Southeast Anatolia. "Jericho" island in Israel. "Çatalhöyük" village in the vicinity of Çumra town of Konya in Central Anatolia.
8 Thousand years ago	Indian-Europeans who came from Central Asia are coming together in the Middle East and Northern Europe. "Cayonu Tepesi" village in the vicinity of Diyarbakir in the Eastern Anatolia.

7 Thousand years ago	**Dravidians (relatives of Sumerians) in the "Anav" cairns in the vicinity of Ashgabat in the East of Caspian Sea and in the "Mohenjo-Daro" and "Harappa" cities near the Indus river; in Asia, "Minusinsk Cult"; Also, Subaryans (Hurris) (relatives of Sumerians) in Eastern Anatolia, Mesopotamia and Iran.**
4500-2000 B.C.	Ural Altaic **Sumerians** and Elamites (Gutis) came to Iran.
4000-2200 B.C.	Sumerians and Elamites (Gutis) invented writing in Mesopotamia.
4000-2200 B.C.	Semites came from Arabia to the upper Mesopotamia and founded **Babylon** city.
4000-3500 B.C.	Japanese people coming from Asia settled in the Japanese Islands.
4000-2750 B.C.	Acadians in Aegean islands
4000-1200 B.C.	Cretans are in Crete
4000-525 B.C.	Egyptians are in Egypt
4000-3000 B.C.	Elamites and Gutis are in Iran. Hittites, Lelegs, Kars came to Anatolia from the North. Pelasks are in Europe and Anatolia. Also Pelasks, Lelegs, Kars are in Greece. Semites came from Arabia to Anatolia. Parts and Seleucids are in Iran.
3000-2000 B.C.	Subaryans (Hurris) are in Iran and Anatolia.
3000-1800 B.C.	Babylonians are in Babylonia. Hittites, Lelegs and Kars are in Europe and Anatolia
3000-1200 B.C.	Palestinians and Aram people are in Asia Minor. Italics came from North to Italy.
3000-2800 B.C.	"Urgakina", the King of Sumerians, collected the first known laws of the world together. Then he wrote that he put an end to the slavery, and he brought the liberty.
2600-640 B.C.	Elamites, (Guttis) are in Asia Minor.
2000-612 B.C.	Hammurabi the King of Babylon, made his own laws written down. Sumerians and Semites mixed together, Assyrians appeared in Asia Minor. Subaryans (Hurris) came to Syria and Mesopotamia from Anatolia. Greeks came to Greece from the North. Arameans who are the Semites living in the deserts of Syria, invaded Babylon and moved the capital to Damascus. They were previously called as "Ahlamu". Persians are in Iran.
2000-VII. Century B.C.	Phoenicians are in Asia Minor.
2000-600 B.C.	Veda religion is in Asia.
2000 -VIII. century	Hittites are in Anatolia around Kızılırmak river.
2000 – IV. century	Persians are in Iran
2000 B.C.	Greeks are in Greece.
2000 B.C.	Indian European Aryans are in North West India.
2000 B.C.	Celts (Cymry) of Indian European Aryans are in Europe.
2000 B.C.	Indian European Aryans and Vedics are in India; Khorezm, Sogds (Massaget), and Alans (Ases) are between Caspian Sea and Aral Sea. They dominated Indus Valley and Mohenjo-Daro, Harappa cities. Later, Alans and Sogds joined Saka Confederation composed of Mongolian Mancu and Proto Turks, which were the other tribes in the Altaic area. This Confederation was composed of three groups:
2000 B.C.	1) Fergana-Qashqai Altaic region, 2) Between Aral Sea and Caspian Sea, 3) Today's South Russia, This Confederation has survived up to Alexander the Great, in 323 B.C.
1800 B.C. -1600 B.C.	Hixoses (mixture of Hemites and Semites) came from Syria and invaded Egypt.
1800 B.C. – 1200 B.C.	Hittites are in Anatolia. Achaeans (India Germans) are in Europe. Hurrians (Sub-Arians) came from Anatolia to the Northern Mesopotamia to found Mitanni Kingdom.
1700 B.C. -522 B.C.	**Sogds (Massagets) are in Asia.**
1680 B.C. -1160 B.C.	Kassite are in Asia Minor.
1600-XIII. Century B.C.	Hebrews are in Asia Minor.
1600 B.C. -1200 B.C.	Ural Altaics are in Asia. Slavs are in Europe and Greece.
1200 B.C. -IV. Century B.C.	**Scythians (Saka) are in Asia Minor.**
1200 B.C. -IV. Century B.C.	Sea people came to Egypt from Mediterranean Sea.
1200 B. C-VIII. century B. C	**Urartians are in Anatolia. Aztecs are in America.**

1200 B.C. -676 B.C.	Phrygians (Phrygian means Venus) are in Northwest Anatolia.
1200 B.C. -546 B.C.	Lydians are in the West of Anatolia.
1200 B.C.	Phrygians came to the Northwest of Anatolia from the North. Dors (Slavs) came to Greece from the North. Italics came to Italy from the North.
1000 B.C. **1000 B.C. -295 B.C.** 1000 B.C. -245 B.C. **1000 B.C. - 20 A.D.** 1000B. C–VI. Century B.C. 1059 B.C. -249 B.C.	Germens came to the Central Europe from the East. **Etruscans (Tuscans) are in Italy and Asia Minor.** Huns are in Asia and Europe. **Mayas are in Mexico. Australians are in Australia. Indian Sioux Nation are in America.** Chinese are in China. Tibetans are in Tibet. Proto-Turks are in China
1059 B.C. -900 B.C. 1059 B.C. -550 B.C.	Ural Altaic Urartians came to the East of Anatolia. Van has been the capital city. Medes are in Iran. Persians are in Iran and Anatolia.
700 B.C.	Lydian Kingdom in the West of Anatolia. Sardis is the capital city. Cimmerians Scythians came to Anatolia from the North.
546 B.C. 500 B.C.	Coming from Iran, Persian King Darius defeated the Lydian King Kroisos who has invented the first gold coin. Lydian Kingdom has come to an end. Cymry (Celts) are in Europe.
476 B.C.	Attila and Huns came to Italy from the North. Roman Empire was demolished.
334 B.C. IV. Century B.C. -106 A.D.	Greeks came to Anatolia with Alexander. Nabateans are in the Northwest Arabia.
200 B.C. -600 A.D.	**Vikings are in Northwest Europe.**
190 B.C.	Romans came to Anatolia.
II -IX centuries A.D.	Celts (Cymry, Avars) are in Central Europe.
III. Century A.D.	India German Visigoths are in South-west Russia.
III. Century A.D.	Ostrogoths are in Southeast Russia.
III. Century A.D.	Anglo-Saxons and Francs are in Northern France.
III. Century A.D. 552 A.D. - 659 A.D. VIII. Century A.D.	Vandals are in Hungary and Spain. Gök Turks are in the Central Asia. Yakut Turks are in the Asia Minor. Karluks are in Central Asia.
-IX. Century A.D. X. -VI. Centuries A.D.	Hungarians are in Hungary Oghuz Turks are in Asia Minor and Anatolia.
1071 A.D. 1206 A.D. -1226 A.D. **XIII. Century A.D.** 1299 A.D. -1920 A.D.	Seljuks came to Anatolia from Central Asia. Mongols are in Asia and Anatolia. **Incas are in the Central America.** Ottoman Empire is in Anatolia, Hungary, Arabia and Northern Africa.
1977 A.D.	Population of the world is 4, 6Billions.
2020 A.D.	Population of the world will be 8 Billions.
2620 A.D.	Population of the world will be12 persons per meter-square. (If calculated correctly)

The first thinking and talking humans "**Homo Sapiens**", in other words today's humans hunted in Central Asia, Asia Minor, in the caves on the slopes and riversides, using the weapons, which they invented of hewn stones, animal bones and horns. They survived by eating the wild plants. Later, they developed weapons like arrow, bows, swords and spears. While taking the agricultural activities from the cave environment to the riversides, they passed to the communal life in the tents made of animal skins and adobe houses. In the meanwhile, they succeeded in lighting a fire, building ox-driven carts, tilling the soil by primitive ploughs, raising grains and grinding the grains by stones to make bread. Humans built villages and cities near the streams and started to breed by the sheltering and feeding opportunities of rich natural environment.

Approximately 10.000 B.C.: In the great flood, before the ice thawed in the Northern hemisphere and the seas in the world raised about 80 meters.

14. 000 B.C.: Migrations from Asia to North America through Bering Strait; and from South, to South America through Philippine Islands. Because it might be possible to pass overland through the Philippines region before the seas in the world raised about 80 meters. Eskimos in the North America and Indians coming from Mongolians proves these migrations. Also after the Spanish people invaded the Central America, there were hundreds of Turkish words in the dictionaries, which Spanish priests prepared to teach Christianity to the native population. It is another example of that: like Europeans, the old America people were also originated from Central Asia. According to one of the researches of National Geographic Magazine; common ancestors of modern humans, were from a tribe which has migrated from Africa to Central Asia and which today still lives in South Africa. This was proved by the DNA tests.

SOME ADDITIONAL PICTURES AND DRAWINGS ABOUT THE HISTORICALEVOLUTION OF HUMANS:

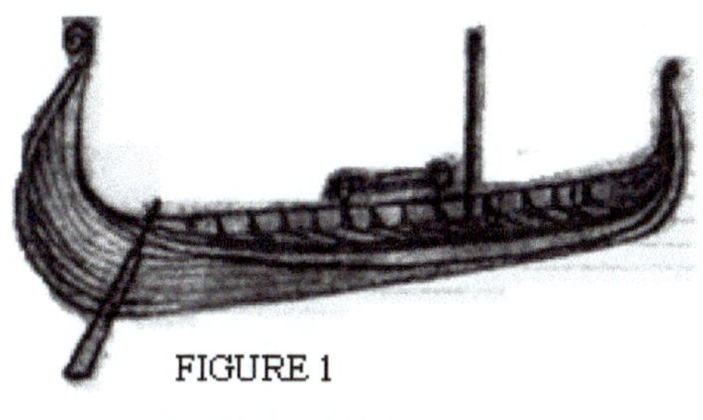

FIGURE 1

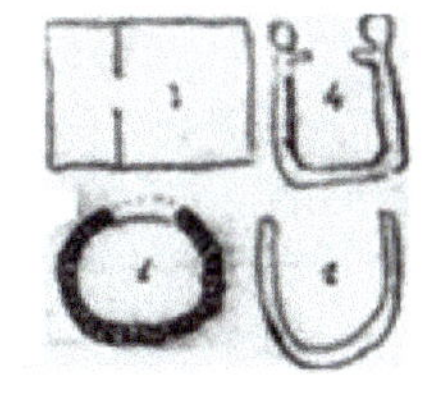

FIGURE 2

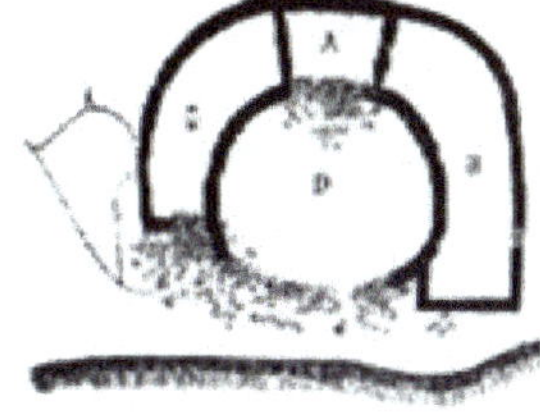

FIGURE 3: THE WHEELED TENT OF DJENGIS KHAN

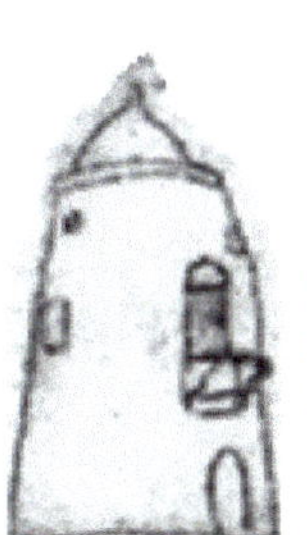

FIGURE 4: ANCIENT INDUS HOUSE IN SINAI DESERT

B)   EVOLUTION OF HUMAN LANGUAGE:

LANGUAGE: It is the system of words, which the humans use to express the things they hears and think. It is the physical form of thought and it exists by means of words. All of the languages determine, store and transfer the knowledge of humans to the new generations. Therefore, as a common language is necessary for the development of a nation; a common language is also needed for the development of humans. As the rhythm of sounds creates harmony, in other words, it creates the music; common language will represent the rhythm and harmony between the humans in the world, it will create the mathematics and thus it will be able to start a new era of the common symphony of humans.

Expression of thoughts of humans by means of sounds, created the languages; and expression of thoughts of humans by means of figures, created the writing. As in mathematics, expression of things by means of sounds creates the numbers; and expression of things by means of figures creates the geometry. Thus, knowledge of humans were managed to transfer from generation to the generation first verbally, then in written. Today, **Computers** widened and globalized the development and accessibility of knowledge, worldwide.

Languages of the world are separated into three groups according to their word structures:

1- Differential Languages: Words are syllable and they do not vary in the sentence. Their single syllables create the difference in meanings by expressing indifferent tones. (Chinese, Tibetan, Siamese)

2- Agglutinative Languages: Meanings and relations vary by adjunctions in the unchanging word roots. (Turkish, Mongolian, Japanese, Hungarian). Turkish is a uniquely regular language. The logical structure of Turkish language is so strong that one feels that this language is planned and developed by a group of mathematicians. If it is really so, then its source could be sought in the civilization where Sogd, that made also the Iranian literature and Persian unique and Uighurs, who were told to be the last empire on the continent of Mu according to the Internet were together. This Sogd and Uyghur civilization could have been the source of the Sanskrit, extending to the Vedas and the Brahman priests. Salute the Brahmans and Tibetan Monks of today.

3- Inflected Languages: For the inflection and production of words, adjunctions can be used**, and the word roots can vary**. Arabic, Persian and German languages may owe their advanced level to the **Sogd and Uighur civilizations**, where they were together in Asia, like Persian is based on the Sogds.

However, in the above classification, some of the languages in the same group may be non-relatives. None of the languages of today, except Turkish, is scientifically perfect. Among the known languages, in India, "**Sanskrit**" which has been prepared and only been used by the Brahman priests is the only perfect language. Besides, in the twentieth century, some studies were performed for a scientific and practical "**World Language**". "**Esperanto Language**" in the old Yugoslavia is an example. It is very shameful that, today the humans could not prepare a scientific language, which Brahman priests could prepare approximately three thousand years ago. Esperanto language was firstly published by Dr. Louis-Lazare Zamenhof in 1887. Likewise, Descartes, Montesquieu, Leibnitz, Voltaire and Condorcet has also conducted Universal Language studies previously. German priest J. Schleyer published "Volapuk" language in 1880.

If a new scientific world language research is to be carried out on the basis of the expression of human thought in both forms and geometry, it can be said that the new language should be based on a formal, semantic foundation and geometry, by analyzing the isolating, agglutinating, fusional languages and perhaps the golden proportions as in the case of the Arabic script. At least, English, the most widely used language in the world today, should be freed of contradictions, the articles and perfected like the examples of Turkish and Sanskrit or the search for a global language that will integrate all of the languages of the world and easy for everyone should be started urgently. Because also linguistics is an intersection and meeting point between human like mathematics..

Expression of thoughts of humans are separated into three:

1) **Verbal language**: Various languages.

2) **Body language**: Behavior properties, expression of mysticism: "**Status**".

3) **Music language**: Universal common language

Only a World Language, constituting the common symphony of the universe and having a musicality can conciliate the communities of the world.

C) EVOLUTION OF RELIGIONS:

I) COLLECTING, HUNTING and WITCHING:

28. 000 B.C. Arrow, bow and bird pictures, witching activities about hunting in the caves of Ergani, one of the districts of Diyarbakir city in Anatolia and Mexico.

23. 000 B.C. – 20.000 B.C. Two aurochs figures of 60 cm long, and footprints of dancing man and woman in "Tuc. D'audoubert" caves of Ariegnée, France.

12 000 B.C. -5. 500 B.C. Family life; arrow, bow and tools made of bones; Mother goddess, deer, aurochs, horse and mammoth pictures in the caves of La Madeleine, France.

II) AGRICULTURE, FIRST VILLAGES AND MOTHER GODDESS:

10.000 B.C.: The oldest settlement and the oldest temple of the world in Göbeklitepe near Urfa city in Southeast Anatolia. Humans of the Biblical Flood.

7000 B.C.: In the Central Anatolia, near Çumra district of Konya city, "Çatal Höyük" settlement, the "Mother goddess" which gives its name to Anatolia.

6000 B.C.: We can see the mother goddess up to the years before Jesus; in Southeast Anatolia, "Çayönü Tepesi" Village in Ergani district of Diyarbakir city**;** also in Jerusalem (**Jerico**) city of Israel;

Crete Island; Europe, Egypt and Iran; "**Mohenjo-Daro and Harappa**" cities in the Indus Valley of India; "**Anav**" cairns in the Central Asia; in many parts of Anatolia, for example in **Troya**.

III) POLYTHEISTIC RELIGIONS:

~20 000 B.C. ? MU LOST CONTINENT (P-193)
~ 15 000 B.C.? ATLANTIS CONTINENT (P-193)
~ 10 000 B.C.- 1500 A.D. MAYAS are in Central America. Jade stone is holy.
~ 9 500 B.C. - 14-15 Century INKAS are in South America.
6-7 000 B.C. ANNAV CULT In Central Asia, in present-day Turkmenistan **(P-164)**. **ANNAV CULTURE (OXUS, OK-UZ)** relative to Göbeklitepe, Scythian and Uighurs. The name ANNAV comes from the sky god AN.

SUMERIAN WATER GOD – "ENKI" 4000 B.C.

4 000 B.C. -3 000 B.C. SUMERIANS: According to the "Epic of Gilgamesh" of Sumerians which have come to Mesopotamia through Indus Valley from **Annav Territory**; **and the creation mythos "Enuma Elish"** (In the sky); first of all there was **Water (Apsu)**. The word "**Sumer**" means "**Water Mountain**". They were perceiving the sky as the lights shining in the water. **Sky God An (Ansar)** and **Soil Goddess Gaia** have been born from the **Water Goddess Nun** (Today, 'Nene' means 'grandmother' in Turkish**)**; and **God Enlil** has been born from their relation. ENLIL has created the **Moon Goddess NANA** to enlighten the sky; and Nana has created **Sun God Utu** and **Goddess of Love and War Inanna**. Nanna's husband **Dumuzi** has represented the "Fertility of Man" **and he** has been the **Shepherd God**. Sweet waters god **Enki** has been born from Mother-Earth **Ki**. Living in the depths, **Enki** is the god of art and wisdom who has protected the humans from extinction in the great flood (**EA** in Semites, **Thalia** in Altaics). Priest Kings of Sumerians were called **Ensi** or **Patesi**. **Kur** is the "Netherworld". **Dilmun:** god of heaven; **Adat:** god of storm and rain; **Antum:** wife of Sky God**; Belit-Sheri:** court clerk of underground lawyers; are important characters of crowded Sumerian mythology family. As the Sumerian gods have later been used by **Egypt, Anatolian, Greek, Etruscan, Roman** religions in different names; they show similarities with the **European and Indian European** religions. As an example; Gilgamesh which has been the divine king in the **Epic of Gilgamesh (Gilgamesh** meaning knowledgeable, holy, noble valiant), was transformed into **Herakles** in Greek; **Sandan** in Tarsus (Cilicia) and Ibreez in Anatolia; **Rustem** in Iran; **Samson** in **Hebrews**.

4. 000 B.C.: Asian people have come to **Egypt** through Syria and Palestine. The people before Semites used to call South Egypt as **Kemet** (black land) (Kem means 'bad' in Persian); and North Egypt as **Desert** (Red land). The first settlers of Egypt have established the small city-state named **Khassep** ['Kasaba' (means small city in Arabic)]. Later Greeks called them as **Nom**. In delta area in the North, 20 cities had united. The falcon shaped one was the good Sun God, in other words, **Horus**. 23 cities had united in the South Nile Valley. The scorpion shaped one was the bad Sun God, in other words, **Thebes**. In Egypt, the Gods, which were born from Nun - the Water God of Sumerians, really exist as they are. This condition has developed in different cities of Egypt in triple Gods as follows:

GODS	THE FIRST THREE	IN NORTHERN EGYPT		IN SOUTHERN EGYPT	
	ALONG THE NILE	IN ABYDOS	IN MEMPHIS	IN AMBOS	IN TEB
SKY	GODDESS NUT	GOD OSIRIS (FATHER)	GOD PTAH (FATHER)		GOD AMON
EARTH	GOD GEB		GODDESS SEKEMET (MOTHER)		GODDESS MUT
SUN	GOD RA	GOD HORUS (SON)	GOD. TMOTEP (SON)	SET (BAD)	
MOON		GODDESS ISIS (MOTHER)			HONSHU

2000 B.C.: The Gods of Arameans (Armenians) of Semites living in **Syria and Mesopotamia** were as follows: **Simegi** was the Sky God, **Hepa** was the Earth God, **Teshup** was the Storm God, **Kusuh** was the Moon God, **Namu** and **Hazzi** were the Mountain Gods.

2000 B.C. -800 B.C.: HİTTİTES: In the polytheistic religion of **Hittites, Utu** (İştanuş, Arinna) was the Earth God; **Hepat** was the Sun God; **Teshup** was the Storm God; **Aruna** was the Sea God. Also Hittites worshipped the gods of other communities.

2000 B.C. -800 B.C. VEDAISM: (**Veda**=Hearsay knowledge) It is the religion of Indo-Europeans coming from North to invade India. They were collected in 4 books after the invention of writing:

1-**Rigveda** (Knowledge on praying and wind).

2-**Samaveda** (Knowledge on religious ceremonies).

3-**Yajurveda** (Knowledge on witching).

4-**Athervan Veda** (Knowledge on Medicine and Witching).

Among the numerous gods of this religion:

a. **Indra** (God of night sky, war, storm, lightning).

b. **Mitra** (God of sunny daylight sky, justice and light).

c. **Varuna** (God of daylight sky, mind and universal order).

Being the mother of all these three, Goddess **Aditi** is the common essence of all universal beings. Besides; "**Dyaus Pitar**" is "Sky Father", "**Mata Prithvi**" is "Mother Earth". There are numerous gods like Sun God "**Surya**", Law God "**Vata**". According to this religion, the human actions creating the gods, are the sacrifices. Actions create the beings. Veda Religion assumes that the numerous gods are various visions of divine Mother Goddess **Aditi**. In one of the verses of **Rigveda,** (the oldest of Vedas) it says: "**Wise men remember the almighty ones with various names. They call them as Agni** (fire and stove god), **they call them as Mitra, they call them as Veda.**". As can be seen in this verse, Vedaism is some kind of a search for a single god.

2. 000 B.C. -1. 000 B.C. ORPHICRELIGION: It is a religion, which gets its name from the musician singer **Orphe** in the Thracian mythology in the Balkans before Greeks. Orphe witches all beings and wild animals with his music and poetries. The instrument named "**Lyre**", prophecy and witching are believed to be invented by him. This religion was born first in Thrace and then passed to Greece and Rome. According to Orpheus; there are "**Dionysian**" for goodness divinity and "**Titanic**" for badness in the humans. Humans may reach to the eternal fortunate by a redemption. Soulis purified by passing from body to the other body. To exist in the other world, suffering is necessary in this world; and living like **Orpheus**, not eating the foods like meat, eggs. **Orpheus** had an understanding, which unites the humans with animals; with the nature. This was similar to some kind of **Pantheism,** Public divinity, later **Stoicism** of Greeks; new **Platonism**; **Kabbalah** of Jewish people; an understanding of Eastern **Unity of Body**, **Hermesians** and **single god, and Islam Mysticism.**

1894-1595 B.C. BABYLONIA: "Bab" means door in the the Sumerian language, while "El" means god. Its first king Nimrud, the fatherless son of Marduk Semiramis, is the chief god of Babylon, the lord of the earth and the sky, and created people from clay.

1. 700 B.C. – 1. 200 B.C. CENTRAL ASIA-BAIKAL REGION: There were idols for the agriculture, stockbreeding, advanced hunting techniques and earth in the Central Asia Baikal region. Jade was a commercial item to be sold outside by means of Nomadic people.

1700 B.C. -1586 B.C. HIXOS: East Anatolian **Hurri (Hurri and Hor**=Sun) and the race of Arabian **Semites**. They came from Syria and invaded Egypt. They had lightning, storm and sky gods.

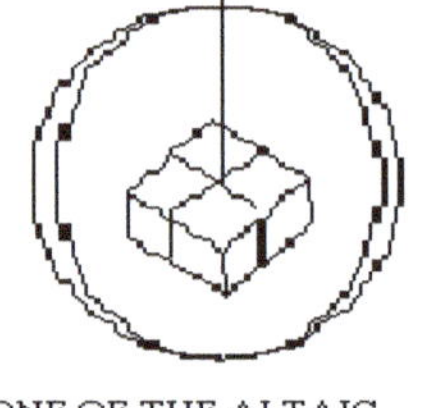

ONE OF THE ALTAIC IDOLS

ALTAIC SHAMAN DRUM

1400 B.C. CENTRAL ASIA BAIKAL OTUGEN (AEDUGEN) REGION: There were religious beliefs about the **sky, sun, fire, stove, souls of ancestors and element** in the Turkish speaking races.

1059 B.C. -249 B.C. PROTO-TURKS: There was a religion opinion, which conceive the universe orders as one in Time and Space in the mostly prototurk races which were called as "**Chou**" (In Turkish: kök, gök) by the Chinese who have come from the West of China and invaded North China together with the races in North. They assumed that the Universe was a dome-shaped tent (royal tent) or a two wheeled (one wheel was moon and the other was sun) car on which there was a tent. 28 slices of the dome of tent represented the zodiacs, which the Moon passed by while travelling the World in 28 days. These were considered the symbol of four directions in four groups of seven zodiacs, in accordance with the time of the appearance in the East in spring. There are the **nine palaces of gods** on the top of sky. In the center there was the palace of Sky God on the **Pole Star;** four of them were in the four main directions and the remaining four were in the secondary directions. These nine divisions were applied to the races and states on the Earth. Provincial organization was composed of nine provinces, one in the center and the others in the eight directions. State was sacrificing a young, male, red colored animal in the mid-summer (in June) for the **Sky God;** young, male, black colored animal in the midwinter for the **Earth God**. In the religious ceremonies, state used to turn its face to the South; esquires and public used to turn to the North in front of it; in the right side of state, small officers used to turn to West; in the left of state, master officers used to turn to East. Religious ceremonies were performed for the souls of ancestors, ground waters at the beginning of four seasons and for the Sun, Moon, mountains and rivers in the four directions.

B.C. 554-480 BUDDHISM: It is the religion which founded by Goutama (Buddha) of Turkish origin, in Nepal. (**Buddha** means: Enlightened, wakened, meeting with heavenly light, knowing, saving). Youth, health and life are all ephemeral.

One day he recognized the four main truths under the fig tree (tree of knowledge):
1- The only truth in the world is the suffering.
2- Resource of pain are desire and the possessions.
3- Getting rid of the pain and reaching to Nirvana are only possible by being purified from the requests.
4- The only way for this purifying is **Dharma**, in other words, a law based on **ethics virtues**.

Divine path has 7 levels, accordingly:
1- Pure belief.
2- Pure wish (will).
3- Pure word.
4- Pure work.
5- Pure effort.
6- Pure memory.
7- Pure consideration.

The holy book of Buddha is **Tripitaka** (three baskets). It includes:
1- Monk rules.
2- Alternatives for a salvation.
3- Philosophical and psychological opinions.

The texts of Buddha are called as **'Canon'** ('Kanun' in modern Turkish). "How nice is the loneliness of the happy one who knows and considers the fact; how nice is the situation of man who doesn't leave his path, who does not give harm to none of the living beings. Beating the obstinacy of the ego is really the highest of all happiness. Among the two extreme ends, in other words, **the lives of beer and skittles which are disgusting and empty; and diet and fasting which are gloomy and empty, should be avoided. The middle path between these two should be selected to reach to the knowledge, quietness, and a nihility full of happiness.** There is no continuity in the **Material** or **Soul** worlds. But there are conditions and the terms of these conditions are created by the previous ones. Thus a false, empty, temporary **Universe** and **Ego** is created. This is symbolized by the wheel rotating around itself. This is the **wheel of existence.** This is **Karman**, in other words, the stuff which creates this special universe rotating around itself. **Mandala** is the place of religious ceremonies in Turkish.

5 negative ethics rules of Buddhism are:
1- Do not kill.
2- Do not steal.
3- Do not indulge in sexual misconduct.
4- Do not lie.
5- Do not take intoxicant.

Positive ethics of Buddhism:

It orders the personal suffering, resignation, thinking about the living beings, sharing the pains and happiness of them, even mentally, being kind, mercifulness, excusing the insults, sacrificing for the others. Kindness results with sacrificing the others. If grudge is responded with another grudge, what would the result of this grudge be? Humans should give not only everything they have, but also their time, lives and personalities. "**Nirvana**" is a relative absence, in other words, absolute reality, **heaven with a golden sky**. There are gigantic Lotus flowers in it. There, humans recover from all kinds of suffering with a pleasant music, which sounds nice, and beautiful belly dancers (Apsaras) in a mellow light. A believer comes here with the grace of **Amida**. The holy remainings of Buddha are kept in Stupa monuments. Tibetan Buddhism is called as "**Lamaism**" (**Lama** means supreme).

VI. Century B.C. JAINISM: Jina (conqueror) who has established this religion in India, was a fighter and from the prince family. According to Jina, there are no creators or creatures. The world is immortal. No harm should be given to any of the living beings, fasting and nudeness are required.

IV. Century B.C. MITHRAISM: Mithra (meaning **friend** in Sanskrit) is the messenger of Indo-Arian Sun Goddess. He first subdued the *Cosmic Bull*, and then sacrificed it, in the image of a light coming out of a rock. All of the living beings were born from the blood of this bull. In the image of a fire in an unbearable brightness, lighting the darkness, enlightened the humans, got up to the sky, united with the Sun to be a God. Then, previously dark **Universe was half-enlightened** (day/night). As a god, **Mithra** helps the humans in struggling with the evil. The ones, who have succeeded in the judgement after the death, reach to salvation and immortality. They go up to the sky to live forever in the **land of the lucky ones**. The ones willing to embrace have to pass the *pubescence ceremony*, and they reach to happiness. Worshipping happens in the caves and underground secret rooms. Women cannot attend.

In Mithraism, there are 7 levels, which can be proceeded by means of examinations:
- **1-** Crow.
- **2-** Gryphon.
- **3-** Soldier.
- **4-** Lion.
- **5-** Pars.
- **6-** Messenger of the Sun.
- **7-** Almighty.

Besides their master is called as **supreme of supremes**. In the seventh degree ceremony, a bull is sacrificed; a baptism ceremony is performed using water and honey; brand is used for searing; meat of sacrificed animal is eaten with bread and wine; thus, the last meal of Mithra before reaching to the sky is commemorated. Besides, in these secret ceremonies, people wear the today known masks, fast and sit down on long stone chairs and eat.

VIII. Century A.D. SHINTOISM: Since Japanese people learned writing very late, they collected the rules of this ancient Japanese religion in the **VIII. Century**, in a holy book named "**Kojiki**". In this polytheistic religion, "**Shinto**" means the **path of gods**. The oldest native population in Japan, **Ainos** in the Hokkaido island, used to worship to the bear, fox and holy souls, in other words, **Kami**; they had **moon, sun, fire, storm and earth** gods, "inviting male" god named **Izanagi**, and "inviting female" goddess named **Izanami**. There is a creation myth in the Japanese Mythology, like Adam and Eva, including the first male **Izanagi** and the first female **Izanami**. **Emperor is the son of Sun God**. All of the dead people become god in the Shinto religion. Their souls rule the nature. They are good when respected, but evil when they are not respected. Gods are the images of them as various natural beings. **Amaterasu** was the Sun God, **Tsukiyomi** was the Moon God, **Susanoo** was the Storm God, etc. It was used for political purposes in the name of "**Koka Shinto**" (National Shinto) after the **XVIII. Century**. It was asserted that Shinto religion named "**Kamino Michi**" in Chinese, was derived from the idiom "**Shin Tao**" which meant "Good Souls". (It is interesting that the word 'Şen' means happy and cheerful in Turkish).

800 B.C. - I. Century A.D. BRAHMANS: Religion in India, which has been founded by putting the knowledge based on Vedas, in other words, verbal information, in writing by Brahman Priests for disciplining after the invention of writing. It is based on the written texts named **Upanishads** (secret notices) and **Brahmanas** (sacrifice formulas). Legitimating the superiority of priests, in other words, Brahmans, this religion is based on three things:

1-**BRAHMAN** (Macro cosmos, basic essence of the universe)

2-**ATMAN** (Deep ego in the micro cosmos, in other words, humans)

3-**KARMAN** (Actions of previous living beings, soul immigrations, Samsara).

When humans believe the sameness of Atman and Brahman, their desires will die, and they will reach to Brahma, in other words, immortality. This is a salvation, which is reached by knowledge, instead of sacrifice.

Three gods of Brahmans are:

1-**Brahma** (Creator, thesis).

2-**Vishnu** (Protector, anti-thesis).

3-**Shiva** (Destroyer to re-build, synthesis).

Shiva objects to the will of Brahman to create a happy world free of matter and troubles. It symbolizes the idea that the vitality and development would realize by means of contradiction. Its symbol is a triple gammadion.

700 B.C. -245 B.C. HUNS: According to the Chinese records, each one of the four tribes constituting the small province could settle on one of the four directions, in the provincial organization of **Tysins**. "**Onguns**" the divine souls in the four directions, were the sons of "**Ogan**" the divine soul of 'center', in other words, town. The element of East was **wood**; the element of South was **fire**; the element of West was **metal**; and the element of North was **water**. The tribe in the East was sacrificing a sheep for **Sky Khan** in spring; the tribe in the South was sacrificing a red rooster for **Red Khan,** in summer; the tribe in the West was sacrificing a white dog for **White Khan** in autumn; the tribe in the North was sacrificing a black pig for **Black Khan** in winter. And in the middle of year, a yellow ox (**Tosun, Tysin, Tahsin**) was being sacrificed to **Ogan**, the divine soul of the 'center'. The tribe in the right was close to the gods in the nine layers of sky; the tribe in the left was close to the gods in the eight layers of underground. In the beginning, there was a foursome classification. Later a 'center' was added and the war ended between communities in the four directions. Peace started. There was also a triple classification like **sky gods** and heaven in the sky; **underground gods** and hell under the ground; **souls** of Earth (like mountain, rock, lake, river, tree, forest)**, humans** and **life** on the Earth. Marriages were happening out of the tribe, in the town, in other words, between the tribes.

660 B.C. MAZDEISM: It's the Iranian religion before Persian religion. **Its three gods are:**

1- **Ahura** (good, beautiful, big, perfect, bright, enlightment, joy, purity; god of fire, nature, sun which plan the order of the world, the perfect scholar, 'Mevlâ')

2- **Mithra** (Sungod),

3- **Anahita** (Kybele, earth mother goddess, fertility figurative young female goddess).

Also the six immortal powers are (Ameta, Spenta, Ulu, Yazata):

1- **Behmen** (good thoughts).

2- **Erdibibist** (good virtue).

3- **Shehrivar** (good empire).

4- **Spedardioh** (good look).

5- **Hurdat** (health).

6- **Murdad** (immortality).

In Mazdeism, animals are sacrificed, dead people are buried, fire is divine. There is a holy water named **Haoma**. Giants are bad devils.

660 B.C. RELIGION OF ZOROASTER: Zoroaster established a religion, which is commemorated by his own name, in Iran. His book is **Zend-Avesta** (text comment). While god **Ahura Mazda** (Hürmüs, Ormus, Ormazd, bright, magnificent, big, beautiful, clear, wise, creator, delight, reality, purity, life, immortality, upper world) represents the goodness; god **Angra Mainyu** (Ehreman, hitting soul, dark, death, dirtiness, temporary force, lower world) represents the badness. Earth and soul of humans are the arena of these two powers. All of the humans should fight in the war which would be the victory of humans in the universe. Humans should feed, take care of and improve themselves. Working is a must. A real believer is a hardworking villager, calculating animal breeder, right and fair landlord or village headman. The main ideal is working, family unity and the life passing by. Fire represents a divine being. A no extinguishing divine fire is lighted in the temples.

VI. Century B.C. YOGA: It is the doctrine, which founded by **Bhagavad-Gita** (song of heaven) with two books with the same title in India. The purpose is; creating a perfect human who will live healthy and happy among the others; who is communed with everything, living or non-living, in the universe; and also with his own existence. **Life is a torment. The reason is that we can see only a part of the reality. If we can understand the reality completely, the pains will disappear.** The realities, which humans can only achieve by knowing himself, are secret and private information. Various ways to know the self, five senses and to stop the functions of brain (In this case logic stops, superior way begins. **Brahma** is contacted. **Schamadhi**) are called as **Yoga**. Yoga is the ordeal of **body** and **soul**. The main ego is **soul** and it is beyond the personality. Unless the humans understand their own spirit, which comes from the God, soul stays connected to the world by means of the body. After the humans die, soul returns to the world in a new body. The main rules of Yoga are as follows:

1-Yama: They are the orders, which forbid violence, stealing, cupidity, rapacity, tortuosity, indulgence.

2-Niyama: Purity, simplicity, growth, working, commitment to God.

3-Samadhi: When achieved, Yama and Niyama should be obeyed. Because understanding the unity of everything is to know that the good of others, the good of humans are the same as the good of ourselves. Yama and Niyama take us understand that. Besides, **Jnana** (knowledge), **Karma** (action), **Bhakti** (love), are the three laws of Bhagavad-Gita.

Tantric Yoga: Adopting the one and only power is a spiritual enlightenment by means of physical ways. Universe and body, which is the microcosmic universe of Gods, are used for a mystic experience. In this experience, symbolic sounds, images, light, motions and sometimes at the peak point of the ritual, sexual intercourse take place. The purpose is going beyond the duality of world (male and female), and reach to the one and only and infinite reality.

Hatha Yoga: **Ha** (Sun), **Tha** (Moon). Mind-affecting and body functions (Bearing, breathing, thinking, etc.) can be extraordinarily controlled by means of body.

Kundalini Yoga: Universal energy and potency are considered to be the same and yoga is performed in this way.

Laya Yoga: It includes the assumption of Kundalini Yoga and **Chakra**. Chakras are lotus flowers with various numbers of leaves. These flowers are related with holy elements like **earth, water, fire, air, ether and awareness**, and all of them have their own symbolic sounds. These **symbols, indicate the bodies of humans as the micro scaled sample of Universe; and thus moon, sun and fire** unite to reach immortality.

YANTRA WITH LOTUS LEAVES

LOTUS FLOWER

II. Century A.D. RAJA YOGA: (**Raja:** Like a magnificent king) He has eight arms. The path of this Yoga is divided into eight steps.

FAKIRISM: In India, the persons who adopted to live by torturing themselves are called as "Fakir"; and this doctrine is called as "Fakirism". Fakirs are generally the persons who can only endure the physical tortures, in other words, hunger and the injuries of fire and sharp objects like, nails and glass; in other words they are the persons who could not complete their evolution to be a Yogi, yet. Because gaining such kind of powers constitutes only a part of disciplines to be a Yogi.

I. century -250 A.D. HINDUISM: It is the continuation of Brahmanism in India.

It is based on two written epics:

1- **Mahabharata** (Big Brahman War)**: It was written in the IIIrd-IInd centuries B.C.** Its most beautiful part is **Bhagavad Gita** (Song of heaven).

2- **Ramayana** (The adventure of God Rama).

They both originated from the legend named **Purana**. It is about the life of **God Rama. God Krishna** takes the place of former god Rama. It is the mother, father, procreator, protector, end of knowledge, cleaner and organizator among all of the divine things. It is the syllable "**OM**" (**O: triangle** and **M**: upper horned circle; in the Brahma alphabet), it is the **word**, farewell, **sound; it is the master and feeder;** it is the house, home, shelter, **everything**. For a real **wise man**, happiness and pain are the same. He does not desire, does not begrudge, **does not become afraid**, does not get angry; he lives being contented about his life. He cut himself off from the world and worldly things. He lives his inner life by eluding from all earthly affairs, and keeping his balance. He is not bound to anything, anyone. Since he is free from all kinds of desires, happiness and unhappiness do not affect him. These are the characteristics of a real scholar. In Hinduism, salvation is not by sacrificing or knowledge; salvation comes by loving the saver God. Transmigration of soul is believed. There are holy trees and waters. The most important Gods are: "**Krishna**" (Creator instead of Brahman), "**Shiva**" (destroyer to re-create, with a crescent on his three eyed head), "**Vishnu**" (he is the Fire God in the form of **Rama and Krishna**, protecting the world. His eyes are like a lotus flower. He holds Indra's sign "**chakra**" in his right hand and **hammer** and **pearl shell**), "**Indra**" (Night sky, air, storm, Rain God), "**Ganesha**" (Elephant headed **wisdom god**). In Hinduism, while **Mandala and Brahma** are the symbols of God or his forces; YANTRA (above picture) is the symbol, which indicates the God and unity. Lower classes and women could also be embraced in this religion.

III. Century A.D. MANICHEISM: Founded by "**Mani**" who was born on the Babylonia border in Iran. It was a universal salvation religion, which was a synthesis of reasonable parts of the orders of Buddha, Zoroaster, Jesus and ancient wisdom. The structure of all beings in the universe is a mixture of **goodness and badness**. As a part of the universe, **humans** are also the same: **goodness** (in other words, **soul, mind, light, knowledge**) and **badness** (in other words, **body, superstitions, ignorance**) are struggling. **Humans** are freed of ignorance and being shackled by the laws of nature by increasing their knowledge and rule the nature, accordingly. Souls, which are trapped by and suffer from the **material things** should be rescued. The world would come to an end when all souls are redeemed and reached to the skies. Mani recommends sharing of the goods, properties and women.

V. Century A.D. MAZDEKISM: Established by Iranian **Mazdak**. He asserted that the good was sensitive and free, bad was blind and ignorant. He stipulated sharing of air, water, fire, money and women. He was executed and this religion disappeared in the X. Century. Orkhon Inscriptions were written on the stone in the name of Göktürk ruler Bilge Kaan. In these inscriptions, it is said that "When the sky was created above, when the earth was created below, the mankind was created between the two".

VI. Century A.D. GOKTURKS: In Asia, at the beginning of winter (November 8th), after the religious ceremony for Earth God, "Hakan" was used to go to the "Budun Inli" caves in the West of Otuken mountain together with the clan leaders ("Başbuğ") to sacrifice to the souls of their ancestors. Orkhon Inscriptions were written on the stone in the name of Göktürk ruler Bilge Kaan. In these inscriptions, it is said that "When the sky was created above, when the earth was created below, the mankind was created between the two". **Orkhon Inscriptions were written on the stone in the name of Göktürk ruler Bilge Kaan. In these inscriptions, it is said that "When the sky was created above, when the earth was created below, the mankind was created between the two".**

IV) MONOTHEISTICRELIGIONS and THEREAFTER:

a) 1370 B.C. -1352 B.C. EGYPT SUN GOD: Amonhotep, pharaoh of Egypt enshrined Aton (evening sun) as the one and only god. For Aton, he founded a new city named **Akhetaten**, which means "Horizons of the Sun" (Today's Tel-el Amarna). He moved the capital city from Thebes to this new city. He deleted the names of sky god Amon, who was the head of the gods in Thebes and the other gods. He changed his name, which means, "Amon is glad" to **Akhenaton**, which means "Glory of Aton". His ideal was establishing a religion for all of the humans. Between God and humans, priests were not necessary any more. Not only the Sun physically, **Aton was representing the resource of all goodness, temperature, brightness, vitality, joy, happiness and freedom; and also nature,**

loving all beings and joy of living. He was the one who radiated these beauties to the world with this light. In the eulogies which Akhenaton has written to the Sun:

"When you rise and spread your lights on Earth, you chase the darkness away. You give light to us. Humans get up happily, they get to work, trees and meadows become green, Birds come out of their nests, goats jump, birds on the ground and flying start to live, ships come and go, fish of rivers also rush into you. Because your lights go in the deeps of the waters. You are the one who feeds the baby in the bosom of mother, you give the reviving breathe in every child you create. When the little bird is in the egg and when it cries, you are the one who gives the air to keep it alive; and thanks to you, it can find power to break down the shell surrounding it. The things you have created are in all shapes and sizes. You created the Earth together with all of the humans, herds, everything living on the ground and flying." After Akhenaton died while he was 29, priests removed this religion and again the old religion was adopted.

b) XIV. Century B.C. JUDAISM: It is the first of the biggest three monotheistic religions in the history of religions. The prophets of all of the three **monotheistic religions has started with prophet Noah**. After the prophet **Noah (10 000 B.C.), Prophet Abraham was born on 2000 B.C.** in Kuşa town of Urfa city in the South East of Anatolia. His father was a Sumerian idol sculptor named Hazer. Believing in monotheism, Prophet Abraham broke down the heads of all God eikons in the ziggurat in Ur city. The King of Babylon Nimrod was going to burn him in Urfa. But water welled out in the middle of the fire. And Nimrod chased him, his family and believers away from Ur. When they went to Palestine, **Canaan people** there called them as "Jew" which means "the humans of the other side of river". Expulsion of **Prophet Abraham** is the **1st expulsion of Jews**.

After Abraham, in the **XIII. Century B.C.,** when Jews were in Egypt, they were chased away by Pharaoh **Tutankhamun** for the 2nd time and they came to Palestine again in the leadership of **Prophet Moses**. Here in Mount Sinai, Prophet Moses was inspired with the 10 commandments. This time, they mixed with the Canaan people. According to Torah, the holy book of Jews, Judaism is based on three fundamentals:

i.   **Body** and
ii. **Soul** are separate beings. Afterdeath should be believed.
iii. **Ethics**: God was not born from anything. He is the almighty; He knows and sees everything; He is potent, he is immortal.

Following are the 10 commandments according to Torah:

1- I am the Lord your God.
2- You shall have no other gods before Me. You shall not make for yourself an idol.
3- You shall not take the name of God in vain.
4- Remember and observe the Sabbath and keep it holy.
5- Honor your father and mother.
6- You shall not murder.
7- You shall not commit adultery.
8- You shall not steal.
9- You shall not bear false witness.
10- You shall not covet your neighbor's wife or house.

Religious ceremonies: The purpose was being protected from the idolaters. Later, music was also included.

Civil rights: The father had unlimited rights on their children. Marriages happen by means of the decision of family. A bride price is paid for the girl. Widow marries with her brother in law.

In 1000 B.C., they were invaded by the Palestinians coming from the sea. But only a small kingdom could be established. The prophets after **Prophet Moses** are; **Prophet Harun, Prophet Yusha, Prophet Ilyas and Prophet Elyesa**.

In 1025 B.C., Prophet Dawud (David) was the second king of the small state, which founded in Palestine. He established Jerusalem. His son **Prophet Sulaiman (974 B.C. -931 B.C.)**had the temple of Prophet Suleiman in Jerusalem (Al-Aqsa Mosque). After Prophet Süleyman died, kingdom was divided into two as Israel and Jews kingdoms.

In 721 B.C., Assyrians ravaged Israel kingdom, they chased the people to Assyria (**III. expulsion**).

In 586 B.C., Babylon king Nebuchadnezzar invaded Jerusalem and chased the people to Babylon (**IV. Expulsion**).

In 536 B.C., Persian king Cyrus invaded Babylon and chased Jews back to their country.

In 339 B.C.: After Alexander the Great ravaged Persian Kingdom, they were invaded by Greeks.

In II. Century B.C. Sani Religion founded. In this religion, personal properties were cancelled; trading and using gold and silver were forbidden. All properties became the common properties of the community.

In 70 A.D. After revolting against Rome, Emperor Titus chased away Jews to various locations. (**V. expulsion**). Later they were distributed to everywhere in the world. In the World War II, millions of Jews were thrown to the furnaces. Later:

In 1945 A.D. They returned back to Palestine to establish Israel.

V-XV. Century A.D. KABALISM: It is the God universe doctrine which is an inspiration older than two written books of Jews and which passed down from disciple to disciple for long years, throughout the middle age.

Kabbalah: Means the tradition of transferring the comments of Torah from generation to generation.

ITS BOOKS ARE:

I- Sepher Yetzirah (Book of the creator).

II- Sepher Haz-zohar (Book of the light). God exteriorized Himself and everything in the universe came into existence by this exteriorism. This exteriorism was realized by 32 circle stages, which were called as Sefirot (circles):

1- The first physical map in which everything is in the form of seeds. **2-** Reviving air. **3-** Water. **4-** Fire. **5-** Direction of the head **6-** Direction of the foot. **7-** Right. **8-** Left. **9-** Front. **10-** Back. (The first four circles of these ten divine circles, show the elements of being; and the last six circles show its location in space.) **11-** Self. **12-** Quantity. **13-** Quality. **14-** Relativity. **15-** Affect. **16-**Passion. **17-** Time. **18-** Space. **19-** Ownership. **20-** Opposition. (In these second ten divine circles, the situation of being, and its possible shapes are shown). **21-** Infinity. **22-** Mind. **23-** Intelligence (This divine three is the first triad and establishes the mental world). **24-** Charity. **25-** Justice. **26-** Beauty (This divine three is the second triad and establishes the ethics world), **27-** Power, **28-** Occupying a space, **29-** Dimension (This divine three is the third triad and establishes the physical world), **30-** Mental world. **31-** Ethics world. **32-** Physical world.

(This divine nine, which is composed of three triads, establishes the last triad kingdom of God). Each of these circles get one of the names, which Torah gives to the God, and the last one is named **Adonai**. All of them form **Adam Kadmon** role model. The first **ten circles** are the creator words, the **Word**. The last **22 circles** represent the 22 letters of alphabet (Jew alphabet) which forms this creator word. Each letter is a number at the same time. **Divine secret is hidden in these letters and numbers.** It opens to the ones who know how to read. Cabalism doctrine presents the soul conditions of which a part is male and the other part is female, and which combine in a single being. These beings, which have separated from each other in the Earth, search for each other to re-unite. When all of the waiting souls complete their pilgrimages on earth, Messiah will come and start the happiness age. In **1648**, **Sabbatai Zevi** of Jews origin announced himself as a Messiah in Izmir. He had 18 commandments. His disciples migrated first to Thessaloniki and later to Istanbul.

c) 570 B.C. -490 B.C. TAOISM: It is a philosophy and religion system which founded by Chinese **Tao-Tzu** (Wise old man). His book, **TAO-TECHING** (TAO: the truest path to follow, ritual, real, seasons, order of the universe, immortal essence from which all beings had come to existence), (**TE: Virtue, perfection, power, appearing of Tao on the humans**), (CHING: name given to the classical books). He is the **real** one and he has an order and a unity principle. He is full of secrets. He cannot be named. He is both transcendent and immanent. He is **Tao**. But this name does not indicate Him completely. Tao, who is being tried to be understood, is not Tao himself. The name, which is given to Him, is not His real name. This is the highest reality. All visual developments disappear in Him. World came to existence by the integration of the existent in other words **Yang** (male) and inexistent **Ying** (female). Everything is relative. To get the unity again, we need to give up reading, living collectively; we need to act with sensations not the intelligence; we need to pick up instead of disorder;

we need to be simplified; we need to neglect everything. Like living just for living. We need to learn the art of living simple and joyful, from the children, animals and plants. Humans should be like a baby who is purposeless and who smiles everything. Humans can be enraptured with dance and drunkenness. The rules of Taoism are:

1- Sparing.

2- Spending a simple life, being modest and unpretentious. (The one, who leaves his mark behind, is not really big).

3- Being merciful. Even the ones, who are rude to themselves, should be behaved well.

Besides, there are some methods, to protect health and to live longer by means of various physiological treatments in Taoism. The one with the highest virtue does not do anything. He has no purpose. The ones with a lower virtue can do everything and they have a purpose. The highest honor is dishonor. The saint persons do not want to be hewn like a jade. They want to be spreaded like pebbles. Tao gives birth to one; one creates two; two forms three; three brings all beings of the world into existence.

d) THE BIRTH OF CHRIST and CHRISTIANITY: It is the religion, which was founded by Prophet Jesus (of Jews origin) from Nazareth. His book Bible could be written after Jesus in IV. Century. Later, Torah was called the **Old Testament**, and the Bible was called the **New Testament**. Announcing that he was the Messiah (In Greek language **Christo** means messiah, the saver. Egyptians used to call the Sun as **CHRIST** or **YESHUA**, which means protector. The name of Prophet **JESUS** in Latin was derived from this word. Besides, its equivalent in Hinduism is God **Krishna**) who has been heralded in Torah, Prophet Jesus ameliorate Judaism. "Do not think that I came to abolish the holy law or the Prophets; I did not come to abolish but to fulfill.", "You have heard that it was said to the men of old, 'You shall not kill; and whoever kills shall be liable to judgment.' But I say to you that every one who is angry with his brother shall be liable to judgment; whoever insults his brother shall be liable to the council, and whoever says, 'You fool!' shall be liable to the hell of fire.", "You have heard that it was said, 'Love your neighbor and hate your enemy.' But I tell you: Love your enemies and pray for these who persecute you.", "For if you forgive humans their trespasses, your heavenly father will forgive you too [your trespasses]", "Judge not, that you be not judged. "Contribution of Prophet Jesus to the Judaism can be summarized as love and equality in poverty. He was explaining his birth from the Virgin Mary without a father, by his father's being the God. Thus the son of God was both a God and a human. After being crucified, he appeared to his associates once again. He promised to help them forever. He ordered all of the races to be educated in his name, and raised to the sky to His Father. The medieval universities were founded because of this reason. He used to say: **"I came not be served, but to serve."** Therefore, church was regarded as a public service. The political power was the subject of justice and was linked to judgement. The state was no longer the only source of law. **It was accepted that the nature could be understood by reason**. The human might understand the nature by means of reasoning, but nature would not be able to understand the human. **Ethics** became the art of reaching the infinity by self-sacrifice. The notions such as existence of God, divergence of **soul** and **body** and **independence** were accepted. **Goodwill**, devotion to god and an understanding a world away from any kinds of **discriminations and concessions**. The holy trinity of Christianity (trinity); **(Father, Son and Holy Spirit)**. Prophet Jesus will come again on doomsday to judge the people. After Prophet Jesus, one of his disciples Paulus (of Jews origin) separated the religion from Judaism in order to establish the real Christianity. He separated the commandments of Prophet Moses from Christianity. **(5 -67 A.D.)**.

e) 571 A.D. -632 A.D. ISLAM: The religion established by **Prophet Mohammad** in Mecca. **The origin of the word Islam in Arabic is the word SILM which means Peace and there are 4 main types of peace in Islam:**

1. **Peace among all creations in the universe;**
2. **Peace among the world nations and states.**
3. **Peace among the family members**
4. **Peace between the selfishness (needs) and the mind of the individual.**

All of those 4 types of the peace are collected in the soul, which is lodged in heart. Thus, the harmony, rhythm, beauty, balance, symphony and melody that we observe in the astral

**and quantum universe are in the hearts of the human, because the heart represents the love
and embraces the entire universe.**

a) **PILLARS OF ISLAM: According to the holy book Quran, the pillars of Islam are:
(Creed: 'Amentü'):**

a1)-6 ARTICLES OF FAITH (Belief):
1- Belief in the existence and oneness of God (Allah).
2- Belief in the existence of Angels.
3- Belief in the existence of divine books (104 Books).
4- Belief in the existence of Prophets (28 known by name and the others).
5- Belief in the existence of the Judgment Day.
6- Belief in the existence of God's predestination, whether it involves good or bad.,

a2) 5 PILLARS OF ISLAM (Ritual Obligations):
1- Declaration of faith
2- Obligatory prayer (Five times a day)
3- Fasting in the month of Ramadan
4- Zakat (prescribed purifying alms, compulsory for rich people to give 1/40 of their
properties to the poor people once a year)
5- Pilgrimage to Mecca (for rich ones at least once in their lives)

**a3) THE MOST IMPORTANT CONDITIONS AND RULES TO ORGANIZE THE RE-
LATIONS AMONG HUMANS:**
1- Equality among humans,
2- Denying the racism,
3- Prohibition of usury,
4- Rights of women (As men have rights on women, women have rights on men as
well),
5- Social justice (alms and solidarity),
6- State type: Republic (All 4 caliphs were elected by the public).

Prophet Mohammad told: **"I was sent to complete the epitomes of ethics"** and **"the ones who
encourage each other for patience and mercy will be judged from the right side; the ones who
deny our evidences will be judged from the left side, flames will rain on them."** The orders in
Qur'an, which are a 'must', are called as 'Religious duty' (**Farz**); the ones, which are not a 'must', are
called as 'incumbent' (**Vacip**). The rituals which Prophet Mohammad has said or done in His life or
which he has forbidden are called as **'Sunnah'**. Since one word can have dozens of different meanings
in the Qur'an, the people tried to understand the Qur'an according to the Qur'anic Commentaries made
by competent people in every era. The concept of ethics expressed by Muhammad will be discussed
in more details on **page 175-185.**

b) **ISLAMIC MYSTICISM (SUFISM):**
b1) **DEFINITION OF SUFISM:** After the foundation of the religion of Islam in the VII.
century, some of the Muslims have developed a mentality and life style by taking the life
and ethics of Prophet Mohammad as a model. In the beginning they have been called as
'Sofi' (In Greek, Sophia: Knowledge, Philosophy; in Arabic **Sofi**: (woolen cloth which
has also been used by Prophet Mohammad like the poor people) or "**Sofu**". Later, this
trend developed as **mysticism** and **philosophy of Islam,** great 'Sufi's were raised. Sufism
could be considered as the effort to be integrated with all beings and the spiritual path to
mystical union with God, by setting **soul** free from the **matter** and **body** albeit living in
accordance with nature of human. In the terms of mysticism, this is "Kesrette vahdet", in
other words, "**unity in plurality**". This movement developed under the term "**tasavvuf**",
which is an "**Islamic mysticism**" (**the secret that the mind cannot grasp**). This term is
the combination of two words in Arabic: "**vâkıf**" (**aware**): meaning someone, who is
aware of, understands and knows something; "**tasavvur**" (envisagement): **formaliz-
ing in the mind;** and "**saffet**" (purity of heart): meaning **pureness, honesty, ethical,
polite.** According to Sufism, "**edep**" (decency) differs humans from animals. In

Mawlana's words; "The **mind** was asked 'What believes in Allah?' mind whispered to the ear of the heart and said, '**Faith** (believing the existence of Allah) is **edep** (decency)". According to Sufism, the **universe** and **human** are composed of the **beautiful names of Allah** [**Esmâ-i Husnâ: 99 beautiful names of Allah**]. Among these names: "**ADL**": **Fair, balanced, symmetry in** geometry; **equality in** algebra; and **justice** in the management. "**MUQADDIR**": **order, fate, pre-calculated.** "**Allah**" (c. c.): The One, who has created the universe, keeps the universe alive and will destroy it. "**NÛR**" (Light): **The one and only source of light for everything.** "**AL-KHABIR**": The All-Aware. He who has knowledge of the inner, most secret aspects of all things. "**QUDDUS**": **The Pure One. He who is free from all error.** In Quran, **Allah** (c. c.) has more than 100 names. Although He is said to have 1000 names according to the Hadiths and some of the Ehl-i Beyt (family of Prophet Mohammad) resources; when we say "**Esmâ-ul Husnâ**" (names of Allah), it means 99 names of Allah. As the human is the trustee of Allah on Earth, and since **Prophet Mohammad** reflects the names and attributions of **Allah, he is the model of "perfect human".** In other words; he is **complete and perfect characteristics in good words, good behaviors, high ethics, good knowledge.**

b2) OBJECTIVE OF SUFISM; is the happiness in the world and afterdeath; keeping the human away evil behaviors, thoughts, words and actions; bring in the good qualifications which Prophet Mohammad had, and raising **perfect** (competent, faultless) **humans** and thus make them gain the eternal happiness when still living.

According to Sufism, there are 4 stages in Quran:

1. **Shariat**, denotes Quran itself, meaning the **laws.**
2. **Tariqat**, denotes orientation, heading towards Quran and Allah in different ways. **Tariqa**, means path and direction in Arabic.
3. **Marifat**, denotes the knowledge learned in life on the basis of experiments. The very first command of Quran is '**Read**' (**Iqra**). It does not mean reading a written book. Everything in the universe is like the verses of Quran. Learn the knowledge in them, **be well-informed,** in other words, be "**wise**".
4. **Haqiqat:** In other words, there is unity (vahdet) in plurality (kesret). **Universe is a whole in the unity of Allah.** Recognizing Allah understands this truth (reality). If this person, reaching that level is a man, he is called "**kemâl**" (perfect man); and if a woman, then she is called as "**kemâle**" (perfect woman). The door of Heaven is always open to '**kemâl**' and '**kemâle**' persons. According to Muhittin Arabi **(P-83)**, one of the ways of being a perfect human is to grasp the "**universe of the supremes in a unity**". The **Rabias** we present in this book (P.9-10,173-178) can make it easier for people to pass through the gates of **Marifat** and **Haqiqat.**

b3) HISTORY OF ISLAMIC MYSTICISM: There are 3 periods in the history of Islamic mysticism:

1. **Period of Zuhd**; the period lasting for 2 centuries after Prophet Mohammad, when only the orders and prohibitions of the religion were followed.
2. **Period of Tasawwuf**; the period lasting for 3 and a half centuries, when the concepts of Sufism were formed after these first two centuries.
3. **Period of Tariqat (Tariqa means path and direction in Arabic);** the period, when the **tariqa** (sects) were formed as the organizations of Sufism **in the XII. Century** and became a part of social life. Since Sufism is not only science but also an attidude and education issue, in the third phase, "**sects**" (orientations) started to be formed as various education organizations. The **ethical, social** and **Islamic** rules followed to take the spiritual (humanistic) characters, the desires (nafs) and innate powers of human under control through training under the supervision of a spiritual mentors, who were called "**sheikh**" in a lodge, known variously as a zawiya or tekke) were called a "**sect**" (tariqat).

b4) SECTS IN ISLAM:

Some of the sects in Islam are given below in chronological order according to their emergence:

BATINIYYA - IX. Century A.D. and later: The term bāṭiniyya derives from the Arabic bāṭin, meaning "inner," "inward," "interior," hence "hidden," "secret," and, most importantly in this context, "esoteric." It is the path of people who interpret Islamic religion in rationalist terms. According to them, external meanings of the Qur'an make no sense for people who attained the internal meanings. For example, the aim of performing the Salaat is to approach to the God, so the Salaat is not necessary for people who had already approached to the God.

They believe in 5 main powers in creation of the universe:

1- The Former (The one who had existed before anything else, some kind of God, but not a creator. The creation does not begin with his will but with himself) 2-**The Subordinate** (the **reason**, **intelligence** that follows the Former) masculine principle 3-**The Predecessor** (**Matter's ability of formation**, some kind of breath and female principle) 4-**The Conquered** (The **space**, location in which the formation takes place) 5-**Imagination** (The **time** which will accompany the formation).

There is a material existence, which is ongoing in the universe. Its mutual contact with the active masculine principle and passive female principle are also ongoing. Esoterics used to eat brotherhood food and Heaven bread, so living in a full community of goods. They previously had 7 books, then this increased to 9. It was an earthly sect. The religious leader of the 7 stages had the entire esoterism knowledge and he was the head of all esoterics. The universe was formed by the rotation of the stars.

BABEKISM 816 A.D.: Mazdekism was continued in the Rey village by Hurreme who was the wife of Mazdek. Then, Javidan and eventually Babek took the lead of Mazdekists and established a political administration in Azerbaijan. Then, they united with Esoteric, Qaramith and İsmaili doctrines.

QARAMITH 889 A.D. (Qaramith means secret keeper): It is an Islamic sect that is Esoteric and Mazdekist established by Hamdan Qarmat ibn al-Ash'ath. This is a socio economic and political conception based on the principle of community of goods. It is assumed to be the basis of craft guilds and universities established in the West later.

MELAMISM IX Century A.D. (also known as Melametism, Malamatiyya, Khorasanism or Khorosan Saints): It is a sect, which unites Sunni Path, which founded by Turks in the Nishapur city of Khorasan with Esoterism. They accept no special knowledge, dress, ceremony and meeting place. Melamet means reproach, disgrace and rustle in Arabic language. The Sura No. V of the Qur'an states *"if any from among you turn back from his Faith, soon will Allah produce a people whom He will love as they will love Him,- lowly with the believers, mighty against the rejecters, fighting in the way of Allah, and never afraid of the reproaches of such as find fault."*. Khorasani Turks had considered themselves as this tribe. One should not see the faults of other people, he/she must condemn and insult him/herself in the first hand. One should not brag about anything; piety should be kept hidden; **if doing a favor for others gives the pleasure of superiority, even this pleasure should be suppressed**. You should do favors and help others, **because tasawwuf is nothing more than favor, but one should not seek the help of others.** The God is necessary, people should imitate the prophets. The melamists do not look for secret meanings in the Qur'an, but try to further explain the explicit meanings. Someone, wishing to participate into Melamist sect, sits in front of the Murshid (that is the spiritual mentor) and answers the question "What do you want? " by saying that "I want the God". The Murshid responds him "The one who wants the God shall drive anything other than the God out of his heart, you should also do this". Then, the candidate tries to purify his/her heart from everything. Afterwards, he begins to live with "Melamet Cheer" by considering that everything comes from the God and he is also a part of this whole just like all other beings.

FUTUWWA - IX. Century A.D. (**Futuwwa** means "young-manliness" or "chivalry" in Arabic); is a craft guild with religious nature arising from the organization of craftsmen by the melamists. Their sheiks are called as "**akhi**". Every profession has a separate founding leader. The founding leader of tailors is the **Prophet Idris (Enoch)**, the weavers is the **Prophet Seth**, and the founding leader of cotton weavers is **Ammar**. If anyone would do a harm, they used to put up his shutters and punished him by throwing his shoes on the roof of the shop. **Futuwwa in Islamic Mysticism** (tasawwuf) **meant being considerate of others' wellbeing and happiness on Earth and hereafter first despite of self** or Futuwwa members would be **people who have no enemies** and **who cover up the shame of his/her friends can avoid the hostility of his/her enemies**. Such people were called as "**fata**" before the Islam, however they were not organized, they used to attach importance to personal virtues and had military qualifications. They were called Alperen (combatant), Cavlak (hairless) in Khorasan and Tur-kistan; jawanmardi, rind in Iran; shatir, futuwwa in Arabs.

ISMAILISM (Fatimism) 908-932 A.D.: The sect of people who accept Ismail, one of five sons of Ja'far al-Sadiq, one of the grandsons of Hz. Ali as the sect leader,. It founded by Abdullah ibn Meimun. They founded the Fatimid State in Northern Africa. 7 is the sacred number of Ismailis. There are 7 days in a week, 7 big stars in the sky and 7 prophets (Adam, Noah, Abraham, Moses, Jesus, Muhammad, Ishmael, Jafar) and similarly they accepted 7 sect leaders.

QADIRIYYA 1166 A.D.: A Sufi sect founded by Abdul Qadir Gilani in the south west of the Caspian Sea. After 25 years of isolation and fasting, Gilani established the sect as a continuance of the Sunni Sufism tradition.

YESEVIYYA 1166 A.D.: It founded in Turkistan by Ahmad Yasawi. (Yasi: Turkistan). He was a member of the Hanafi School. He placed emphasis on **intelligence,** agility, honesty and discretion when serving the sheikh in his followers.

RIFAIYYA 1183 A.D.: A Sufi sect founded in Arabia by Ahmed ar-Rifa'i, who was a member of the Shafii School. There are two kinds of seclusion (ordeal) which last for 40 days to clean and enhance ethics. One includes fasting in a secluded place. There is another way of seclusion other fol-lowers that includes performing an ablution, not sleeping with their spouse, not eating meat and keeping silence. Rifais used to wear a black turban, sit on prayer rugs, play *tambourine* and *frame drum* in their ceremonies, stick needles in their body, step in fire and chew glass.

SHADHILIYYAH 1196 A.D.: A Sufi sect founded in Tunis by Abul Hasan Ali ash-Shadhili. Followers wear clean clothes. They do not hesitate benefitting from earthly wealth.

SUHRAWARDIYYA 1234 A.D.: A Sufi sect founded in Baghdad by Abu Hafs Umar al-Suhrawardi.

CHISTIYYA 1236 A.D.: A Sufi sect founded around Bukhara Samarkand by Muʿīn al-Dīn Ḥa-san Chishti. They do not care about private possession and tangible property. They believe in the need for generosity like the sea, sweetness like the sun, humbleness like the earth. It spread in India and Pakistan.

AKHISM XIII Century A.D. (Akhi means **my brother** in Arabic and **generous** and **open handed** in Turkish): It was founded by **Nasreddin Mahmud (Ahi Evran)** in Kırşehir, Anatolia, when Turkish artisans were organized as an ethical art institution. It is follow-up of Futuwwa. Artisans are taught the secrets of the profession in a hierarchical order of **assistant, apprentice, craftsman, mas-ter**. In the evenings, ethical teaching was provided at zawiyahs, guests were welcomed, **semah ritual** was practiced with them. They used to wear futuwwa shalwar, welcome strangers, prompt to serve food and to satisfy the wants of others, and ready to suppress injustice and kill those who persecute people. The one entering the organization was called *yiğit* (valiant), *akhi* on the way of the sect, *shaikh* when reaching the destination, and the shaikh on top was *Ahi Baba*. **Nasreddin Ahi Evran** was the master of leather craftsmen and akhism. Those who had not any profession would not be accepted in the organization. Towards the 18th century, akhism transformed into economic institutions and guilds rather than religious ones.

BEKTASHISM (Bektashi Order) XIII Century A.D.: The sect founded by Turkish Sufi Haji Bektash Veli. It is based on Malamiyya, Yeseviyya, Babaiyya, Ithnāʿashariyyah. It aims to associate Islam with the old Turkish religion. It is a mystic and confined sect, which carries a socialist, Esoteric character focused on **equality, brotherhood and collectivism on property,** and has traces of Akhism. Not every human has acquired necessary knowledge, those who have acquired necessary knowledge and maturity are called the **pole.** Similar to rotation of the grindstone around the center bar, such the universe rotates around the poles. The most powerful of many poles is the **pole of the poles.** On his right side sits the "**right imam**" and on the left side sits the "**left imam**". These are followed by **four pillars** ruling across the universe and then come pleiad and tricents. The universe is ruled by these organs. Bektashism maintained its integrity and did not divide into groups for seven centuries and melted various Turkish sects, such as Qalandariyya, Hayderiyya, Abdalism and Hurufism in itself in the course of time.

MAWLAWIYYAH 1207-1273 A.D.: The sect founded in Konya by Sultan Veled, son of Maw-lana Jalal ad-Din after his death. Mawlana was born in Balkh, Khorasan, his father was Bahā ud-Dīn

Walad one of the most important scholars of the era. Mawlana, growing up in an environment of science and culture since he was educated by and discussed with major scholars of the period as of his childhood. He had been equipped with all the contemporary knowledge and culture during his education in Aleppo and Damascus. When Mawlana was about 50 years old, after the arrival of Shams-e Tabrizi in Konya, they developed the conception of the world and life style which formed the basis of the Mawlawi sect. Mawlawi sect does not accept **dhikr**, **seclusion** (suffering) and **esma** (dhikr calling 99 names of Allah) as a rule. It is based on **ecstasy** (rapture with the love of God, being under ecstasy of God). The dervish, who completed 1001 days of suffering (seclusion) earns the right to become a **dede**. The power of human belongs to Allah. Allah created motion. It is required to try to advance in moral as well as tangible things. It is required to be dedicated to the *ihtiyar* (**believe in predestination and fate), be decent, take the blame and see the good from Allah in this struggle. The one who fails to correct himself/herself, desires to be occupied with the faults of others. The follower who speaks much in the presence of his Shaikh resembles a person whose cleanness is disrupted during prayer**. It can be concluded that Mawlana elevated Sufism to the most inaccessible level. To understand him, it would be useful to remember he was from Khorasan and to think about saints of Khorasan.

What can really describe him is actually his own words:

Seven Advices of Mawlana

1. In compassion and grace, be like the sun...
2. In concealing other's faults, be like the night...
3. In generosity and helping others, be like a river...
4. In anger and fury, be like dead...
5. In modesty and humility, be like the earth...
6. In tolerance, be like the sea...
7. Either appear as you are, or be as you appear..

Mawlana was asked about decency; he replied "The **mind** was asked 'What is believing in Allah?' Mind whispered to the ear of the heart and said, '**Faith** (believing the existence of Allah) is **edep** (decency)'" (in Sufism **heart** represents love, therefore it is called **soul** (in spiritual sense) in mysticism. Mawlana does not say "Do not misbehave" since he does not want to break the heart of the atheists. Also in this book, though we cannot find the word to explain the infinite magnificence of our Universe in astral and atomic scale, what could the ignorance about the owner of our Universe could be other than crudity?) Let's continue with Mawlana: "Religions are the same, the difference lies in the path. Moreover, the religion, that is the belief and motion are the means. **The one who reaches the truth is not a member of any religion or any belief**, and worships are a restraint for this kind of man. Intelligence is the shadow of God, God is the Sun, can the shadow ever resist to Sun? The one who is really mad is the one who says he is smart. What is **intelligence**? The candle of the worthless world. Nothing is either absolutely good or absolutely bad. Everything has both benefits and harms depending on the conditions that is why the **knowledge** is necessary. The knowledge should not be a purpose but a tool. **Knowledge has two wings, suspicion has one; presumption has not any, so it cannot fly.** Even there are one hundred books, they all have one chapter; all revolve around a single mihrab; **all roads lead to one home. Thousands of spica originate from a single seed**. One should be mature on the way to the desire. One should keep hands off from this world, treat his/her own eyes, otherwise the world is entirely made of it but you need an eye to see it. Souls descend because of the body, bodies ascend because of the souls. We are the doctors of work and word, you are inspired by the light of the glory of greatness. We think that there is ascedism (fasting, breaking the desires of the self). Your path is entirely the path of living, peace and tranquility. **Justice is putting everything in its own place, cruelty is putting them where they do not belong to"**. According to Mawlana, **love** is **one of the qualification of the Creator. Mawlana strictly opposes to unnatural love.** According to him, the real love is commitment to a mature human or seeing own maturity in the other so that after ecstasy, this love spreads to all people and the world. **He targets for good, beautiful, truth and unity: Mawlana had this high and universal joy in every stage of his life. According to Mawlana, everything is nothing but beloved. One who is not in love is like a bird without a wing, this love is the only medicine to release people from greed, arrogance, possession and ego. The soulful dignity of Mawlana filled with poem and wisdom accepts all people equal on the way to the right and the**

truth. His statement is written at the entrance of his tomb in Konya: "Come, come, whoever you are, wanderer, worshipper, lover of leaving. It does not matter. Ours is not a caravan of despair. Even if you have broken your vows a thousand times".

BADAWIYYA 1276 A.D.: A Sufi sect founded in Morocco by Ahmad al-Badawi. It is similar to Alevis. Their cardigans, ensign and flags are red.

DASUQIYYA 1277 A.D.: As Sufi sect founded in the Southern Egypt by Ibrahīm Bin ʿAbd-El-ʿAzīz 'Abu al-Magd Dasuqi. Commitment to Sharia, dhikr and fasting, adoption of social ethics and commitment to Allah are important; they wear green clothes.

NAQSHBANDIYAH 1389 A.D.: A Sufi sect founded around Bukhara by Baha-ud-Din Naqshband Bukhari. In this sect, time is like milestones on the spiritual path. It is necessary to discuss every moment experienced, care for breathes and be grateful. They look at their toes to avoid sin. Naqshbandis were members of Hanefi School; they used to put emphasis for confidentiality in rituals and did not let non-members in.

BAYRAMIYYA 1429 A.D.: A Sufi sect founded in Ankara by Hajji Bayram-i Veli. He routed his followers to make a living of hand labor like himself. Thereafter famous Sufis like Akshamsaddin and Aziz Mahmut Hudayi became shaikhs of this sect. However, Aziz Mahmud Hudayi founded another sect called Jalwatiyya in 1628.

BEDREDDINIYYE XV. Century A.D.: The sect founded in Anatolia by **Shaikh Bedreddin Mahmut** of Simavna. **They wanted to establish a state order, which removes all kinds of discriminations of religions and sects, considers people equal and partner in property with a religious, political and social approach.** Shaikh Bedreddin says in his book titled Varidat "God created the world and gave it to people, thus the earth and all products of this earth are common properties of people. It is against the rules of God if one accumulates property and the other remains hungry. I should sit in your home as if it is mine, you should use my property as if it is yours, because all of them are for all of us and belong to us". Shaikh Bedreddin strictly objected to the idea of partnership in women and defended **the concept of family**. Every good is heaven and every bad is hell. Worshipping and praying are all for correcting ethics and purifying the inner; real worshipping has no conditions, limits and shapes.

ḤURŪFIYYAH XVI. Century A.D.: A Sufi sect founded by Fazlallah of Iran. It is based on Pythagoreanism, Cabalism and Batiniyya. It says, **"Creator is the letter"** and tries to derive meanings from letters. Human is the speaking God. Letters have unlimited possibilities of combination. La ilahe illallah (Allah is the only God to worship) is written with three Arabic letters, **lam, Elif, be**, they respectively represent **intelligence, self and fate**. Its formation of four words indicates four natures of human, 7 syllables refer to two eyes, two ears, two nostrils and one mouth in the human head, 12 letters stand for 12 organs of humans. **Thus, this statement gives voice to human** and proves the occurrence of God in human. According to Hurufis, existence is explained with letters. **The purpose is human, explaining human explains God**.

BABISM (Bahá'iyya) XIX. Century A.D. Islamic sect founded by **Siyyid Muhammad Bab** of Iran. God is one, Muhammad Bab is the mirror of God, everyone who looks at him can see God in him. The number 19 is sacred. The year is comprised of 19 months and the month is comprised of 19 days in their special calendar. Co-religionists are administered by a council of 19 members. Every year, one month, i.e. 19 days, fasting is required. Number 19 is the number obtained by calculation of words **Vahid** and **Vucud** reflecting God in accordance with abjad. 19 pages of Beyan, the book of Muhammed Bap should be read every day. (Abjad calculation: the sequence 8 words indicating the former sequence of the Arabic alphabet, every letter has a numeric equivalent).

c)   SUFISM IN OTHER RELIGIONS:

~1000-2000 BC, **in Egypt** at the end of the era of the pharaohs, priests believed in a single god. According to **Hermes Trismegistus**, who was the first Greek learning the knowledge of the priests in the Egyptian temples, none of our ideas could depict Allah; a shapeless being cannot be perceived by our senses. The one out of the time cannot be measured with time. Nevertheless Allah gives some of the selected humans the ability to have certain appearances from its supreme maturity. Those who deserve it, cannot find any word to tell what they see and feel to the public. People reach this level after a long and exhausting suffering stage.

c1) According to BUDISM in 600 B.C.; it was required to get away from earthly pleasures and spend time with fasting and thinking to prevent passion and desire, the source of all bad, trouble and sorrow, from being the purpose of life and to release the soul from being the slave of the self.

c2) In VI. Century B.C., Pythagoras and then Socrates and Plato were the founders of the mystical movement **in Greece. According to Pythagoras, the human body and the soul are a minor model of this world. What is dominant in this world is disorder and intrigue. When peace and welfare is established in the society, Allah comes down to the conscious of people.**

c3) IN CHRISTIANITY: Bishop Denys L'arepagite says in his book **"The Divine Names"** in order for the soul to leave the material world, the individual must leave *masiva* (everything but God) and to consider themselves non-existing, this is the only way to reach God. He suggests mystical belief is not acquired by **intelligence** but **love**. Realization of **observation (God in the heart)** is possible with **pieous life** (excluding anything but God from the heart). According to a Christian mystic **Saint Victor**, **reading**, **begging to God** and **control of self** are three levels of mystic activity. In the 12th century in Byzantium, people in a Christian sect named **Hesukhia** tried to get close to God with methods such as **dhikr** (invocation) and **ecstasy** (losing consciousness). Especially the monks of Mount Athos believed they combined with God with **introspection** (introspective thinking) in silence.

c4) BRAHMANS (Priests) between I- VIII centuries A.D.; In India, when Aryans began settled life on agriculture, they left all wishes and desires and lived a secluded life away from the public to think about and reach the god named Brahma.

c5) Between V. and XV. Centuries A.D., kabbalist mystic thinking of the **Jews**, can be summarized as **ilm-i ledün** (innate science), **tecelli** (appearance), **suffering**, **seclusion** (isolation). Appearance of God in the form of fire to Moses and friendship of Moses with Khidr led to adoption of knowledge about God. According to Jewish mystics, it is required to have a **form** (substance) to have an appearance. The most beautiful form of the world is **fire**. It suits the best to the glory of Allah. Suffering that was inherited from Prophet Moses during the seclusion was adopted by Jewish mystics.

d)   OTHER ORGANIZATIONS SIMILAR TO SUFISM

<1723 A.D., FREEMASONRY: (Mason means a **bricklayer** in many western languages). The term is derived from the teachings of **Hermes, Pythagoras, Kabbalah, Akhis and the Guilds**. The first freemasonry association was founded in England in 1723. Others were founded in Germany the same year, in France in 1725 and in USA in 1760. Masonry associations have spread all over the world today. Freemasonry is attributed to the period, when **Hiram, the architect** of the Solomon's Temple, broke the workers into three categories, namely **apprentice, fellowcraft and master** under the reign of **Solomon**, the king of Israel and a prophet, who lived between the years **973-923 A.D.** The degrees of Freemasonry retain the three grades of apprentice, fellowcraft and master. Their organization is made up of Lodges of at least 7 members. The chair of the lodge is called worshipful master. He has two surveyors, 1 spokesman and 1 secretary. There also treasurers, marshals and wardens. Lodges combine to form Grand Lodges. Today's superior degree system has 33 degrees in Royal Arch in England (symbols originated form Christian beliefs) and Ancient Accepted Scottish Rite. 33 degrees are divided into four groups. Degrees 31-33 are the highest administrative levels. Their president is the Inspector General. Degrees in Turkey also follow this Rite. Operating principles of freemasonry are brotherly love, relief and truth. Their fundamental rules are mutual tolerance, self-respect and respect to others, freedom of conscious. They leave the religious preference to the discretion of their members. They try to improve financial and moral status and maturity on intellectual and social terms of humanity. A freemason is a free person who values truth and justice above all, has freed himself from superstitions of the public, loves the virtuous rich and poor ones and has high ethical manners. Humanity has three enemies: **ignorance, bigotry** and **greed**. Number 7 is sacred. Salt and sulfur used at masonry membership ceremonies remind Hermetics and wearing leather aprons reminds wrapping loincloth in Akhi order and the lodges.

V) RELIGIONS OF TODAY'S WORLD

Religions other than **Hinduism, Buddhism, Shamanism, Taoism, Confucianism, Shintoism, Judaism, Christianity and Islam** have been completely extinct or have survived at insignificant levels. Some examples are:

1) RELIGION OF THE NATIVE AUSTRALIANS: Tribes settle in circles. The circle is divided into the number of clans in the tribe. The right hand is close to holy objects, the left to unholy ones. Religious paintings and sculptures (mammoth, reindeer, aurochs, horse etc.) are available in caves and sacred places, during **intishuyuma** (a totemic) ritual, clan members collectively go to a cave with many symbols of their ancestors in rocks and dust powders around for wealth and fertility (The association of animal reliefs on rocks in Göbeklitepe, near Urfa, in East Anatolia with the oldest temple in the world is worth investigating).

2) RELIGION OF THE NATIVE AMERICANS: In Hopi Indians, living in the plains in the northeastern Arizona, tribes settle in four directions, each direction has a color, marriages between tribes occur among tribes of different directions.

3) SHAMANISM: More precisely, Toyonism; this religious belief called by this name today is encountered in slightly different forms in Tunguska, Mongolia, Manchuria in Asia, Laponia, Eskimos, Voguls, Ostyaks and Samoyedics, Caucasia, India, China, Indonesia, Greenland, Melanesia, Polynesia, Australia, Pacific Ocean Islands, Alaska, Iceland, North America, Guyana, Amazon region and numerous territories in Africa

VI) CONCLUSIONS ABOUT THE RELIGIONS:
1)DEFINITIONS OF RELIGION:

RELIGION: In accordance with the metaphysics system of thought, many Western philosophers state the historical beginning of religions in a different way. According to **E. B. Tylor** Animism; according to **P. W. Schmidt** Monotheism; according to **R. R. Marett**, a universal, immanent force or energy (Dynamism); according to **Spencer**, the worship of the spirits of deceased humans: ancestor cult (Manism); according to **Auguste Comte** Fetishism; according to **R. Smith**, belief about the relationship between people and nature (Totemism); according to **R. Thurnwald**, worshipping animals (Zoolatry); according to **N. Söderblom**, belief in a hidden and secret power (Spirit); according to **J. G. Frazer and many others Witching** was the first type of religion. According to **E. Durkheim** who is one of the most popular metaphysics oriented philosophers, religion is composed of two basic categories: Beliefs and ceremonies. Religion is a system of belief and actions about separated and forbidden divine things. All of these belief and ceremonies constitute a religion. Religion is a system of symbols, which makes the community reach to its consciousness.

2) UNIVERSAL RELIGION:

A new belief design above all regional beliefs and various religions has been proposed from time to time by a variety of thinkers in order to unite all beliefs, that discriminate people and in some places, antagonize each other. The proposals of **J. J. Rousseau, Auguste Comte and Felicien Challaye** are among the most popular ones. At the end of his work "History of Religions", **Challaye** says: "Historical evaluation of special religions takes us to a universal religion. One day, this religion will find its own language, which will take it from heart to heart. As a result **"no religions can be separated from each other with strict boundaries"**. For example, **"the first spring festivals"** which we can assert to go back to the first ages of humans; **"Hıdır Ellez"** (Khidr- Elijah) in Muslims; **"Easter"** in Christians (while its name was **Mitra or Saturnalia in Romans,** Christians made it before the new year); **"Peshah"** in ancient Semites and Jews; **"Mitra"** (Sun) in India, Mitanni Kingdom in South East Anatolia and Hurrians; **"Saçu"** in Mongols; **"Saçılga"** or **"Islah"** in Yakuts; **"Mihrimah"** or **"Nevruz"** in Iran Persians and Parsees; also in China and Japan, they have been celebrated and are still celebrated. We can give many other examples for the common points from the first religions to the most developed ones.

In Mawlana's own words: **"Even if there were a hundred books, they all have only one chapter. They turned towards a single mihrab in these hundred directions. All of these roads lead to a single house. These thousands of wheat spica are made of a single seed".** There are the verses on the door of his tomb in Konya*: "Come, come, whoever you are; wanderer, worshipper, lover of leaving. Ours is not a caravan of despair. Even if you have broken your vows a thousand times, it does not matter. Come, come yet again, come".* It cannot be denied that all religions, cults and sects are related with the level of knowledge and living conditions of humans in that particular era and region. Almost 400 sects, which emerged after the last religion Islam, came to rise by the efforts of some Islam philosophers to incorporate previous religions and knowledge into Islam.

According to Mawlawi sect and many other Islamic orders, none of the members of any religion can consider the members of other religions as enemies.

On the last page of the fourth part of this book, a synthesis is made according to our knowledge level in the 21st century. The **"one and only seed"** in His holiness Mawlana's verse, **"These thousands of spica grow up from one and only seed"** is assumed to be the **Big Bang** and the "**Bose-Einstein Condensate**". We do not know, yet. Thus, humbly; **the target of Mawlawi dervishes to get closer to Allah mystically was achieved in a scientific way.** In one sense, I think that we reached to the **Divine Light** which Allah has written down everything on before creating (or as in the name of our book "**the Master Plan of Universe**" as also told in Quran, and which Mysticism defines as the "**ancient divine light**" (never ending divine light), in religions and in science, as in His holiness Mawlana's verse "**All of these roads lead to a single house**". Apologize for any slip of the tongue!

However, **Challaye's** longing for a new "**Universal religion**" stated above is wrong. The world has experienced enough holy wars. It is enough to look at Sufism and His holiness Mawlana, who had been inviting and accepted in Anatolia and worldwide for 700 years. Otherwise, if we would make religions materialistic or make them serve to materialistic purposes, while trying to unify them, it would become the greatest evil for humanity. Humanity can enter a desperate path that is catastrophic and irreversible.

ADDITIONAL PICTURES AND DRAWINGS FOR HUMAN'S EVOLUTION OF RELIGION:

C IN PART TWO
ADDITIONAL FIGURES AND DRAWINGS ON THE RELIGION EVOLUTION OF HUMAN BEINGS:

CHINESE COSMOLOGY (DURING THE CHOU FAMILY 1059-249 B.C.)
AS PER THE ROTATION OF MOON AROUND THE SUN
IN 28 DAYS, THERE WERE 28 SIGNS OF ZODIAC,
7 FOR EACH SEASON. BUT IN THIS FIGURE
WE CAN SEE TODAY'S COSMOLOGY WITH
12 SIGNS OF ZODIAC. ACCORDING TO
THAT AGE'S KNOWLEDGE, EARTH
WAS IN THE CENTER AND SUN,
MOON, PLANETS AND SIGNS
OF ZODIAC WERE AROUND IT.
7- SATURN
6- JUPITER
5- MARS
4- SUN
3- VENUS
2- MERCURY
1- MOON
EARTH IS IN THE CENTE

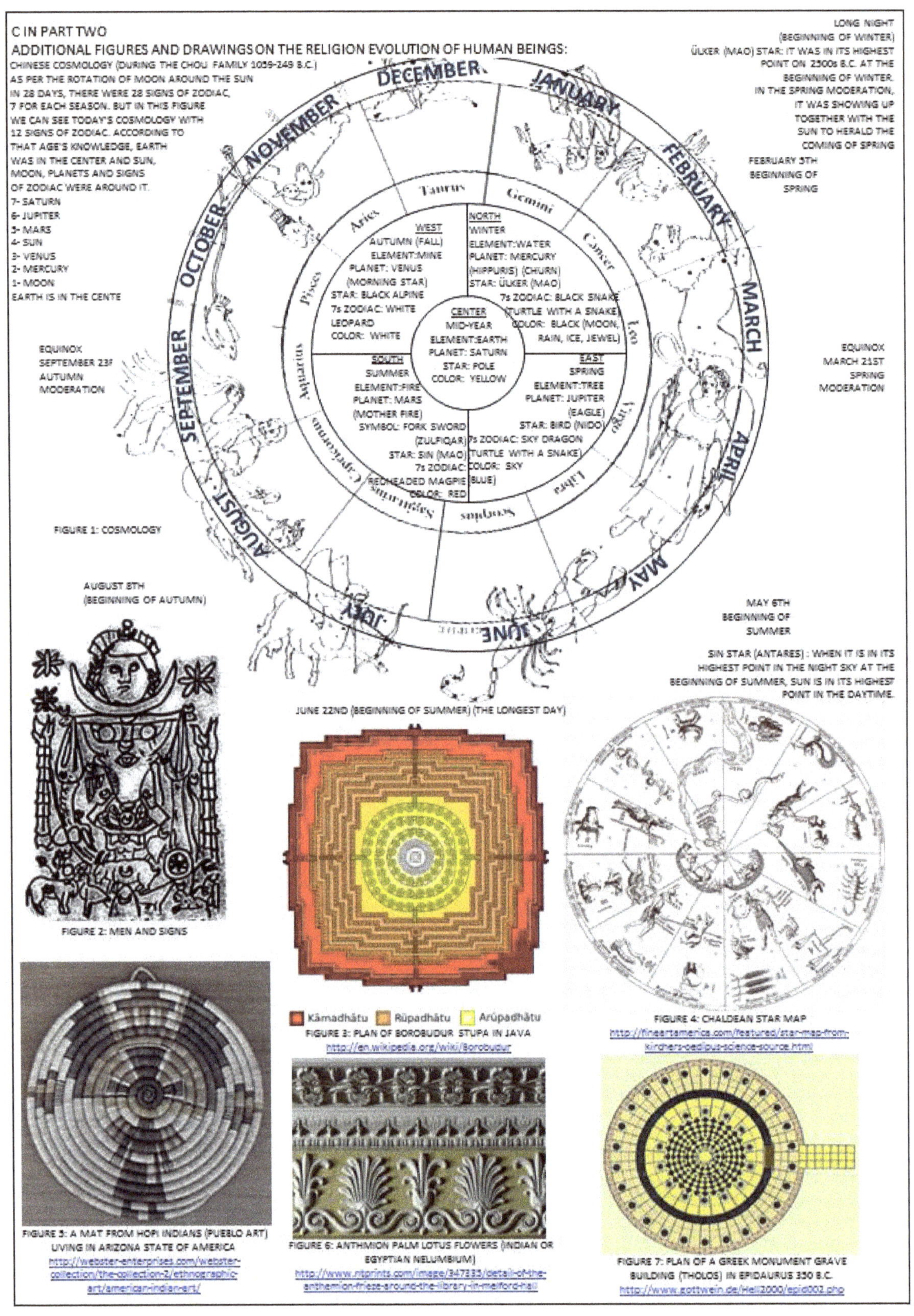

EQUINOX
SEPTEMBER 23F
AUTUMN
MODERATION

FIGURE 1: COSMOLOGY

AUGUST 8TH
(BEGINNING OF AUTUMN)

LONG NIGHT
(BEGINNING OF WINTER)
ÜLKER (MAO) STAR: IT WAS IN ITS HIGHEST
POINT ON 2300s B.C. AT THE
BEGINNING OF WINTER.
IN THE SPRING MODERATION,
IT WAS SHOWING UP
TOGETHER WITH THE
SUN TO HERALD THE
COMING OF SPRING

FEBRUARY 5TH
BEGINNING OF
SPRING

EQUINOX
MARCH 21ST
SPRING
MODERATION

MAY 6TH
BEGINNING OF
SUMMER

SIN STAR (ANTARES) : WHEN IT IS IN ITS
HIGHEST POINT IN THE NIGHT SKY AT THE
BEGINNING OF SUMMER, SUN IS IN ITS HIGHEST
POINT IN THE DAYTIME.

JUNE 22ND (BEGINNING OF SUMMER) (THE LONGEST DAY)

FIGURE 2: MEN AND SIGNS

FIGURE 3: PLAN OF BOROBUDUR STUPA IN JAVA
http://en.wikipedia.org/wiki/Borobudur

FIGURE 4: CHALDEAN STAR MAP
http://fineartamerica.com/featured/star-map-from-kirchers-oedipus-science-source.html

FIGURE 5: A MAT FROM HOPI INDIANS (PUEBLO ART)
LIVING IN ARIZONA STATE OF AMERICA
http://webster-enterprises.com/webster-collection/the-collection-2/ethnographic-art/american-indian-art/

FIGURE 6: ANTHMION PALM LOTUS FLOWERS (INDIAN OR
EGYPTIAN NELUMBIUM)
http://www.ntprints.com/image/347333/detail-of-the-anthemion-frieze-around-the-library-in-melford-hall

FIGURE 7: PLAN OF A GREEK MONUMENT GRAVE
BUILDING (THOLOS) IN EPIDAURUS 330 B.C.
http://www.gottwein.de/Hell2000/epid002.php

C IN PART TWO
ADDITIONAL FIGURES AND DRAWINGS ON THE RELIGION EVOLUTION OF HUMAN BEINGS:

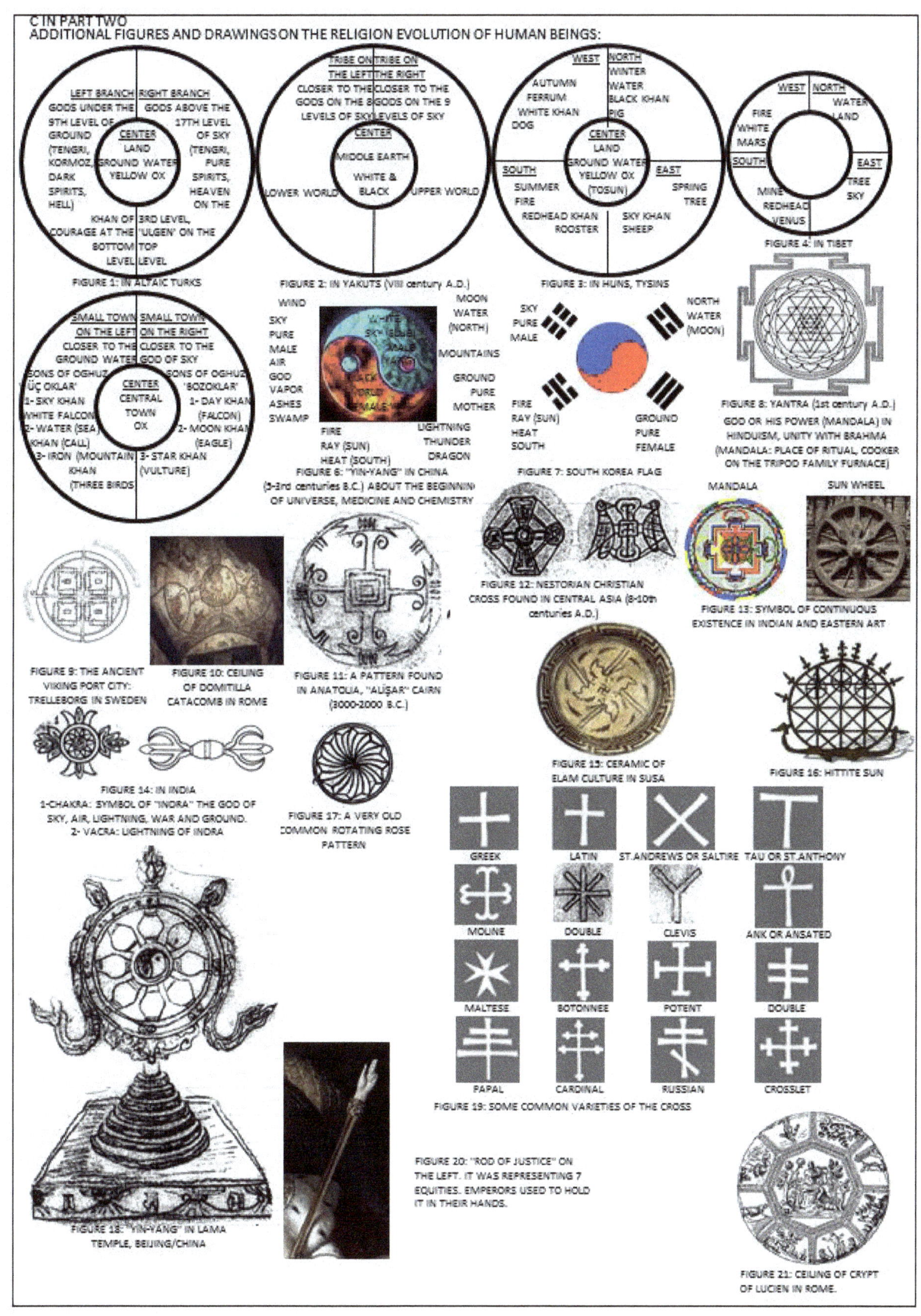

GOD OR HIS POWER (MANDALA) IN HINDUISM, UNITY WITH BRAHMA (MANDALA: PLACE OF RITUAL, COOKER ON THE TRIPOD FAMILY FURNACE)

FIGURE 9: THE ANCIENT VIKING PORT CITY: TRELLEBORG IN SWEDEN

FIGURE 10: CEILING OF DOMITILLA CATACOMB IN ROME

FIGURE 11: A PATTERN FOUND IN ANATOLIA, "ALIŞAR" CAIRN (3000-2000 B.C.)

FIGURE 12: NESTORIAN CHRISTIAN CROSS FOUND IN CENTRAL ASIA (8-10th centuries A.D.)

FIGURE 14: IN INDIA
1-CHAKRA: SYMBOL OF "INDRA" THE GOD OF SKY, AIR, LIGHTNING, WAR AND GROUND.
2- VACRA: LIGHTNING OF INDRA

FIGURE 15: CERAMIC OF ELAM CULTURE IN SUSA

FIGURE 16: HITTITE SUN

FIGURE 17: A VERY OLD COMMON ROTATING ROSE PATTERN

FIGURE 19: SOME COMMON VARIETIES OF THE CROSS

FIGURE 20: "ROD OF JUSTICE" ON THE LEFT. IT WAS REPRESENTING 7 EQUITIES. EMPERORS USED TO HOLD IT IN THEIR HANDS.

FIGURE 18: "YIN-YANG" IN LAMA TEMPLE, BEIJING/CHINA

FIGURE 21: CEILING OF CRYPT OF LUCIEN IN ROME.

C IN PART TWO
ADDITIONAL FIGURES AND DRAWINGS ON THE RELIGION EVOLUTION
OF HUMAN BEINGS:

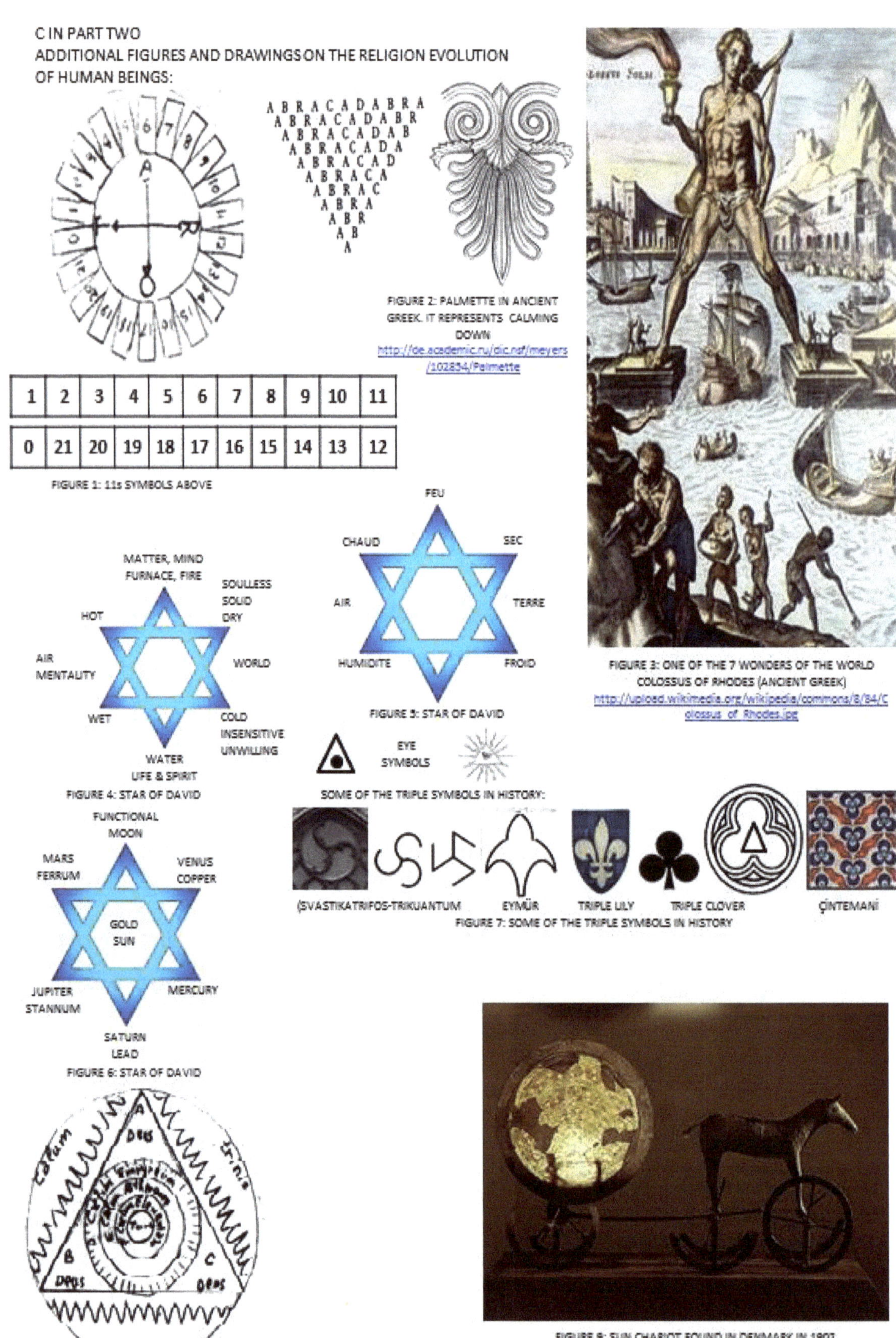

1	2	3	4	5	6	7	8	9	10	11
0	21	20	19	18	17	16	15	14	13	12

FIGURE 1: 11s SYMBOLS ABOVE

FIGURE 2: PALMETTE IN ANCIENT GREEK. IT REPRESENTS CALMING DOWN
http://de.academic.ru/dic.nsf/meyers/102834/Palmette

FIGURE 3: ONE OF THE 7 WONDERS OF THE WORLD COLOSSUS OF RHODES (ANCIENT GREEK)
http://upload.wikimedia.org/wikipedia/commons/8/84/Colossus_of_Rhodes.jpg

FIGURE 4: STAR OF DAVID

FIGURE 5: STAR OF DAVID

FIGURE 6: STAR OF DAVID

SOME OF THE TRIPLE SYMBOLS IN HISTORY:

FIGURE 7: SOME OF THE TRIPLE SYMBOLS IN HISTORY

FIGURE 8: GOD OF SKY, AIR, CLIMATE AND HEAVEN: "CIE" WHICH HAD BEEN PREPARED IN OPPENHEIM/EUROPE IN 1619

FIGURE 9: SUN CHARIOT FOUND IN DENMARK IN 1902
http://www.pinterest.com/pin/460000549410599087/

D) HUMAN'S EVOLUTION OF KNOWLEDGE:

I) KNOWLEDGE IN THE COLLECTING - HUNTING PERIOD:

2.200.000 years ago: The first stone axe in Ethiopia.

400.000 years ago: **Fire was utilized** without extinguishing in China.

30.000 years ago: The immigrants from Europe to the South, know how to cover themselves by sewing leathers; and how to throw stones by means of slingshots.

II) KNOWLEDGE IN THE AGRICULTURE AND STOCKBREEDING PERIOD:

15000 B.C.: Primitive plough is known in Central Asia. The human, knowing **how to make bread from flour, live together.** These people who emigrated from Asia to the Mediterranean region, have brought flour, bread and primitive plough to Europe.

15 B.C. -10000 B.C.: In Central Asia, in the region of Altaic Mountains, people knew **how to melt and mould the minerals in furnaces graved in the rocks; how to use wheel, carts and horses; and how to light a fire.**

III) KNOWLEDGE IN THE FIRST VILLAGES AND LATER:

9500 B.C.: There are animal reliefs on the obelisks in **Göbeklitepe in the vicinity of Urfa province in Southeast Anatolia, which is the oldest known settlement and temple in the world and also in which agriculture has first been done in the world.**

8000 B.C.: There are earthenware and stone figurines, unicolor pots and pans, necklaces, stamps and seals in the Central Anatolia near of Burdur.

7500 B.C. -5000 B.C.: In Central Anatolia; "Can Hasan" settlement in the vicinity of Karaman; **"Aliler"** and **"Alacahöyük"** settlements in the South of Çorum, **"Tilki Tepe"** settlement in Van, Eastern Anatolia.

7000 B.C.: People **in "Çatalhöyük"** village, in Çumra town of Konya in **Central Anatolia** used to enter their houses from the chimneys. **"Jericho"** village in today's Jerusalem city in Israel.

6000 B.C.: In Eastern Anatolia, in the vicinity of Diyarbakır, **"Çayönü Tepesi"** village, there were ventilation systems and stores in the grill shaped houses. **People were using flour-grinding stones.**

5000 B.C.: In the East of Caspian Sea, Anav Culture, in two of the cairns in the vicinity of Ashgabat, people knew **stockbreeding, agriculture, how to make bread with flour. People were building adobe houses, colored clay potteries.** They knew how to ride horses, how to process mines and used **ox-driven carts.** Also **"Minusinsk cult"** in Asia.

IV) EVOLUTION OF KNOWLEDGE FROM THE FIRST CITIES UP TO DAY:

5000 B.C. -4000 B.C. INDUS VALLEY: In "Mohenjo-Daro" and **"Harappa"** cities, they had two storey brick **Ziggurat type houses, intersecting roads and streets, waste water and potable water canalets, wells, public baths, wheat depots, mills, multi storey temples and palaces, cotton and weaving, hieroglyphs on the bricks, sculptures of Mother Goddess and bearded Gods, a three headed** God (to be later transformed to **Shiva**) and water cult.

4000 B.C.: Having come **to EGYPT** via Syria and Palestine, Asian people brought an advanced civilization and founded **city-states.** After this date, hieroglyphs were known in Egypt. There were a temple and state palace in the center of cities surrounded by city-walls.

4000 B.C. -3000 B.C.: Sumerians. (Sumeru: The name of **Golden Mountain** in Buddhist cosmology) in **Mesopotamia:** They came with an advanced culture with thousands of figures and pictographs. They founded 11 city-states in a place named **Sinear**, (In **B.C. 3000,** these city-states united) they built water channels. **They used adobe brick and applied dome and arc system for the first time in the structures. They made sculptures and reliefs. They knew how to make primitive plough, wheels, carts, weaving loom and metallic tools. They had astronomic knowledge. They sorted the stars, zodiacs; divided one month into 30 days; one year into 360 days, 12 months and 13 months; they divided the months into weeks and weeks into the days, they gave a name to each of them. They invented sundial, and discovered that the lunar eclipse and solar eclipse happen periodically; dealed with mathematics; prepared multiplication and division tables; were able to measure area and volume; divided the circle into 360 degrees; had length, weight measurements and medical knowledge; used to prepare medicines; had granaries, depots, shopping areas, schools and workshops in their 7 storey towers (ziggurats), temples and around them. They were trading with far countries.**

2980 B.C. EGYPT: Memphis became the capital city. The oldest written works related with **philosophy, astronomy, medicine and mathematics, which could survive up to today. They invented a system to count ten by ten, and hundred by hundred.**

2800 B.C. -2050 B.C. EGYPT: As in the ziggurats of Sumerians, **various information and pictures have started to be doctrine in the schools around the pyramids.**

2200 B.C. BABYLON: A city-state was founded. Among all other gods, the biggest god was *knowledge* **and at the same time Sun god, "Marduk".** All of the eight entrances of the city, which was surrounded by walls, were given names. ("**Sin**" was the God of Moon; "**Enlil**" was the God of Wind; "**Shamash**" was the God of Sun; **Venus "Ishtar"** was the Goddess of Love, War, Fertility and Sexuality; "**Adad**" was the God of Storms; and there were Gods named "**Uras**" (the Goddess of Earth) and "**Shabaka**"). Entering at "Ishtar Gate", ran the Processional Way, which was lined with walls covered in lions on glazed bricks (about 120 of them). On the left, there was "**Temple of Marduk**" which was 20 meters high; and on the right there was the "**Tower of Babel**" which was 90 meters high, and which had 7 floors. Every floor of the tower represented a planet and they were in different colors. Seventh floor was covered with blue tiles. On the lower floors, there were temples of priests, granaries and guest rooms.

2000 B.C. In EGYPT, "Osiris", the God of measurements and mind, appeared in the Gods. Osiris was teaching **art and handicrafts. "Ptah" was the God of Memphis city, artists and miners.**

2000 B.C. VEDAS IN INDIA (means verbal information) information constituting the basics of religion were written down. In this religion, **VARUNA** was the God of daytime sky, mind, universe and ethics.

2000 B.C. IN GREECE, there were **"Muses",** the nine daughters of **Athena, God of Mind and Intelligence** and Zeus (Father of Gods), **the Goddesses of Art and Science.** They were; "**Clio**" the Goddess of **history, "Euterpe"** the Goddess of **Song and Elegiac Poetry, "Thalia"** the Goddess of **Comedy, "Melpomene"** the Goddess of **Tragedy, "Terpsichore"** the Goddess of **Dance, "Erato"** the Goddess of **Lyric Poetry, "Polyhymnia"** the Goddess of **Hymns, "Urania"** the Goddess of **Astronomy**, and "**Calliope**" the Goddess of **Epic Poetry**. Muses have also existed before the Greeks.

2000 B.C. IN EUROPE: Handicraft in Celts; **god of technique** was "**Mercury**"; **handicraft teaching god** was "**Minerva**".

1300 B.C. -800 B.C. HERMESTHOTH: "**Hermes Trismegistus**" (Thrice great Hermes) or "**Mercure Trismegistus**" (Thrice great Mercure). There are some knowledge in Clement of Alexandria's (150-216 A.D.) book "Stromata" on Hermes Thoth. According to this book, **Hermes Thoth has written 42 books:**

10 of them were about religion,

10 of them were about religious ceremonies,

2 of them were about hymns for gods and the rules, which the kings had to obey,

4 of them were about Astronomy and Astrology,

10 of them were Cosmography, Geography and their ceremonies,

6 of them were about medicine, human body, organs and illnesses, medical instruments, drugs, ophthalmology and feminine manners.

The original of Clement's book is in the Museum of Athens; and its replica is in Topkapi Palace - Istanbul. Among the Greek Gods, Hermes was the youngest, smartest, downiest and master of the thieves. As **Hermes** was also the name of a coterie, which had the secret truths; at the same time it was the name of planet Mercury which has been known as a god. **Thoth** was the Moon God, which has managed the births, deaths and thoughts in Egypt. According to Torah, Hermes Thoth was a prophet from the sixth generation, he was **Enoch**. According to Quran, he is the third prophet (Idris) after Adam and his son Sheath. **He is the first human using a pen to write and who used stitching. In Islam, he is introduced as the one who had known the secrets of skies. According to Hermes, light is soul and darkness is matter. Souls come down to earth from the sky to unite with the matter**. This is an examination for them. If they can pass, they rise to the holy light again. Human soul is the child of God. If the human succumbs to the matter, the divine light in his/her soul will retreat. The soul is left in the dark, dissolves and disappears. The rising soul becomes all beauty, all power, all mind by means of sense of clearness and this is the immortality. In the science of medicine, it is asserted that everything in nature is composed of **salt, sulphur and mercury** and that the illnesses arise when they become unbalanced. The right medicine is the one, which can bring them into balance again. In his speech to the public, Hermes Thoth indicated that: "**There is only one supreme, divine light and it is hidden everywhere. The divine light is closer than anything to human. It is hidden in the soul, conscience and heart of humans**. Everybody can search and find it in self. **If the divine light in heart becomes acquainted with the divine light in the sky, the human incorporates with the divine light, which is the only divine power and becomes immortal.**" In his book, addressing Pharaoh, **Hermes Thoth said; "Search the divine light. Because a ruler can manage well only if he can see the divine light in the hearts of his people". Then, addressing to the public**, he said; "**You are the divine light, may this light shine forever". The divine light,** the supreme light is the Egyptian god **Osiris**. Previously, that was the one and only God "**Amon Ra**" which the gods of sun and sky had formed together.

Because they believed that the planets and the sun were rotating around the earth; according to Hermes Thoth, the ranks which the souls of humans would rise by means of *knowledge* (the path of souls coming down to the world and rising from world to the divine light, 7 paths of Nur'u Kelâm (the divine light of word)) are:

7th planet Saturn (Zuhal): Holds the golden of globe of universal wisdom; it hides the great brightness, it is the immortality. It hides all of the secrets of mind.

6th planet Jupiter (Customer star): Holds the divine scepter of most supreme power. It is the genius of science.

5th planet Mars (Merih): (Aton, Atun, Ra): Holds the whip or sword of justice in its hands.

4th planet Sun: Rises the victory torch of immortal beauty. It spreads lights of success.

3rd planet Venus (Zuhre): Holds the mirror of love in its hands.

2nd planet Mercury (Utarit): (Hermes, Anubis): Shows the way to the souls which comes down to the world and which rises up with the snaky scepter of science,.

1st planet Moon (Thoth): Holds a silver sickle. It is the doctor god, who manages the thought, birth and death.

World (Fire): The dark universe of matter. It is the divine fire, which hears, sees, moves. It is the creator, the ore, the divine word.

HERMETICISM: It is a secret monotheistic sect that the priests in the temples of Egypt achieved as a result of their studies during the previous 3-4000 years, at the end of the age of Pharaohs. While managing the prayers of people to the old polytheistic religions, the priests were believing in that one and only god now.

In **31 B.C.,** when the Romans conquered Egypt, 600.000 books were found written on papyrus rolls in the temples of priests. In this last period, there were 36.525 books according to **Stromates**; 20.000 books according to **Diodorus Siculus and Julius Firmikus Maternus**; and according to **Iamblichus**, Hermes Thoth was said to have 1.200 books only about Gods. These were most probably the books belonging to the priests in the temples. It is probable that Hermes Thoth had 42 books as Clement of Alexandria wrote. But the important thing is that; knowledge of 3000-4000 years of the priests has been the resource of monotheistic religions and today's world of science by means of Greek and Islamic scholars. **Hermes Thoth** is the first Greek who managed to enter into one of the temples of Egypt as a priest and learned the knowledge in the temple before **Pythagoras**. At the end of the age of Pharaohs, the doctrine, which we call as Hermeticism today, was the biggest divine secret of the priests of the **Temples of Thebes and Memphis in Egypt**. Therefore, they were not written by the priests on papyrus. Later, the secrets, which were revealed by Hermes Thoth, were verbally told to the priest candidates on 22 pictures (11 on the right and 11 on the left) along a corridor in the temple of Memphis. First the priest candidates were taken to the temple of **Isis** (wife of Osiris). After serving here for a while, they were left to enter into the temple of Memphis through a small hole, where they could proceed only crawling on their knees and elbows, between snakes and scorpions. A priest, they used to meet in the small rooms on the way, used to ask if they would like to go back, or not. Later, while they went on crawling, they used to hear sounds and screams advising them to go back. While they going on down from a steep slope in hunger, thirst, with wounded knees and elbows; they used to face with a precipice. If they could save themselves from falling, they would see a small door on their left, behind the door, climbing the long stairs, they would reach to a corridor. **This was the enlightened temple of God Osiris**. A priest would tell them the secret meanings of the 22 letters and pictographs. Every figure had three meanings; being one **divine**, one **mental and one in the physical world**. The letter 'A' corresponded to the number 1; to the **god,** the source of everything **in the celestial world**; **to unity in the mental world**; and to the senses and **human in the physical and natural world**. The sign 'A' was a clairvoyant holding a scepter in his hands and having a golden crown on his head. White dress used to symbolize the **soul and clarity of mind; scepter** used to symbolize the **potency, dominance; golden crown** used to symbolize the **light**. After telling the secret meanings of these 22 letters and numbers, the priest candidate would pass through a hot furnace. After succeeding in the **fire test** and the **water test** later, if he was not discouraged or did not lose his passion, he would enter into a **lust test**. In a bedroom where a young, beautiful and half-naked woman served alcohol and foods; which was enlightened with red lights, color furnished; and where music of lust can be heard. He would succeed if despite his hunger and thirst, he would not eat, drink anything and he had an intercourse with the woman. Otherwise, they would be lifetime captives and servants in the temple; and if they tried to escape, they would be killed. After each test, priest candidate used to be left in a room made of stone for months on his own to think. The last test was the **grave test**. In a ceremony, they were buried alive in a grave. In the grave, they would see a bright pentagonal star with a mystic music in the background. Later the star would turn into a flower. After the grave test, the disciple would attend to a feast with the other disciples who passed the same tests and got the same rank. Thus, 7 divine lessons: **grammar, logic, rhetoric, geometry, mathematics, music and astronomy** were taught to the priest who passed the first degree. Second degree was called "**Serapis**", and the third degree was called "**Osiris**". These levels were not the highest maturity and perfection levels. Education and instructions would go on until the death to find the **secret light** and to be alone together with it. These tests were examples to the tests, which the souls enter into, on the earth, after separating from the only creator Soul.

    XIII-III. Century B.C. In Italy: In Etruscans, "Minerva" was the god of handicrafts.

    XII-VII. Century B.C. In North-West Anatolia: In Phrygians, "Sabazios" was the Priapus god and the god of handicrafts.

    625 B.C. -545 B.C. THALES of Crete: Greek poet, mathematician, politician, nature scientist, musician, astronomer, tradesman and philosopher. His ancestors were from a family that migrated from Phoenicia to Crete. He visited Egypt. In **B.C. 585,** he predicted the solar eclipse. He divided the year into 365 days. He accepted that the source of nature was water; and the earth was a piece of wood on that water. He said that "the world is held up by water and rides like a ship, and when it is said to 'quake' it is actually rocking because of the water's movement." He believed the existence of a mind made of water. He invented some of the poetic measures and Thales' Theorem in geometry. Besides, he is said to invent the magnifying glass.

    615 B.C. -545 B.C. ANAXIMANDER of Miletus. He was a pupil of Thales, but he had a different perspective and was a materialist. He learned astronomy, mathematics and geography from Mesopotamia, and he taught to the Greeks. He tried to establish a systematic design of universe. He defended his opinion that the first reason was an unclear essence surrounding the world completely.

    600 B.C. -300 B.C. In India, BRAHMANS declared in their Upanishads that salvation could no longer be reached by means of sacrifice, but by means of knowledge; all kinds of evil were arising from ignorance.

B.C. 585-525 ANAXIMENES: Philosopher of Miletus, pupil of Anaximander. None of his works survived today. According to him; **air** is the essence of everything. The first reason is **cosmos, which** surrounds the world from four directions. The mother of all of the beings is the biggest god. **Fire, water and earth** come into existence by means of condensing air. Everything comes from air and goes back to the air.

570 B.C. -490 B.C. PYTHAGORAS: He was born in Samos. He has no written works. In the **III. Century B.C., Euclid** wrote the things which he collected about Pythagoras. Pythagoras stayed in Egypt, in the temples of Memphis for 20 years. After leaving Egypt, in Croton (in the South of Italy) he built the temple of Delphi and established a sect, which is known, with his name. There were various trials and sufferings, which the ones who were willing to attend the sect had to pass. According to him, **all materials and intangible beings are symbols, which are represented by numbers. Number** is the most dominant thing in the universe. The most beautiful thing in the **universe is harmony among the beings**. All of the moving objects, including the stars, have a voice. **In the universe, there is a divine music, which is composed of harmonious combination of various sounds. The purpose of science is to find the corresponding number of each being.** In the Pythagoras triangle, the numbers from one to ten, which are represented by points, are also indicated by some other terms and they are divine. **Even numbers cover an infinite element, whose principles are confusion and plurality; odd numbers cover a finite element, whose principle is harmony. Everything in the universe are 10 objects like moon, sun, planets, fixed stars, rotating from east to west, with a fire in the center.** In the temple, they are represented by 9 Muses sarcophagus on the sides and in the middle sculpture of Goddess "**Hestia**" who points to the sky with one hand, and protects a furnace with the other hand. Explanation of numbers are as follows:

1 is monad. It is the central fire under the ground. It is the absolute unity. It is the resource of everything. It is God. It is the **point.**

2 is dyad. It is the source of nature. It is the general energy. It is femininity. It shows the appearing of God in the matter, in other words, in time and space. It is the Earth. It is the **line.**

3 is triad. It is the symbol of triples in the universe (Divine universe, inner universe, natural universe) (fire, earth, water, air). It represents the divine unity, integrity, harmony, and besides, all creatures, creative thinking and star trends, which are absolute and mandatory in the nature. It is the only number, which has a beginning, middle and end. It is the first perfect. It is the surface. It is the **triangle.**

4 represents the **divine power**. It is square. It is justice. It is a **tetrahedron.**

5 is the symbol of marriage. (2+1+2). It represents the equality of one on two sides. Besides, it indicates the variety in objects. **It is the sensation** (sense, cognizance).

6 is various types of organic and vital beings. (2+1+3). Since the femininity principle (2) and masculinity principle (3) are combined with absolute (1), it shows the continuity of generations. Besides, it is the **mentality** (intelligence).

7 is Minerva. It is the Sun. It is both the dangerous times, and symbol of **mind, light** and **force.** It shows the immortal change of the nature and that everything will return to a unity. It has not been born by any of the numbers in the first ten numbers; and it has not born any of them.

8 is mind, ethics and virtue. Because it is composed of two evens composing the **cube.**

9 represents the justice and divine perfection. It shows that everything is an evolution since it is the square of the first odd number.

10 is the tetractys (divine four). It is equal to divine square. It is the sum of odd and even numbers (**1+2+3+4=10**). It is the most perfect number. It signifies the **virtue the most**. It realizes **everything. It is the principle and guide of the life.** It is **rather celestial** than internal. **Everything is unclear and dark without it.**

Besides, the letter **Y (epsilon)** is like two paths opening in front of a **human:** the paths of **badness** and **virtue**. The principle of **unity is the reason.** Dark powers, which are factors of passions and anarchy, object to that.

10 Ethics commandments of Pythagoras are:
1- Respect to gods. Resigning oneself to their will.
2- Endure the deeds of the gods in life with fortitude.

3- Thrusting out the firm hand of laws to the instigators.
4- Being faithful to friends; agree to share everything with them.
5- Showing respect and love to the parents. Loving the country as much as to protect it. Abiding by the laws blindly.
6- Accepting the absolute power of the state.
7- Keeping the promises. Tasting the organic pleasures in a moderate and balanced way.
8- Not to reveal the secrets of community.
9- Behaving passive, sensible and gentle in using the wealth.
10- Being ashamed of oneself when an evil is done. Not to swear vainly.

Also, opinion, loyalty and toleration are the requisitions of this ethics. In the temple, first of all it is thought to love **mother and father**. Father represents the god; mother represents the nature. The doctrines of Pythagoras on mathematics, physics, astronomy and philosophy have become research subjects for his successors, philosophers and scholars. Pythagoras ascertained the ratio between musical notes.

B.C. 551-479 CONFICIUS (K'UNG-CH'IU) (meaning the respectable master): Chinese philosopher, politician. He declared that he wanted to convey the institutions of ancient age. **He asserted that China had been governed by wise rulers**; and tried to bring individuals to perfection in terms of ethics, to live that age again; he addressed the reason. He said "**One should know that he knows what he knows; and that he does not know what he does not; this is the real knowledge.** You do not know life, how could you know death". He was a humanist and positivist. Exceeding any kind of metaphysics, he aimed to establish a **logic**, which prompts human to **think fair** and to **express the purpose**, and an **ethics**, which prompts human to **live well. A good order is up to a good elocution. Therefore, the definitions should be corrected. An ungrateful child is not a real child. An unfaithful spouse is not a real spouse.** They are inveigled on definitions. **If you are a father, then be a father; if you are a child, then be a child; if you are a prince, then be a prince. Money is necessary to live. One should earn money, not live to earn money**. Differences between the humans come from their acquired culture, more than their natures or characters. **The first rule of ethics is 'filial piety'. Mother and father are representatives of ancestors. They should get full obedience and love from their children and grandchildren. The respect and love of the younger ones towards the elderly; of the wife towards her husband; of citizen towards the ruler; are like the respect and love for the mother and father. A citizen should behave like a child; and a ruler should behave like a father; he should bring peace, ease and knowledge to his citizens. "Justice" makes all humans get along with each other. Goodness should be replied with goodness; injustice should be replied with justice. Do not do unto others what you do not want others do unto you."**

? -490 B.C. **EMPEDOCLES** of Agrigentum-Sicilia. Greek philosopher, politician, poet, physician, astrologer, lawmaker. According to him, four elements, which have created everything, are **water, air, fire and earth**. Two principles rule these elements; **love (connecting them)** and **hate (separating them)**. There is only one world. It is created neither by a human, nor by god.

540 B.C. -480 B.C. **HERACLITUS** of Ephesus. Greek philosopher. According to him, the main element of universe is (and will always be) the **fire** igniting and extinguishing as per its own rules. **Water, soil** and everything come into existence from the first factor: fire, and then everything transforms back to the fire. The important thing is the being, not the existence. Universe is the chasing of creation and extinction each other in a universal harmony, rhythm and unity which will last forever. Everything is in a continuous process of *being* by transforming to the contrary of itself. Since badness cannot exist without goodness, and goodness cannot exist without badness, goodness is some kind of badness, badness is some kind of goodness. Thus Heraclitus is respected to be the founder of dialectic method.

540 B.C. -440 B.C. **XENOPHANES**: He was born in Ionia. He believed in the unicity of god and that the world was a reflection. His philosophy is an idealist pantheism. All creatures have a particle of god inside.

540 B.C. -480 B.C. **PARMENIDES**: He went from Ionia to Athens. Here, he met the young Socrates who was the disciple of Xenophanes and Anaximenes. He believed that the existence could be understood by thoughts, not experiments. Universe is a system of radiant circles. On the uppermost,

there is "**Olimp**" and it is solid. In the middle, there is again a solid circle, central fire and the earth. Parmenides accepted the absolute unity of existence and said that the world is spherical.

497 B.C. -406 B.C. SOPHOCLES: Athenian tragedian, physicist, politician. He developed the theatre technique. He introduced the third actor for the first time.

490 B.C. -? ZENO of Elea. According to the Greek philosopher, the infinite divisibility of space makes motion impossible. The slightest motion consumes the infinity. Therefore, the concept of motion contains a conflict and is not real. There is no space, because another space is required for the space to exist.

485 B.C. -411 B.C. PROTAGORAS: According to the Thracian Greek philosopher, we cannot distinguish the good from bad. Because both are have their own measures. There is no general customs or ethics.

468 B.C. -399 B.C. SOCRATES: A philosopher of Attike. First, he was engaged in physics and natural philosophy, later he was involved in the issues of **inductive method, ethics and virtue** in the humans. He dignified the spiritual sciences. According to him the "spiritual world" gathers around the thoughts, and the "material world" gathers around the sun. **All evil comes from lack of knowledge. He directed everyone to believe in the knowledge, which they get by their own thoughts, not prejudices**. Athenians sentenced him to death because of trying to break down the traditions. Socrates asserted that knowledge is present in every **mind**; and that alleged that he could reveal the individual minds by directing them to the right thinking by means of dialog method (rationalism). Knowledge will also give the happiness (Eudemonism). **The correct knowledge performs the correct action. Virtuelessness arise from ignorance.** The wise man has to be virtuous (determinism). **The state is necessary for the social order. Morality is the foundation of social order. Differences in human are only in appearance; the tendency to goodness is the same in all of them. This common tendency in people can only be revealed by means of knowledge.**

460 B.C. -350 B.C. DEMOCRITUS: He is from Thrace. According to him, all objects are coherent. They the formation or extinction of things by uniting or separating. The matters in infinite numbers have indivisible characteristics (**atoma**). Motion exists spontaneously; it is in the essence of the objects; it is not for a purpose, but a necessity of objects. Matter is non-incipient and infinite. Nothing comes out of nothing. He stayed with geometry scholars for five years in Egypt. He is mechanist, atomist and materialist.

460 B.C. -377 B.C. HIPPOCRATES: He is from Chios (Istanköy); the biggest physician of ancient ages. Later he went from Chios to Thessaly, and died here. Nowadays, the doctors still repeat his oath to become a doctor. He attributed the illnesses to the liquid disorder in the body.

430 B.C. -360 B.C. ARCHYTAS: He is from Tarentum. He had knowledge on all of the sciences and he was a general, statesman, scientist. He was a Pythagorean. By means of a rational calculation system, he balanced the relations between the individuals in community. **Preventing the dictatorship of an individual or a social class by laws, elections and draws, he tried to establish the unity by means of justice.**

430 B.C. -347 B.C. PLATO: He is from Athens. He was the pupil of Socrates. He visited Egypt, Kyrenia and Southern Italy. He classified the "**justice**" idea of the state as "**management**", "**defense**" and "**economy**". "**State is a magnified symbol of functions of an individual. "Justice" idea is a must in every subject of the state. Otherwise, state retrogrades**. Plato describes the reality as 'idea'. Political reform of **Plato was; establishing a site in which the ratio between its parts and classes gets as much closer to the fixed and immortal ratios between the ideas.** (Idea: thought) He is against the personal properties. He suggests some kind of communist system for the upper classes. This is a must for them to serve the people. In his book "Republic", **he asserted that geometry knowledge was a must to be able to rule the people.**

413 B.C. -327 B.C. DIOGENES (archetype of the Cynics)**:** He is from Sinope. According to him, the most eminent 'good' is **virtue** and merit. Science, fame, reputation and wealth are 'made up goods' to be despised. The essence of **philosophy is denigrating the imitation everywhere, and putting nature instead of imitation**. A wise man should keep himself off from desires and emotions, decrease the needs to a minimum.

IV. Century B.C. MO-TZU: A Chinese philosopher. He advocated the community to be managed by scholars.

384 B.C. -322 B.C. ARISTOTELES: A philosopher from Athens. He is a mathematician and biologist. He was not a theoretician in politics, but an observer. He passes from **metaphysics** (science of beings) on to **physics** (science of motion). "All knowledge, taught or achieved, base on the previous knowledge; our observations show that this is the same in all sciences, including mathematics in the first place. **He considers the synthesizing of acquired knowledge in such an age, when science advances so fast, as a problem, which philosophers should solve.** According to him: "**Time is the number of motion.** Motion is the basic necessity of the nature; **matter is the existence in the form of power**. Impairment of a being is the breeding of another being. He tried to achieve his goal with a wide encyclopedic work by means of deduction method. According to him, there are two types of economic activities: **Family economy and exchange economy**. He considers the exchange economy as immoral since it has an income purpose. "Every object has a *matter*, and a *form*. After this existence takes a form, in other words, after it is formed as a wooden crate or bronze statue, it becomes an existence in action. The 'complete' compared to the 'incomplete' is the same as 'power' compared to 'action'. That is, existence is composed of the combination of a form and a matter suitable to be shaped. **Every object is the form of the one on the lower step**; **and it is the matter of the one on the upper step** (kike soil-brick-building)" According to him, **philosophy is the science of principles and first causes**.

Aristoteles is the founder of logic**.**

Aristoteles says that "soul has three layers":
1- **In the plants**: Assimilation, reproduction.
2- **In the animals**: The sensation is the source of imaginations based on pleasure and pain.
3- **In the humans**: Mind and desire transform into conceive and will.

341 B.C. -270 B.C. EPICURUS: From Samian (or Athenian). His philosophy is based on Democritus. He accepts that the nature is made of matter, and that it came to existence by means of fusion of atoms. But he also accepts the coincidence in this formation. He is a materialist based on an immortal matter idea, which does not require the intervention of gods. "Happiness can be achieved by using the pleasures wisely. The required happiness can be achieved by peace, conforming with nature and avoiding prejudices. Feeling pleasure is the essence of ethics. But pleasures show an earnest simplicity, which is suitable for a calculative wisdom. **The only science required for humanity should be the science of living happy.** The way of preventing ambition is avoidance of pain. **Philosophy is the science of life.** It is an actional system, which is designed to have a happy life. **Natural needs should be satisfied, the others should be prevented."**

336 B.C. -264 B.C. ZENO of Citium: A Cypriot Greek philosopher. He is a stoicist. The basic principle of stoicism is to behave in conformity with the **nature** (pantheism, unity of existence). Compliance with the **nature is compliance with *reason***. Four cardinal virtues of Zeno are; **"right choice"** (phronesis), **"enduring patiently"** (andreia), **"being temperance"** (sophrosyne), **"fair share"** (dikaiosyne). Happiness is in the **wisdom; wisdom is in behaving in accordance with the nature.** Nature is matter. The gospel truth is the solid and physical one. Because only something physical can be active and passive. **Since first reason should be active and passive as well,** it should be material. Wisdom can be reached by **obtaining the theoretical and practical virtue. Theoretical virtue,** means obtaining the right knowledge on the objects. On the other hand, **practical virtue** is to behave in compliance with the mind. These two virtues are closely linked to each other. Only a mind free from designs and assumptions can acquire knowledge of nature. Stoicism brought some religious institutions among the aristocrats but it had not attained the interest of the public. The religionists used to grow beard and wear long coats. **All kinds of pains, illnesses and death are natural. We need to be patient, unconcerned. But humane mistakes and badness should be disgusted and should be fought. Natural criterion take human to the independence and equality. Differences among the humans are contrary to the nature and artificial.** The humans, the children of nature are siblings. **Depending on the same nature takes the individuals to humanism and cosmopolitism.** The doctrine of Zeno is based on this principle of Socrates: **"Behaving well is only possible by thinking right. Philosophy is a life science, which teaches how to think right. The purpose is living a good life. Knowledge is necessary to live a good life."**

IV-III. Century B.C. ANAXIMANDER of Miletus: A philosopher from Lampsacus. He is the master of Alexander.

III. Century B.C. EUCLID: A mathematician, who founded the most famous ecole of its age in Alexandria. In his book "Elements", he discusses the definitions, general notions, as well as the evolutions of these notions in logical order. (Golden Ratios)(P-152)

B.C. 106 B.C. -43 B.C. MARCUS TULLIUS CICERO: A Roman orator. He was engaged in rhetoric, politics, ethics, philosophy and the nature of gods.

23 A.D. - 79 A.D. GAIUS PLINIUS SECUNDUS MAIOR: Latin natural scientist and author from Comum, Italy. **In contradiction to the animals, human should learn to secure his life.**

4 B.C. – A.D. 65 A.D. LUCIUS ANNAEUS SENECA: Latin philosopher from Cordoba. He was especially engaged in ethics.

55 A.D. -120 A.D. P.C. TACITUS (GAIUS CORNELIUS TACITUS): Roman historian and orator. He was the best describer of ancient history.

203 A.D. -270 A.D. PLOTINUS: He was born in Egypt. He wanted to establish the site, which Plato had always imagined in Rome. He was engaged in astronomy, astrology, mathematics, mechanics, language, soul and ethics. He made trilogy like **Plato. He tried to generalize the knowledge of plurality to raise to the unity.** He accepted that god was showing itself in the form of a trilogy as **soul, intelligence and unity**; and that **the knowledge was advancing step by step to the unity.** He believed that he could reach beyond the mind by means of ecstasy (mystical). **"Ethical understanding is the condition of raising the soul to the unity. There is duality under unity, in other words, a last notion that the mind could apprehend. The longing of soul is the desire of joining with all living creatures. Everything that separates, distracts and disperses, is bad".** He accepts a divine principle, which is named one, or good in which everything is mixed; and a mystic way beyond the consciousness of the soul.

354 A.D. -430 A.D. AUGUSTINUS: He was from Tageste and the most famous among the Latin church fathers. His thought involved a **universal perspective**. He was a rhetoric instructor; founded a brotherly community in his house. He had two fundamental thoughts: the fate of human, who was ruined by his/her sins, and rescued by the blessing of God; and God. In his writings, he was engaged in philosophy, principles of religion and ethics.

780-850 A.D. AL-KHWARIZMI (Muḥammad ibn Mūsā al-Khwārizmī or Mehmet of Horzum): He was born in Horzum (Khwarazm) in the east of the Caspian Sea. He went to India in 825 under the patronage of al-Ma'mun, the Caliph of Baghdad. In his work "*al-Kitāb al-Mukhtaṣar fī Hisāb al-Jabr wal-Muqābala,* The Compendious Book on Calculation by Completion and Balancing", which he wrote after his return from India in 830, he explained for the first time how the 9 number system, i. e. the decimal number system and zero, to be known as Indian-Arabic numerals subsequently, is used in arithmetic operations in the present meaning. The Europeans left the Roman numerals and started to use numbers with only after the XII. Century, when this work came to Europe through Andalusia. Al-Khwarizmi is the founder of **algebra** and **algorithm**. According to him, the Universe has been established on the basis of symmetry and he expressed symmetry as equality (equation = equivalence = equilibrium) with numbers. In his work, all algebra solutions have been conveyed to a systematic form based entirely on Geometric considerations. **Algebra** was transformed into **algorithm** in the west. His work "Siddhartha", which al-Khwarizmi translated from Sanskrit to Arabic, contained tables expressing the angles with Trigonometric Functions like sinus. He was also an astronomer and a geographer. They asked Harezmi what a human is. He said;

If a person has good ethic, he gets "1".

If he's handsome, add 0 to it, it's "10".

If it is rich, add another 0 to "100".

If he is noble and has ancestry, add another 0 to "1000".

But if the **"ethic"** is gone, zeros are left with no value.

801-872 AL KINDY: He was the first Arab philosopher from Basra. He said, "**Mathematics is the foundation of sciences**".

841-926 ZAKARIYA AR-RAZI: An Iranian (or Turkish) philosopher and scholar. He was educated on philosophy, mathematics, natural sciences and astronomy in the city of Rey. He learned medicine and he wrote an encyclopedia of medicine.

870-950 AL-PHARABIUS (AL-FARABI): A Turk from Turkistan; a Muslim judge, philosopher, doctor and musician, mathematician. According to him, God makes the person virtuous. **Virtue is cooperation. The highest goodness is obtained by living in a metropolitan city.** A city that cooperates in the things to take them to the real happiness is a **virtuous city.** A cooperating community **is a virtuous community**. A **cooperative community** to make all towns and cities achieve the happiness is a **nation**. If all of the nations help each other, the **earth becomes virtuous. A virtuous city seems like a body having all organs complete. Al-Pharabius classified the sciences as well. The most perfect state is a world state embracing all of the human race.**

934-1020 FERDOWSI: Persian poet. He is the author of the Shahnameh, the history of humanity.

980-1037 IBN SINA (AVICENNA): He is a Turkish Muslim philosopher and physician from Bukhara. He classified the **sciences.** He was engaged with natural sciences, geometry, logic, grammar, syntax, Islamic law and literature. He wrote a book titled "The Laws of Medicine". He combined the rationalism of Al-Pharabius and experimentalism of Zakariya Ar-Razi. He believes that mind, experiment and observation are inseparable whole in the development of a **Scientist.** Besides, he is broadminded about Mysticism, Metaphysics, Mind, Psychology, Logic, Knowledge and Music. While he was 17, he saved the Prince of Bukhara from a dangerous illness. The library of palace was opened to him.

He classified the science:

i. **Natural sciences or lower sciences**. These are the science of forms that are not separate from their matter.

ii. **The sciences of forms separated from their matter.** These are metaphysics, logic or high sciences.

iii. **The sciences of forms separated from their matter only in the human thought** and forms sometimes with their matter and sometimes separate. Mathematics or medium sciences. This third type of science features a passage and a bridge between higher and lower sciences, and connects them in a particular subject. It links the mathematics especially with logic and metaphysics.

XI. century YUSUF KHASS HAJIB: An Uyghur scientist from Central Asia. In his book "**Kutadgu Bilig**" (knowledge on happiness), he wrote that only **goodness and ethics** would bring the happiness. Goodness is **knowledge, peace and righteousness**. The elements of badness **are; ignorance, illness and going astray. One can steady the ignorance through doctrines; illness through treatment; badness through education.**

His book consisted of four parts:

I- **Justice** ('Gün doğdu' (sunrise), Ilik Khan).

II- **Power** ('Ay doğdu' (moonrise), vizier, gives happiness, bliss).

III- **Mind** (Öğdülmüş, son of the vizier, science).

IV- **Sobriety** (Otkurmuş, brother of vizier, end of life, future, balanced, temperate) (in Turkish, settled, matured). The book includes the conversations between these four persons.

? - 1077 MAHMUD AL-KASHGARI: Turkish linguist. He wrote the book "**Divanu Lugati-t Turk**".

1044-1136 OMAR KHAYYAM: Persian poet, scientist. He was engaged in logic, philosophy, mathematics and astronomy. He wrote books in the fields of physics, metaphysics, astronomy and poetry ("**Rubaiyat**", "**Akıl Bahçesi**", etc.)

1058-1111 AL-GHAZALI: Islamic scholar from Tus (Afghanistan). He classified the **sciences.** He said: "You cannot trust in the knowledge of the ones who do not know logic". Ghazali accepted that body and soul were different. "Because a dead person has a body, too. **The human soul is nothing, but the waves of the divine soul on the ocean."**

1093-1166 AHMET YESEVİ: He is a Turkish-Islamic Sufi. He was born in the town of Sayram in Turkestan. He became a source for Hacı Bektaş-ı Veliye and Mevlana. His Works of Wisdom, i.e. Hizmet, are Fakrname and Divan-i Wisdom. He is the founder of the Yesevi sect. However, he defines it as the states and words of people who are at the level of perfection in knowledge and wisdom. He wrote his poems in Turkish quatrains with syllabic meter.

1126-1198 IBN RUSHD: Arabic philosopher from Andalusia. He tried to explain the philosophy in terms of principles of the mind. "The main subject of philosophy is trying to explain the existing and what is given to the humans, as a whole. The Supreme Being on top of all other beings, God, can only be reached through the existing things, which can be perceived through five senses, explained, and interpreted through the principles of reason. **Philosophy should be based on universals.** Universe is the whole of beings, which find their essence in the matter. **State should be managed by old and wise people, and philosophers.** Reason is the basic principle providing knowledge. People need to be educated, raised, trained. There are no differences between men and women except the structure of their bodies. **Community is a whole, composed of all of the humans. Humans need to live together in a community, not separately.** Solidarity, peace, cooperation, division of labor in specific areas are inevitable necessities. These are general and indispensable rules." It is said that Leibnitz and Descartes had benefitted from him.

1135-1204 IBN MAIMON (MAIMONIDES): Arabic philosopher of Jewish origin from Andalusia. According to him, "**universe is a whole.** God cannot be known completely". He was engaged in medicine, astronomy and philosophy. He was a rationalist philosopher.

1149-1209 FAKHR AL-DIN AL-RAZI: An Arabic theologian from Rey.

1155-1191 SYIHABUDDIN ABUL FUTUH YAHYA SUHRAWARDI: Persian philosopher. He is the founder of "Ishrakiye" (enlightenment) age in Islamic philosophy. He was engaged in Islamic laws and mysticism. **According to him, "knowledge and fact are conceived by the heart as a light** (Keşf)". Like the neo-platonics, he considered god as a light (**Nur**) and explained the physical beings as that light descending and spreading and then condensing on the world.

1162-1240 MUHYIDDIN IBN ARABI: Andalusian Sufi. According to him, **the beings that link the hidden and visible universes are the humans (perfect human); the universe (nature) is not identical with the god.**

The ways to reach to the perfect human are:
1- Certainty arising from and based on intelligence.
2- Divine enthusiasm.
3- Deep joy, in other words, the joy that the mysticism puts emphasis on.
4- Perception.
5- Grasping the universe of nobles as a whole.
(He connoted the Theory of Everything)
All of these paths lead to god.

1201-1274 NASIR AL-DIN AL-TUSI: Persian philosopher, astronomer. He was engaged in words, mathematics and hadith. According to him, all rules of ethics and education should be based on mind. "Foundations of ethics is based on goodness (charity). **Children should be supported in choosing a profession in accordance with their tendencies and capabilities.** Shame in the childhood is the first indication whether his/her abilities will tend to good or bad. Besides, **they should be kept away from behaviors, which would cause a show off, wealth, deception, gluttony, messing someone over; and from the miscreants. In the choice of profession, they should be supported to work in the field, which they show tendency, and in which they are skilled.** Humans are the carriers of an ethics as reasonable creatures. "While he was a minister in the palace of Mongolian ruler Hulagu Khan, he established the Maragheh Observatory which was one of the biggest science centers of the Islamic World. He tried to associate the principles of mind and Shia sect. According to him, all rules of ethics and education should be based on reason.

1207-1273 MAWLANA JALAL AL-DIN AL-RUMI: Great Turkish Sufi from Balkh (Khorasan). When Alā ad-Dīn Kayqubād, Sultan of Anatolian Seljukis, invited Baha Al-din Walad (Sultan of Scholars), who was Mawlana's father, they settled in Konya-Turkey. He established the Mawlawi Sect. He was very competent on astronomy, mathematics, medicine, alchemy sciences; and an expert of his age on Quran, Hadith, Islamic law and kalam (science of words). He stood absolutely for the **unity of existence.** He used to treat lovefully, with fidelity and respectfully regardless whether they were children, women, criminals or ill. **He gave importance to the notions like modesty, politeness, manners, patience, tolerance, understanding and amnesty.** However, since he considered that the main issue was **humans; and that religions, philosophy and ethical systems were means to make humans happier and more precious,** he did not want to insist on any of these means. **According to him, creatures were the appearance of creator. Hence, there were no guilty, innocent, beautiful, ugly, good or bad. The most perfect way was the love for God. An infinite love distributed to the**

universe, and a toleration supporting this love, were fidelities inherited from ants to Suleiman (the magnificent). **Mawlana reconciled the best poems of his age with music. According to him, music was the essence of soul unity of humans. "Even if there were a hundred books, they all have only one chapter. They turned towards a single mihrab in these hundred directions. All of these roads lead to a single house. These thousands of wheat spica are made of a single seed". There are the verses on the door of his tomb in Konya**: *"Come, come, whoever you are; wanderer, worshipper, lover of leaving. Ours is not a caravan of despair. Even if you have broken your vows a thousand times, it does not matter. Come, come yet again, come"*. He made a big change in Islam by bringing **Sema** (Mawlawi ceremony) into the religious ceremonies. He wrote his books in Persian language. But he has also a few verses in Turkish, Arabic and Greek.

1209-1271 HACI BEKTAŞ-I VELİ: He was born in Nishapur, Khorasan. He is the founder of the Bektashi order. After Ahmet Yesevi, the Sufi movement, which aimed to unite Islam with the old Turkish religion, continued in Anatolia with Hacı Bektaş-ı Veli Yunus Emre and Mevlana.

1210-1274 SADREDDIN KONEVI: A Turkish Sufi from Malatya. According to him, **creator and created universe are the same. Plurality is in appearance. In reality, there is unity. Humans reach god by means of thought.** God, the creator of all of the universes, is an unknown for the humans. Humans can reach god only by means of thought. Death is not an extinction, but a return to the source, to the **"immortal essence"**. The divine light is an infinite and limitless appearance, surrounding the whole universe. The human can see just one side of it, which comes into appearance. The reality is not in this appearance but in the core that is hidden behind the appearance. This core may neither be reached by means of reason nor by rising through knowledge step by step. The creator of the whole universe, namely god is unknown for human.

1235-1316 RAIMUNDUS LULLUS: French (Catalan), mystic encyclopaedist, poet, missioner. Later he became a Muslim. He tried to express all of the facts with triangles, circles and statements, etc. in his **'Ars Magna'**. **"Alchemy symbols are the sources which reflect the alchemy science of the middle age".**

1238-1310 YUNUS EMRE: A Turkish Sufi from Sivrihisar (Anatolia). The main subject of his poems is the love of God and humanity. He describes humans in his poems in a whole integrity with their past, future and the situation in the time they are living. Moreover, the relations of the human with the God, among humans and the humans with their own functions according to mysticism and Islamic philosophy and in humanist sense. He says in a poem "The heart is the throne of the Oldest God (Çalab), the Oldest God looked at the heart, the one who has broken the heart will be miserable both in the life and after life". Çalab is a word originated from the Central Asia, meaning the oldest God in the history. Heart (soul) means the center of the body in the Islamic mysticism. Not the reason, namely mind but the heart represents the love. The poem implies that: Soul (heart) is the crown of the creator, the creator looked at the crown, and anyone who has broken the heart of someone, namely hurt someone, he/she is the miserable person both in this World and the afterlife World, namely he/she has shamed him/herself. Some other statements of him are as follows: "The words originate from sound of the Creator, not from the book or individuals" namely the words come from the creator, from the God, not from the Scripture or from person. Furthermore, he also said "The only thing which dies is the body, soul never dies". He means that if the one dies, the body of this person dies but the soul that represents the love never dies. Yunus Emre has integrated the doctrine of the Islamic mysticism philosophy with the public language, delicacy and heartiness of public sense and national elements and so become the leader of a humanism conception considerably beyond his era.

1340-1384 GERARDUS MAGNUS: A mystic Dutchman. He founded the sect "Brethren of the Common Life".

1359-1420 SHEIKH BEDREDDIN: Turkish Sufi from Simavna (Edirne). He was engaged in Logic, Philosophy and Theology. He revolted to annihilate the religion differences and he was executed. According to him, "all of the world belongings are for common usage of the humans. There is no afterdeath." He was a Materialist.

1469-1527 NICCOLO MACHIAVELLI: Italian politician, philosopher and historian. According to him, one should do evil, violence, oppression more than doing kindness, everything which makes someone to achieve his/her objective is good regardless of whether it is good or evil. Religions should be protected provided that it serves for the goals of politics and state regardless of whether it is composed of nonsense beliefs. Human is in general evil in the sense of morality, someone who wishes to be good is condemned to suffer oppression. Evilness should be done suddenly and intensely but kindness should be done gradually and slowly so that it could be enjoyed.

1472-1543 NICOLAUS COPERNICUS: Polish Astronomer. He proved that the planets do not rotate around the world, but they rotate together with the world, both around themselves and around the Sun at the same time.

1478-1535 THOMAS MORE: English Economist, Lawyer. He is against the personal properties. In his book "Utopia", there is an island named Utopia. In this island, everybody deals with agriculture and another occupation. Everybody gives their products to the shops in the city; and takes everything they need from the shops: There are no poor people in the island. Everybody is prosperous.

1483-? MARTIN LUTHER: He was born in Saxony, Germany. He was educated on Law, Literature, Music and Philosophy. He joined into the clergy, and became a philosophy professor. When he was excommunicated by the Pope, he translated the Bible into German. When most of the cities joined him, Protestantism came into existence. In Europe, the absolute dominancy of church started to lose control for the first time, a reformist motion was started.

1509-1564 JOHN CALVIN: He was born in Picardie France. After graduating from the university, he studied Luther. Later he went to Switzerland to build Calvinism. He was more a religion reformer than Luther.

1561-1626 F. BACON: English philosopher and politician. He created the "English Positivism". According to him, "the purpose of a real scientist is not the thought. It is conquering the nature, and modifying it to supply better living conditions to the humans. All kinds of inventions, all of the results are means to reach to a new result. In Physics, success comes with;

1- Observation.

2- Experiment.

3- Induction. In his book "Atlantis", he proposes an 'Island of Virtue'".

1546-1642 GALILEO GALILEI: Italian Astronomer, Physicist, Mathematician, Musician. He invented binoculars, trigonometry and hydrostatic balance. He is deemed to be the founder of experimental method and dynamics. He found that a pendulum could be used for timekeeping. Also he found the laws of falling objects, and that the music intervals are proportional to their frequencies. He is also supposed to invent the microscope. He has various inventions. He indicated that: **"The big book of Nature can only be read by the ones who know its language, and this language is Mathematics."**. Since he was supporting Copernic's theory, he was sentenced to death by Pope in the Inquisition. But he was not executed since he was very old. He has been in custody in his own house. At the end of his life, he became blind.

1568-1639 TOMMASO CAMPANELLA: Italian philosopher and author. He longed for the City of the Sun (Civitas Solis) in which all of the humans would live with a single thought and equally. A team of 4 persons manage his City of the Sun.

1571-1630 JOHANNES KEPLER: German Mathematician and Astronomer. He proved that planets were rotating on a plane around the sun. He is best known for his laws of planetary motion. Kepler's laws of planetary motion are three scientific laws describing motion of planets around the Sun. He indicated that: "The whole created universe is composed of an astonishing symphony as in all matter beings in the order of thoughts and spirits, everything is interdependent with a mutual and indecomposable relationship and everything constitutes a harmonious whole, everything is living and the laws of harmony of the music determines the Motion of planets." ¼ of a rotation period in the motion of planets is proportional with the cube of the long side axis of the ellipse. The more the celestial bodies approach to the center of gravity, the higher their velocity is, this is inversely proportional with the square root of the distance between each celestial body and the center of gravity. The area drawn by vector radius, which connects the center of the Sun and the center of the planet, is proportional with the time that passed to draw this area. The square of the time that passed for conjunction cycle of the planets introduced that it is directly proportional with the cube of the long axis of their orbit.

1588-1674 THOMAS HOBBES: English philosopher. He founded the Materialist Philosophy, which makes the mathematical and mechanical explanation of nature. He indicated that: "As we can steer a machine, we can steer the universe, in the same way. Mathematics and Mechanics principles are enough to identify the universe".

1596-1650 RENÉ DESCARTES: French philosopher. He accepts that the living being is composed of **Matter** and **Soul**. According to him, "mathematics is the model of **science.** We may learn the power and effects of fire, water, air, stars, skies and all other objects to make use of them. Thus, we may be the owners, masters of the nature. **One idea brings the other. If we follow these ideas consecutively, there is no information, which cannot be found. "He suggested a general science and sign language (Scientia Generalis).**

1600-1654 JOHANN VALENTIN ANDREAE: Utopian.

1611-1677 JAMES HARRINGTON: Utopian.

1613-1680 FRANÇOIS DE LA ROCHEFOUCAULD: According to him, "Nothing can be known completely. Because we should know the details to know everything. Since there are infinite number of details, our knowledge would always be partial".

1632-1704 JOHN LOCKE: English idealist philosopher. He classified **sciences semiotically.** According to him, all of the thoughts come from experiments. Experiments are divided into two as outdoor experiments and indoor experiments. He suggested the **Semiotic term, Science of Signs and Symbols Theory.**

1632-1716 RICHARD CUMBERLAND: According to him, "humans, community, state and universe are going forward to the purpose of getting **to the best. All of them are as one."**

1632-1677 A. SPINOZA: Spanish Jew. Physicist and idealist. He shows an approach to every question in integrity or fact. According to him, "God is the root of all beings. As when there are no places geometrical shapes will not mean anything; when there is no God, objects are nothing. He solved the problem of the difference of **physical** and spiritual essence; from the God essence of living **beings, in accordance with the idea that it has been formed by a mathematical necessity. He combines the matter and soul in a divine essence and determines their effects on each other.** Philosophy is the mathematics in general. Humans are the prisoners of what they do not know. They become free when they reach to the knowledge. What develops the good mind, **blurs the bad mind. Virtue is measured by the power and effort of humans to protect their own existence.** Humans get stronger if they come side by side. **So, there is nothing more useful for humans except another human.** Therefore **humans need to unite".** He wrote the book 'Ethica' (Theoretical Ethics) applying the **Analytical Geometry** method of Descartes. The opposite of **Ethica** is 'Morale Social Tore' (**applied ethics).**

1638-1715 N. MALEBRANCHE: French philosopher. According to him; **"minds of humans" are part of an infinite mind (words of God). It is based on the idea of ETHICALORDER, in other words, the existence phase from simple to complicated (from humans to God).**

1642-1727 ISAAC NEWTON: English philosopher, mathematician, astronomer. He found the **infinitesimal calculus, structure of white light, universal gravity theory; he invented the reflecting telescope. We can say that he laid the foundations of Modern Science; and prepared its methods. He also has other various inventions.**

1646-1716 G. W. LEIBNITZ: German philosopher. He tried to create a universal scientific language, the language of signs, a **"Characteristica Universalis" by applying mathematics for philosophy. Besides, he tried to explain the language systems in linguistic, mathematically. According to him, evil is in the details. But, whole is perfect in the highest degree; VIRTUE is the love for god; Happiness is in this love.** He made studies on **Universal Mathematics (Mathesis Universalis), Algebraical Linguistic, Statistical Linguistic and Linguistic Computer.**

1694-1878 F. M. VOLTAIRE: French philosopher. He said: **"World should enter into the domination of mind".**

1701-1780 ERZURUMLU İBRAHİM HAKKI: One of the last Islamic Sufis. He was born in Erzurum in north east Anatolia. In his famous encyclopedic work Marifetname, he brought together his knowledge in the field of Geology, Astronomy, Physiology and Psychology. In this work, he aimed to inform people first about the belongings around them and then about themselves and finally about Allah. According to him, both the universe and the beings seen in the sky are spherical. He wrote more than seventy works.

1712-1778 J. J. ROUSSEAU: Swiss author and philosopher. According to him humans have been good. It was civilization, which has turned them into bad. "The Greek art remains passed to Italy with the collapsing of Byzantine. Thus, the unhappy pedantries and virtueless civilizations of Europeans have started. The more science and art advance, the more virtue disappears. **There are principles of 'JUSTICE'** (law) and **'ETHICS'** (tradition) in the depths of peace."

1714-1762 ALEXANDER GOTTLIEB BAUMGARTEN: German philosopher. He separated the **Science of Beauty** under the name of **"AESTHETICS" from the other fields of philosophy and from** Leibnitz and Wolf philosophy ecole. **He defines aesthetics as "the science of Knowledge Coming from Senses or Lower Gnoseology".**
There are two types of perfections:
1- "Mentally Competence" creating the "goodness" which is related with ethics.
2- "Competence in Senses" creating the "beauty" which is related with aesthetics.
A triple compatibility is the compatibility:
1- Between thoughts and objects.
2- Between thoughts and thoughts.
3- Between thoughts and external signs.
These three compatibilities create the competence of knowledge coming from senses; in other words, they create the "**Beauty**".

1723-1790 ADAM SMITH: English philosopher, economist and liberalist. His book "**Wealth of Nations**" is deemed as a milestone in the economy doctrines. "The wealth of nations are subject to the amount of goods which are essential and indispensable. Labor creates the wealth. The difference between wealths of nations come from the difference between labors. **The efficiency of labor increases with the sharing of work.** Market Price depends on the supply and demand. The condition of economic development is accumulation of capital. Capital appears by means of savings. Personal benefit **brings all** economic activities. Competition is necessary. There are three types of incomes; "**WAGE**", "UNEARNED INCOME" and "**PROFIT**". INTEREST **is** the income of capital. Unearned Income is the amount which the users of land pay to the owner."

1724-1804 IMMANUEL KANT: German philosopher. According to him, **knowledge** comes from the shaped activities of the mind of human. The Law of **Ethics** seems to us as **Respect** in the shape of senses. But respect is not a tendency to lead us to the object like love or desire. Respect is a reviewer emotion, which stops our desires. It limits our tendencies and makes us look at the object from a distance. Respect is the negativeness of tendencies, and a preparation for a behavior coming from a non-sensual world. It makes a combination of **Nature** and Freedom (**Soul**) in terms of having the characteristics of an artwork organism. Virtue is not related with instinct, it is related with mind. We should not take offence at virtuelessness. Because these virtuelessnesses will take the humans to the 'virtue' ambition.

1743-1704 ANTOINE LAVOISIER: French chemist. "We should say 'to unite' instead of 'to be'; we should say 'to leave' instead of 'to die'. There is no other change except the change of appearances. Essence does not change. **Nothing is lost, nothing is created, everything is transformed.** "He dealt with the subjects like amount of elements in the universe.

1747-1826 J. E. BODE: Based on the observation of the German astronomer **J. TITIUS** in 1766; he explained the mathematical code in the radius of the orbits of the sun and planets.

XVIII. Century ÉTIENNE GABRIEL MORELLY: Communist and utopian. He is the author of "La Basiliade".

1762-1814 JOHANN GOTTLIEB FICHTE: German philosopher and idealist. According to him, the basic emotion of existence is the 'Consciousness of absolute freedom'. This consciousness occurs in the will in accordance with the ethics laws. Making philosophy is to know that existence is nothing, and the duties are everything. In this regard, it is the knowledge of ego. **Philosophy is a** 'Doctrine of Science'. "The basic idea in all of his writings is that "'I' creates himself/herself.". He considers **God equal with an ethical World order**. "State is an artificial institution. Someday it will disappear."

1770-1831 GEORG WILHELM FRIEDRICH HEGEL: German philosopher, idealist and utopian. According to him, 'absolute' is not an ore which is impossible to apprehend or another world or a God. It is the subject of scientist. In this context, we cannot talk about God. We can talk about only the absolute knowledge. He accepts that humans have right to interpret their own philosophy under new circumstances in various styles. 'Absolute' is not over the **Nature** and **Soul**; it is 'in' them. **Humans develop by reconciling the opposites**. "VIRTUE" is a reconciliation between **goodness and badness**." According to him **CULTURE** is everything humans do in consideration of creatings of the nature.

1771-1858 ROBERT OWEN: English socialist. According to him; "since the difference of humans is a result of the environment and coincidences, it is not good to differentiate between **humans.** The incomes of humans should not be different. Humans should be able to obtain their needs under equal conditions. In other words, everybody should be imparted as per their needs". Owen founded a colony named "New Harmony" in Mexico. 2500 Europeans came to the colony. But since most of them were lazy, they fell apart two years later. Owen is against **Profit** and **Money**. Since the values of goods can be quantified by means of Labor, he advises substitution of LABOR BONDS which are expressed with labor instead of money. He applied this idea. But his idea of annihilating the profit, later provided the development of retail societies.

1772-1837 CHARLES FOURIER: French socialist and utopian. He accepts the competition. He is against the **profit.** He suggested the establishment of unions named "Falster" including 810 women and 810 men. In Falster, everybody resides, eats together. They obtain their own needs. Everybody works as they wish. 5/12 of income should be divided for labor; 4/12 for capital **and** 3/12 according to the qualification and capability. Fourier accepts the ownership and capital. But he is against the gainful occupation. Worker should be the partner of entrepreneurship. **He should participate in the management and get share from the profit**. His aim was to remove the contradiction between the producer and consumer, between the worker and employer. Although he tried to apply the Falster system, he could not succeed. But these ideas were useful for the development of the cooperative system.

1772-1823 DAVID RICARDO: English economist and liberalist philosopher of Jewish origin. According to him the value of money is determined by its amount. Since the owners of fertile lands get product with lower costs, the profit, which they get more than the other proprietors, **is called as "Income". The values of goods are subject to the production expenses and amounts. Since capital is the production expense and capital is made of labor, the values of goods are formed by the labor used for their production.**

    There are three types of incomes:
1- Income of the landlord: "**Unearned Income**",
2- Income of the worker: "**Wage**",
3- Income of the capital: "**Profit**".

1773-1831 JOHANN WOLFGANG von GOETHE: German poet. According to him, life is not good or bad itself. It is what it is. Humans should face and accept it. Humans, who are able to show the power to call the badness for help in order to reach to the great truth, are not the enemy, but the mediator of goodness. Because virtue is the essence of humans. Whatever the devil (Mephisto) do, it will not be able to separate humans from this essence. At the end humans will be elicited, they will manage to preserve their humanity. In his 58 years of life, Goethe tried to reduce the dual reality that **HUMANS ARE THE BATTLEFIELDS OF GOODNESS AND EVIL, into one**. The reality he found out was: 'PEOPLE SHOULD SAY YES!'. The happiness of individuals are related with the happiness of **community.** Living is beautiful, provided that we join to the common power of **community. **The duty of genius **is to direct this power to the most useful purpose.** What makes the **humans live that joy to open the gates of the heaven on earth is that they feel this purpose in their hearts. Humans understand that with the consciousness, which they reach despite the mephisto. And they can say 'Stay! Thou art so beautiful!' to the life**. In the written work of Goethe, Dr. Faust turned his back on the happiness for years, and spent his time reading his dusty books. He understood that he was wrong. He was too old to devote himself only to the life. He was also too young to be unwilling. From then, he was disgusting from knowledge and thought. What they call intelligence was nothing but stupidity, according to him. Why does innocence not know its own divine value? **The only thing to do is to feel happiness** (Love, God, Heart, whatever). **Fame** is a sound smoking the brightness of skies, but **emotion is** everything. Dr. Faust was one hundred years old. With the opinion that he has wasted his life, at the moment when he was about to commit suicide, Mephisto came along. Mephisto told him that the life was beautiful, and there could be times when he could say: 'Thou art so beautiful. Stay!'. They placed a bet. Dr. Faust accepted to give his soul to the devil in the other world, if he would say that. At the moment when he would die, he said 'Stay!'. But he would not be losing the bet. Because he reached to the clearness, he learned how to say 'yes!'. But not with the happiness, which Mephisto unfolded for him; with his own existence… The bright truth was that: **Life is of whom, who makes it theirs everyday over and over. I want to be in a free land, between free people.** At that time, I can say 'Stay! Thou art so beautiful!' to the time passing by." According to Goethe, the more a poem is hard to understand, the more beautiful it is. In Faust, there are still unexplained symbols related with time. Next generations will solve these symbols one by one; they will believe in their own greatness; they will deem the greatness of Goethe necessary in their own greatness. Both progressivists and reactionists should be pleased with Goethe. Because each of them can comment on Faust in accordance with their own purposes.

1777-1855 CARL FRIEDRICH GAUSS: German Mathematician, Astronomer and Physicist. While he was 16, he presented a method, which calculates the orbital elements of Uranus, and allows to find the orbital elements of a planet by means of measurements from a point on earth. He dealt with electricity, especially magnetism. He invented Magnetometry. He formulated the mathematical theory of Magnetism.

1788-1860 ARTHUR SCHOPENHAUER: German philosopher. He told that: "Universe has come into existence, from the will to exist. Universe is the place of who goes at it hammer and tongs, eternally. So, what really matters is virtueless ness, not virtue. Compare the sorrow of shredded animals and the joy of shredding animals, it will be understood that free will is the basics of existence."

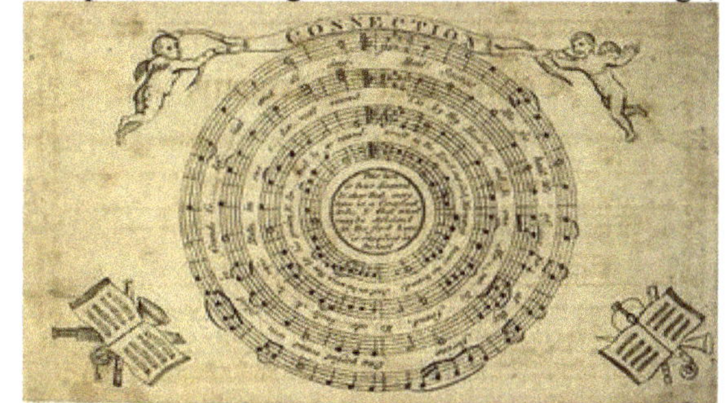

1794- ? WILLIAM BILLINGS: He composed a world harmony named "The Continental Harmony" in America on a nested orbit including 4 circles as can be seen below. "Heil sacred Music heil".

1798-1857 AUGUSTE COMTE: French philosopher. According to him, positivist Western communities have passed from 3 phases like; 1-Theological. 2-Metaphysics. 3-Positivist. He used the Law notion instead of Positive Condition, Reason notions. He classified the sciences as **Mathematics, Astronomy, Physics, Chemistry, Biology, Sociology and Ethics**. According to him, community was not an ideal being, but reality. "Social facts arose only in the Western European communities. All of the sciences are divided into two parts: static and dynamic. The first steps of **Social union are prepared in the family.** Humans should live for the others. Christianity ethics order to love the ones similar to us, as much as we do for ourselves. This means to love each other for our own sake. It is not social, it is immoral. The purpose of ethics is to prefer humanism instead of animalism; to live for the others, not for ourselves. Social life has its own physics as the physical sciences have. And it is the essence of humanism. **Individuals are a part of humanism organism.** *Duty* notion precedes the *rights* notion." In the religion, which he has established, the most high being to worship is "**humanism**". "**world**" and "**space**" comes later. He describes the humanism as a 30 years old woman who holds "love for principles" in her right hand, "basically order" above her head and "advancement as a purpose" plates on her forehead. State was a popery based on science. He suggested the **community to be organized with science.** Besides, he tried to establish the **science on various triads.** He divided the human life in 7 parts. For every part, he planned a unique education. "To live for the others requires to love the others. **Virtue is love. Love the humanism like you love a human**. There is only one absolute principle. Everything is related to each other. **Philosophy** is the **science of sciences, which** unites all of the sciences.

1804-1872 LUDWIG ANDREAS FEUERBACH: German materialist and philosopher. According to him, religion is the self-alienation of humans, it is the consciousness. He proposes a method of education and a new conscious religion. Love is the reality of religion. **The commitment and love for another human will put the** humanism in order. Religion is the self-respect of humans.

1806-1673 JOHN STUART MILL: English philosopher. "Happiness is the benefit. Philosophers have always looked for what **the top best (Summum Bonum)** is, from the beginning of philosophy. And here we give the response of this question, together with Bentham. **Superior good is the PROFIT.** Neither rationalists, nor the psychologist could determine what separates the good from bad. The measure, which separates the good from bad, is the amount of **Profit**. Profession of a doctor is extra good, because its has a **health profit**. Music is good. Enjoyment (Hedonism) brings the **pleasure profit** and the happiness of humans, in reaction to loathing. **Enjoyment** is a result of beneficialness. Humans like the beneficial ones. They do not like useless things. Useful ones are the liked ones."

1809-1882 CHARLES DARWIN: English biologist. According to his evolution theory:

1- Species of the same origin change based on various effects (environment, feeding, other changes). Body and germ cells also change.

2- He reached to the principle of "struggle to live" from the idea of Economist Malthus that "Population increases geometrically, and feeding opportunities increase arithmetically, in other words, feeding opportunities increase less than population". Superior ones have the chance to live and breed by means of **Natural Selection**.

3- Males are in a fight to get the females. The strong one wins. Females choose the handsome and the strongest male. Thus, a sexual selection happens.

4- "**Virtue**" is a product of evolution, which came into existence because of the necessity to adapt to the environment and conditions. It starts from the protection instinct. It is a must for the humans to survive.

1813-1883 RICHARD WAGNER: German musician. He built his compositions on love, salvation and waiver, human independence issue and glorifying the public virtues. Subjects of his compositions were from German Mythology. He played his works in the high acoustic concert hall, which he made constructed for his orchestra, instruments (164 musical instruments) and himself. Some of his works are: "Twilight of the Gods", "Art and Revolution", "Artwork of the Future". They were friends with Rockel, Bakunin and Nietzsche. He transferred some parts of the same theme to various musical instruments in his compositions.

1814-1876 MIKHAIL ALEXANDROVICH BAKUNIN: Russian utopian and anarchist. He believed in atheism, abolishment of social classes, equality of women and men, common usage of all capital goods, abolishment of the states and any kinds of authorities. He was against the state idea of Karl Marx.

1817-1894 WILHELM GEORG FRIEDRICH ROSCHER: German economist. He applied the Economic History Method by denying the Natural Economic Laws. He asserted that immigrants bring more capital to the accepting country, than the capital they have taken from their own countries.

1818-1883 KARL MARX: German philosopher of Jewish origin. According to him, basics of community, its 'substructure' is its production method. Its 'superstructure' is composed of religious, ethics, political and legal organizations. Substructure dominates the superstructure.

History of humanism can be arranged according to the production method as follows:

1- **Primitive** community (production type is collective and there are no personal properties).
2- **Slaver** community (production depends on the personal properties of slaves).
3- **Feudal** community (production depends on the personal properties on the ground).
4- **Capitalist** community (production depends on the personal properties on the means of production).
5- **Socialist** community (production is shaped by collectivizing of the means of production).

Marxist Science is natural, conscious and communal; a science of sciences which covers and integrates all sciences dependent to each other.

1820-1895 FRIEDRICH ENGELS: German philosopher. He worked together with Marx in line with the same ideas. They both determined the fundamentals of any kind of developments in **Dialectical and Historical Materialism. (Dialectic or Dialectics:** Thinking the notions together with their opposites and reaching the reality). Down the ages, a prejudice continues on the main problem of philosophy (**materialism** and **idealizm** problem). Hegel is on the side of **Idealism.** "Economic substructure" is a factor, but "Ideological superstructure" is also a factor. All these factors affect mutually. "Marx and me, we have the responsibility of youngsters' giving more than necessary importance to the economic aspects. We had to resist on the main principle, which our enemies tried to deny. Therefore we could not get a chance to give all factors contributing to the mutual effect, what they deserve".

1820-1903 HERBERT SPENCER: English philosopher. Some of his books are: 'Education: Intellectual, Moral and Physical', 'Classification of the **Sciences**'. "Science is the knowledge only one part of whom has been unified. **Psychological events arise from the biological events; and biological events arise from the physical and cosmic phenomenons**". He is an evolutionist.

1821-1881 FYODOR MIKHAILOVICH DOSTOYEVSKY: Russian author, musician and astronomer. He admired the European civilizations. But he believed that spiritually West was in a period of regression **and that they would never understand Russia. According to him, Russia is obliged to unite the cultures of West and East. "Humanism will tend to reconciliate in liberty."**

1844-1900 FRIEDRICH NIETZSCHE: German philosopher. He first promoted the knowledge and designed a social ethics based on knowledge. Later he criticized the knowledge and at last indicated that the scientist would not be able to provide any rules of life, so the thing to do would be to accept the life in high spirits. Nietzsche never tried to reconcile his ideas logically. He said that "I am writing with the lightnings, not with words. His ideologies, which are not based on logic, but violence, effected Fascism, National Socialism and Anarchism. He lived for 11 years after going mad.

1847-1922 GEORGES EUGÈNE SOREL: French author and engineer. He read Karl Marx, Proudhon, Bergson, W. James and Nietzsche. He propagated his supporting thoughts on violence, in his book named "Reflections on Violence". He indicated the necessary conditions for the Proletarians to come into power. According to him, this class should attempt to struggle on the way of establishing its own civilization without compromising with parliamentarians or intelligentsia. Working class will resort to violence by sustaining its own fight with an uninterrupted and indirect action. If the Myths, which will guide the masses (the expression of collective subliminal in the form of view), do not exist, people may continuously talk about revolution but they never achieve a revolutionary movement. They just try to constitute in the development of a general science of signs by analyzing the characteristics specific to the concerned Myths. **(His book: Elements of Semiology**). He had a role in the education of Lenin, Mussolini and Hitler.

1856-1939 DR. SIGMUND FREUD: Australian of Jewish origin. The founder of **Psychoanalysis**. According to him, the pressure of our souls is what makes our body sick. Our virtues were making us sick by keeping the door of our conscious open and by engraving in our subconscious much of the natural needs, which are requested to be realized. In that case, **virtue** was some kind of microbe in terms of medical science. Previously, 200 years before Freud, **Leibnitz** has proved that there is a Subconscious in the humans apart from **Conscious**.

According to Freud, humans are being directed by three instincts:
1- **Protection** instinct.
2- **Sexuality** (sex) instinct.
3- **Socialization** instinct.

In the civilization of humans, **protection** and **socialization** instinct have lost their importance and today, the only important instinct left is **Sexuality** instinct. This is present even in the newborn babies. Suppressing it under moral ethical constraints is the reason of many illnesses (like hysteria and neurosis in the beginning; later heart, stomach, liver and intestinal diseases). Psychoanalysis method takes the delayed, suppressed, imprisoned excitements out one by one and educates (nurtures), puts in order. We should not underestimate the animal in us. We need to make some of our tendencies real for our health. Today's civilization keeps many people under pressure unnecessarily. Tending to **high purposes are not always sufficient.** As all of the heat given into a machine cannot be transformed into power, all of our tendencies cannot be transformed to high purposes, either. The results of his research are as follows:

1- Unconscious periods have dynamical effects on conscious and behaviors.
2- Mental oppositions play an important role, not only during the illness, but also during the normal development. [An introspection directed to various protection mechanisms in which instinctual tendencies are excluded from conscious or behaviors, or are changed (as in the sublimations), is also a part of this.]
3- Personalities.
4- Power of instinctual impulses (sex and aggression) on motivation (reason to move).
5- Existence of sexuality in the childhood and its importance.

1858-1947 MAX KARL ERNST LUDWIG PLANCK: German Physicist. After some researches on the black body radiations, in 1900, he suggested his assumption about **Conservation of Energy** and he defined the **Quanta Theory**. This theory is the basics of modern physics. "**Planck's Constant**" is the universal constant of Quanta Theory. (**h**=6. 625. 10^{-27} Erg/Seconds) The energy of **n** frequency light particle or quantum is **E=hxn**.

1867-1934 MARIE CURIE: Polish chemist and physicist. She discovered the radioactive properties of thorium; won the Nobel Prize in physics in 1903 and the Nobel Prize in chemistry in 1911. She died of cancer due to the radioactivity because of Thorium contanimation, and became the first martyr of Thorium. The secon martyr of Thorium was the lady Prof. Dr. Engin Arık who died in a plane crash in Turkey in 2007. May they rest in peace.

Some words of Madam Curie:

If half of the millions of manuscripts burned in Andalusia had survived, we would have been traveling among the galaxies for a long time.

We cannot expect a better world without making people mature and moral.

1869-1962 Dr. SUPHI ZÜHTÜ EZGİ: One of these who gave the final form of Turkish Music System.

1871-1935 RAUF YEKTA: Musicologist, composer, music writer, *neyzen* (flute player), calligrapher. Turkish music is the final form.

1871-1959 RUDOLF CARNAP: He was a major member of the Vienna Circle and an advocate of logical positivism. He studied on the value of **Languages** and their functions. He was engaged in **Semantics**. (**Semantics:** It is the idea movement, which believes that words cannot completely describe the communal facts, misunderstandings, happen since they sometimes misrepresent, and there will be no problems if these words are not used.)

1873-1938 MUHAMMAD IQBAL: Pakistani Muslim poet, philosopher and author. He read Western philosophers like Goethe, Nietzsche, Dante and Schopenhauer, and reached East-West synthesis. He wrote Urdu and Persian books. He was influenced by Mawlana, J. M. E. Mc. Tangart, Nietzsche and Bergson.

1878-1960 SAİD'İ NURSİ: One of the last Islamic Sufis. He was born in Nurs village of Bitlis in eastern Anatolia. When he was still a pupil, he memorized 90 books on basic Islamic sciences. When he was 15 years old, his teacher named him "Bediuzzaman" (unique in his time). He wrote the Qur'anic commentary complex Risale-i Nur. Founder of the Nur community. Throughout his life, he tried to establish a university in eastern Anatolia where both religious and natural sciences would be taught together and ethnic languages would be free.

1879-1955 ALBERT EINSTEIN: German Physicist of Jewish origin. He proved the basic logical assumptions in the physical field, which he has derived from the Historical and Dialectical Materialist Philosophy and from the dielectric process of nature by means of various theories which he precisely suggested in **the special relativity, general relativity and unified field theories which verify the dialectical materialism philosophy, matter, time, space, motion, light, gravity, essence etc.** fields. No matter what the **motion conditions are, laws of nature are the same.** *Space, time* and *motion* **are not independent from the other laws of physics. Conditions of objects depend on the gravity field; and the gravity field exists by means of matter. It means that** *space, time, motion, matter* **and all of the natural events are the same.**

He made researches on the structure of atom and Planck's **Quanta theory**. He established his theory by applying the probability calculations to the **Brown motion. He calculated the value of Avagadro number, and applied on the Quanta theory and radiation energy. This calculation made him establish the light particle and photon hypothesis. He explained the photoelectric effect, and founded their laws. Later he laid the foundations of Relativity Theory, which he prepared for time and space.** He was awarded 1921 Nobel Physics prize. In 1940, he became an American citizen.

Relativity Theory is separated into three parts:

1- Special Relativity (1905), which modifies the laws of Newton Mechanics and which proposes that Mass and Energy are equivalent.

2- General Relativity (1916), giving the 4 dimensional gravity theory belonging to the universe which has been presumed to be curvilinear and finite.

3- A theory trial combining Electro/Magnetism and Gravity in the same field. The specialty of these a priori prepared theories were proved by bright experiments on Atom physics and Astronomy, especially when big masses and speeds come into question.

1879- JAMES CLERK MAXWELL: Scottish Physicist. He tried to establish the **Kinetic Theory of Gases** in accordance with Gauss' **Law of Error** by examining the fraction of gas molecules moving at a specified velocity at any given temperature. He showed that the average kinetic energies of molecules in the same temperature, do not depend on their structures. By means of the measurements he made on internal friction, he found the value of mean free path. In 1862, he suggested the "**Displacement Current**" notion, which is found in the Di-Electric (Piezo/Electric) objects under the influence of a variable electric field. He founded the **Electromagnetic Wave Theory, which** occurs with the replacement of a current in a magnetic field. (1865) This theory was derived from the identicalness of speed of light between the propagation of first electromagnetic shock and light. He checked by means of experiments that, the ratio between the electric units in absolute two systems was equal to the speed of light. In **1873, he presented the general equations of Electro/Magnetic field**. He found out the **Magnetic contraction**. Beside she explained the rotation in Crookes Radiometers.

1880-1955 H. SADETTIN AREL: Musicologist, composer, lawyer. He specified the 24 unequally spaced system of Turkish music and described Arel-Ezgi-Özdilen-System used today. He is one of the three musicians who explained Turkish music in the systematic scales in cooperation with Rauf Yekta and Dr. Suphi Ezgi.

1883-1945 BENITO MUSSOLINI: Italian fascist leader

1885-1962 NIELS BOHR: Danish Physicist. He formulated the calculation of orbit radius of atoms, which can be seen under the electrons of atoms. Besides, he developed the knowledge on Hydrogen spectrum and Periodical Table of elements. He explained the reason of expressing the electron orbits in accordance with the Quanta conditions.

RADIUS OF 1. ORBIT	0.5×10^{-8} cm (0. 5 ANGSTROM)
RADIUS OF 2. ORBIT	$2 \times 2x \, 10^{-8}$ cm (4 ANGSTROM)
RADIUS OF 3. ORBIT	$3 \times 3x \, 10^{-8}$ cm (9 ANGSTROM)
RADIUS OF 4. ORBIT	$4 \times 4x \, 10^{-8}$ cm (16 ANGSTROM)

1885-1954 THEODOR FRANZ EDUARD KALUZA: German Mathematician and Physicist. He developed the 5-dimension equation, which is known as the Kaluza-Klein Theory together with Oskar Klein. Thanks to this study, "**String Theory**" was developed later.

1887-1961 ERWIN SCHRÖDINGER: Austrian Physicist. He made studies on Quanta Theory using the physiological examination of Colors. Concerning his main field of study "**Wave Mechanics**" and its application on an atom application, he deepened the parallelism between the **"Wave"** structure and "**Particle**" structure of physical phenomenons. He showed that Wave Mechanics of Louis de Broglie was not different from the Matrix Mechanics of Heisenberg. Establishing a **propagation equation (Schrödinger equation)** which is used for calculating the wave function of a particle displacing in a specific area, he prepared **Today's Methods on the Quanta Mechanics**. He was awarded 1933 Nobel Prize together with P. A. M. Dirac.

1889- MARTIN HEIDEGGER: German mystic philosopher. He suggested the **Existentialism** idea, which was based on a secret mystical thought. According to him, in the universe, humans were the only beings who create their existence themselves. Existentialists reduce the objective existence to the humane existence; and this humane existence to the personal property and personal thought; and they reach to a subjective idealism like Fichte. Since humans create themselves, they have to be free and responsible. The responsibility emotions of existentialists are in fact as a result of fear of death. This fear may push the humans to the action of their personal interests and life. According to the existentialists, world exists thanks to the thoughts of humans. Universe is against the humans. It is inapprehensible and ends with a metaphysical incoherence like death.

1889-1970 ANTÓNIO DE OLIVEIRA O. SALAZAR: Spanish fascist leader.

1889-1945 ADOLF HITLER: German Fascist leader. The founder of National Socialism. He read Sorel and Nietzsche. He threw 7 million Jews into the furnace with the racist discourses. He caused 5 millions of German, 25 millions of Russian and millions of humans from other countries to die.

1889-1951 LUDWIG WITTGENSTEIN: Austrian philosopher. He affected the Vienna movement.

1891-1953 HANS REICHENBACH: German philosopher, new positivist. He is one of the founders of Vienna movement. He wrote the book "**Elements of Symbolic Logic**".

1891-1967 Prof. Dr. SALİH MURAT ÖZDİLEK: Musicologist, Physicist, Mathematician, Science historian.. One of the founders of the system Arel-Ezgi-Özdilek, which is used today in Turkish music.

1892- LOUIS DE BROGLIE: French Physicist. He wrote a thesis named "**Research on the Theory of Quanta**". He tried to reconcile **two light theories** (Emission and Wave theories). Beside she laid the foundations of **Wave Mechanics**. Proposing that the electrons and particles in motion have wave particles, this theory was proved with experiments by Davisson and Germer. This invention allowed obtaining useful results like **Electronic analysis and Optics**. Later, he examined the Magnetic Electron Theory and Spin Particles Theory.

1894-1977 OSKAR KLEIN: Swiss theoretical physicist. He studied on the "**String Theory**" and "**M theory**". He was awarded the Nobel Prize together with Theodor F. E. Kaluza.

1901-1954 ENRICO FERMI: Italian Physicist. He studied on **breaking of X-Rays**. He presented a statistics, which is based on Pauli Principle and applied on **electric particles,** together with Dirac. He defined a new field, which provides the stability of atom nucleus stability. He also developed the **Neutrino Hypothesis,** at the same time with Pauli. He showed that **a Neutron could be separated into one proton and one Electron by giving energy.** He was awarded 1938 Nobel Prize in Physics. He attended in the Atomic Energy production group, during the II. World War. In 1942, he produced the first battery including Uranium and Graphite, in Chicago. According to him, **Quanton is Boson or Fermion**.

1900-1958 WOLFGANG PAULI: Swiss Physicist of Austrian origin. According to the principle, which is called with his own name, **"in an isolated atom no two electrons can have the same four quantum numbers".**

1901-1976 WERNER HEISENBERG: German Physicist. In the II. World War, he made studies on Atomic Energy, in Berlin. He made important contributions to the Ferromagnetism Theory. He discovered the Allotropic Forms of Hydrogen. He founded the theory, which shows that **Atom nucleus is composed of only Neutron and Protons**. In 1926, he found the "Perturbation Forces" in his researches on the helium atom and explained the stability of this Atom. He designed the atom as a table of numbers, which was abstracted from matter, but in compliance with the matrix calculations, by applying Quanta Mechanics on Atom (1925). He explained the Rejection of Unobservable Quantities and the "**Uncertainty Principle**" which changed all theories of Micro Mechanics.

1902-1986GEORGE GAYLORD SIMPSON: American biologist. He researched the Paleontology fossils. He classified the animal types in a circular table.

1902-1987 PAUL DIRAC: English Physicist. He succeeded in linking the Relativity Theory with the Wave Mechanics; and in compliance with the Pauli's Exclusion Principle, he made a statistical explanation of **Quanta Mechanics**, which he was, one of the founders (Fermi-Dirac Statistics). In 1930, he theorized the existence of **Positive Electrons before these electrons have been found**. He was awarded 1933 Nobel Prize of Physics together with Schrödinger.

1903-1957 JOHN VON NEUMANN: American Mathematician and **Computer Scientist of Hungarian origin**. He wrote the book **"Mathematical Formulation of Quantum Mechanics"**. He developed the Game Theory. **He pioneered in the production of cutting-edge computers.**

1915- ROLAND BARTHES: French author. He says that "The myths of **modern industrial society** are the indication of **collective unconsciousness as an image, and comes before the individual unconsciousness, and makes it accept its deep exciting symbol**". He tried to develop a **General Language of Signs** by examining the specialties of Industry **community.**

1915- ARNOLD SOMMERFELD: German Physicist. He developed the studies of Bohr and Pauli. He suggested the 'elliptical orbit' notion for the atoms. **He found the law on number of correlations between the wavelength and frequency of spectral lines.**

1926-1996 MOHAMMAD ABDUS SALAM: Pakistani Physicist, Mathematician, Theoretical Physics Professor. He developed a theory on **Electro/Magnetic interaction and weak interaction of**

elementary particles. He was awarded 1979 Nobel Prize together with Steven Weinberg and Sheldon lee Glashow.

1929- MURRAY GELL-MANN: American Physicist. He started to classify all of the nuclear particles similar to Table of Chemical Elements of Mendeleyev. He wanted to add the recently found new particles to the classification. He showed that some of them were symmetric and that's why they could be examined together. Such that two or three of these particles could be taken as a single object, which could be found in many various conditions. These superior particles have a significant magnitude named "**Isospin**". But in some of the reactions, particles lose their specificity. Gell-Mann called these varieties as "**Strange**" and he suggested a new numerical specialty for them: "**Strangeness**". Suggesting a theory named **"SU-6" (Unitary symmetry in the sestet order) together with the** Israeli Physicist Yuval **Ne'eman, he predicted the existence of UNKNOWN particles**. Among them, a particle named "**Omega Minus Particle**" was found in 1964 by means of "Darkroom" in Broohaven. Based on these results, **Gell-Mann suggested the assumption that the creator element of all Particles were Quarks.**

1933- STEVEN WEINBERG: American Physicist. He worked in the fields of **Quantum Field Theory, Symmetry Breaking, Pion Scattering, Infrared Photons and Quantum Gravity, Higgs Boson, Z Boson,** unification of weak and electromagnetic interactions. **He worked with** Abdus Salam and Sheldon Glashow on **particle physics and quantum field theory, earth gravitation, super symmetry and String Theory**. He wrote articles on Superconductors and Supercollider.

1942- MARTIN REES: English Cosmolog, Astro Physicist, Mathematician. He developed theories on **Black holes, Quasars, Cosmic Micro Wave Background Radiation, Steady State Theories. He studied on Gamma Ray Bursts.**

1946- GERARD'T HOOFT: Dutch theoretical physicist. He shared the 1999 Nobel Prize in Physics with his thesis advisor Martinus J. G. Veltman "for elucidating the quantum structure of electroweak interactions". Besides, he developed theories on the basics of **Gauge Theory, black holes, Quantum gravity and Quantum Mechanics and massless Yang-Mills Fields.**

His studies are in three subjects:

1- Gauge theories in elementary particle physics.

2-QuantumGravity and black holes.

3- Foundational aspects of quantum mechanics.

PART THREE: SYNTHESIS OF UNIVERSE and HUMANS

A) PRELIMINARY PREPARATION TO THE UNIVERSAL SYNTHESIS ACCORDING TO THE MODERN KNOWLEDGE:

I) PHILOSOPHERS RELATED WITH THE UNIVERSE:

1)   PHILOSOPHERS RELATED WITH SEMIOTIC

1632-1704 J. LOCKE: English Idealist. He organized sciences semiotically. He suggested the term 'Semiotic', science of signs and symbols theory.

1235-1316 RAIMUNDUS LULUS: He expressed all of the facts with triangles, circles, words, etc.

1891-1953 H. REICHENBACH: He wrote the book "Elements of Symbolic Logic".

1891-1959 R. CARNAP: He was dealing with semantics.

1915- R. BARTES: He tried to develop a general science of signs.

2)   PHILOSOPHERS RELATED WITH MATHEMATICS:

B.C. 485-411 PYTHAGORAS: "All tangible and intangible beings are signs which can be expressed by numbers. The most dominant thing in the universe is 'numbers'; and the most beautiful one is 'harmony' which is present in between the living creatures. There is a 'divine music' composed of harmonious combination of various sounds in the universe. The letter γ (epsilon) looks like two directions in front of the humans, evil and virtue directions. The principle of unity is the mind."

A.D. -872 EL KINDI: He says that "Mathematics is the fundamental of all sciences".

A.D. 1596-1650R. DESCARTES: He accepted that the living creatures are made of matter and soul. Mathematics was the model of Science. An idea would bring another idea. There was no knowledge, which we could not find by chasing in the right way. He suggested a science and sign language, which would contain all of the sciences.

1632-1677 SPINOZA: "Philosophy is the generalized type mathematics. Humans are the prisoners of what they do not know. They will be free when they reach to the knowledge. The thing, which develops the positive intelligence, is the thing, which blurs the negative intelligence. Virtue will be measured by the will and power of humans to protect their own existence. Humans will be stronger when they come closer to each other. In that case, there nothing beneficial to humans other than humans. Therefore, humans want to unite."

1646-1716 LEIBNITZ: He applied mathematics into philosophy and tried to found a universal science language, the "Language of signs". He also tried to explain the language systems in linguistic, mathematically. Evil is in the details. In fact, whole is perfect in the highest level. Virtue and happiness is because of the love of God.

II) PHILOSOPHERS RELATED WITH THE HUMANS:

1) THE ONES WHO THINK WRONG ABOUT HUMANS AND THEIR SYNTHESIS:

1469-1527 N. MACHIAVELLI: Italian politician. According to him, we needed to do wrong, use violence and torture, instead of favors. Everything, which take humans to the purpose, were good, even if they were bad.

1724-1804 IMMANUEL KANT: German philosopher. Emotionally, moral laws appear to us as the "**respect**". However, respect is no a tendency to direct us to an object like love or desire. It is a judicial emotion, which prevents our desires. It limits our tendencies. It makes us look to the object from a distance. Virtue is not an instinct; it is a matter of mind. There is no need to take offense at the virtuelessness, because this virtuelessness will take humans to the big target, to virtue. Kant does not believe in God. He says "There is no God. There is only power".

1789-1860 A. SCHOPENHAUER: German philosopher. According to him, universe emanated because of the will for being. Universe is the place of eaters of each other endlessly. So, the main thing is the virtuelessness, not virtue. Compare the sorrow of ruptured animals with the joy of rupturing animals. It will be found out that the fundamentals of existence is the will.

1809-1882 C. DARWIN: English biologist. He ascertained all of the animals, which have lived up to-day by examining their fossils. He found out that weak animals have extinct, strong ones have survived. According to his Evolution Theory:

1- Species of the same origin shows changes, which can be attributed to various, affects (environment, feeding, etc.). It changes the body and germ cells.

2- Economist Malthus reached to the 'struggle to survive' principle from the opinion that the population increases geometrically, but feeding opportunities increase arithmetically, in other words, lesser. Superiors have a chance to live and breeding as a result of "natural selection".

3- Males fight to get the females. Strong ones win. Females choose the most handsome and the strongest males. Thus, a sexual selection happens. **Virtue** is a product of evolution coming out of necessity for adapting to the environment and conditions. It appears from the self-defense instinct. It is a must for the survival of humans.

1844-1900 F. NITZCSHE: German philosopher. As a result of Darwin's evolution theory, "A superhuman" thought. "Remain faithful to the earth. Do not believe in the ones who talk about the hopes for other world. God is dead. The biggest evil is needed for the biggest goodness. The biggest goodness is the creativity. The one, who wants to be creative in goodness and evil, should first be destructive and demolish the values. Demolition badness is necessary for the creation goodness. And then this creation goodness will lead to the demolition badness for a new creation goodness. A new race is necessary to become the master of the world. No privileges are possible unless being born from a good family. Poverty of the whole nation is less important than the suffering of a superhuman. **Virtue** is the effort of humans to reach to the "Superhuman". Wherever there is a death, here there is a self-immolation for a birth. Every development arises from an aristocratic population. Molding the humans of future to annihilate millions of idiots, is the target of superhuman." Nietzsche used to tall that he has been writing by storms, not words; and that he has been burning in the fire of his own thoughts. He could not get rid off illnesses throughout his life. He lived for 11 years after he became ill with his last madness crisis.

1889-1945 ADOLF HITLER: He studied Sorel and Nietzsche. He tried to apply the master race theories of philosophers on the world. He threw seven million Jews into the furnace. He caused 5 millions of German, 25 millions of Russians, Asians and millions of humans from other countries to die.

1899- M. HEIDEGGER: German philosopher. He suggested the Existentialism doctrine. "Human is the only creature which has created its own existence in universe. Since humans have created themselves, they should be free and responsible". Responsibility emotions of Existentialists is a result of "fear of death", and this fear may drive humans to the act of their personal benefits and life. According to the Existentialists: "world exists with the considerations of humans. Universe is against the humans, incomprehensible; and ends with a metaphysical conflict like death.

SYNTHESIS OF THE ONES WHO THINK WRONG ABOUT HUMANS: It will be good to take the above motion of thoughts up to the **Vedas**, which were the religion resources of Indo-Europeans who have invaded India on **2000 B.C.** In Veda religion, "Sacrifice creates the Gods. Actions create the existence". **After 800B.C.,** there were 3 Gods in Brahman religion: **Brahma** "creator", **Vishnu** "protector" and **SHIVA** "destroyer to re-create". The symbol of these three gods **Trimurti** (Triple swastika)

There are effects of Veda and Brahman religions on the motions of thoughts, which are listed above, in the last centuries. After the renaissance, while enriching by the industrial revolution, European people who were freed from the pressure of church in the middle age, started to be democratized. In the meanwhile, as the people were directed to the atheism as a reaction to the church which anathematized the kings and judged the scientists, the developments like Darwin's idea of "Since the strong animals are able to survive, it means that a strong human race will destroy the other" and **Karl-Marx** and Engel's **Dialectics Materialism, Communism, Materialism** which gave kind of scientific support to the fight, caused the world live two World Wars.

We cannot say that all claims of these scientists were wrong. However, everything owe its existence to the *balance* and this balance creates the *order*. Balance is the thing which supports the universe and the third "must" which Da Vinci has learned from Andalusia in the XIIIth Century; and neglected after the soul and matter. It is the **wise** and **perfect** man who will provide the balance between matter and soul, which is represented by Mawlawi dervish in Islamic mysticism. It is undeniable that all of the above philosophers, who are defined as wrong, were idealists; and that they were pining for the perfect humans. Who would refuse the purpose of achieving a world ideal including free, responsible and perfect humans? But, superior races do that. And we are that superior race. It was a wrong idea to savage the ones who were not one of us. Because there are no races, in which all of the individuals are perfect. The one who behaves fondly, respectfully and indulgently to the others; who lives as per the ceremonies; is the superior human, no matter if he is Black, European, Chinese or Indian. There is no need to look at his skull. **The most superior man in Islamic Sufism, both wise (Arif) and living in accordance with the knowledge, perfect human is defined as eugenical and being modest is one of the main characteristics**. Because none of the humans are abiding in the world. This is indicated in the Middle East with an idiom like **"World wasn't left to Suleiman, either"** (No man can live forever). A lower level of perfect humans is wise humans. It means the well-informed humans with the true knowledge. The better may be left to the DNA studies. "Great minds think alive" for all of the humans. Solutions by killing or having difficulty are against humans' nature. It would be an inhuman treatment. The problems will disappear if the truth is taught by means of education. Attributing the problems, to the fight to the racial superiority is causing their continuity forever; and it may lead us to a very tragic end. The secret of maintaining the peace by the Ottomans for 600 hundred years and that Westerners have defined as **Pax Ottomana** lays in the Islamic mysticism.

2) THE RIGHT PHILOSOPHERS RELATED WITH HUMANS:

468 B.C. -399 B.C. SOCRATES: "True knowledge makes the right action."

384 B.C. -322 B.C. ARISTOTELES: "In the age of fast advancing sciences, philosophers should make the synthesis of acquired knowledge. Motion is the fundamental necessity of nature. Time is the number of motions."

23 B.C. -79 A.D. C. S. PLINIUS: "Humans need to learn to secure his life, unlike the animals."

1162-1240 MUHYIDDIN ARABI: "Human is the link between the hidden universe and visible universe (**Perfect humans**)."**Comprehending the Universe of the Supremes in an integrity"**, is one of the ways to reach the **"EUGENICAL"**. All the paths reaching the **Eugenical** end in the **Creator**. **(In the 21st century, now we can define comprehending the universe as a whole as the "Theory of Everything".)**

1638-1715 MALEBRANCHE: Mind of a human is a part of infinite mind (Word of God). "**Ethics**" depends on the **order** idea, from simple to perfect; from humans to God.

1509-1564 BACON: "Every invention is a means to come to a conclusion. In physics, success comes with observation, experiment and induction. *"*

551 B.C. -479 B.C. CONFUCIUS: Humans should know that he knows what he knows; and that he does not know what he does not. This is the real knowledge. "Logic" which prompts humans to think fair and to express the purpose, needs to build up an "Ethics" which prompts humans to live well. A good order is up to a good elocution. Therefore, the definitions should be corrected. Ungrateful child is not a real child. Bedswerver is not a real spouse. If you are a father, then be a father; if you are a child, then be a child; if you are a prince, then be a prince. Money is necessary to live. To earn money, it is necessary not to live. The first rule of ethics is 'filial piety'. Mother and father are representatives of ancestors. The respect and love of small child for the big one; of wife for her husband; of citizen for the sovereign; are like the respect and love for the mother and father. Citizen should behave like a child; sovereign should behave like a father; he should bring peace, ease and knowledge to his citizens. "Justice" makes all humans get along well with each other. Goodness should be replied with goodness; injustice should be replied with justice. Do not do unto others what you do not want others do unto you.

430 B.C. -347 B.C. PLATON: "Justice" idea is a must in the military, administrative and financial issues of the state (Idea=Reality). Otherwise, the state would be degenerated. State is the magnified sign of an individual.

330 B.C. -264 B.C. ZENON: He is Stoic (**Stoa**=who behaves suitable to the nature). **He indicated that "Suitability to the nature is suitability to the mind. In other words, it is the suitability of humans to themselves.**

There are four main virtues:
1- **Selecting the right one.**
2- **Abiding with patience.**
3- **Being restrained.**
4- **Sharing fairly.**

Happiness is in the wisdom, wisdom is in the behaving suitable to the nature. Wisdom is reached by getting the **Theoretical and Practical Virtue. Theoretical virtue**: Obtaining the true knowledge on the objects. **Practical virtue:** Behaving reasonable. Pains, illnesses and death is natural. We need to stand them, we need to stand idly by. In fact, **humane mistakes, evil and murders are the things, which should be abhorred and struggled against. Natural measure takes the humans to independence and equality. The discrimination among the humans is unnatural.** Being the children of nature, all of the humans are brothers. Being tied to the same nature, takes the person to humanism and the universal citizenship." Zenon also indicates like Socrates that "behaving well is possible by thinking right; and philosophy is a life science which teaches to think right. The purpose is living in good conditions. **Knowledge is necessary to live in better conditions**".

870-950 AL FARABI: Virtue is cooperation. The town whose citizens cooperate in the things, which will convey to the real happiness is a **virtuous town**. Cooperating community is a **virtuous community**. A nation with all towns and cities helping each other, is the **Virtuous nation**. In case all nations help each other, it is the **Virtuous land**. Virtuous city looks like a body with all organs complete. **The most perfect state is a world state, which cover all humanity.**

XI. Century YUSUF KHASS HAJIB: In his book named "Kutadgu Bilig" (Knowledge on happiness), it is told that happiness can be obtained only by **Goodness** and **Ethics**. "Goodness" is knowledge, peace and honesty. "**Badness**" is ignorance, illness and to estray. **Ignorance can be put right by doctrines**; illness by treatment; **badness by education.**

The book is in four parts:
1- Justice
2- Power
3- Intelligence
4- Moderation

1238-1310 YUNUS EMRE: "The lover's heart is the Creator's throne; God admires and accepts it as his own; The man who breaks a heart shall groan and moan; In both worlds, suffering sorrows and cares." (Gönül Çalabın tahtı, Çalab gönüle bahtı, iki cihan bedbahtı, kim gönül yıkar ise). "**Çalab**" **is the oldest word of** Central Asia history, which means God. At the same time, it is the number 160,000, which is the biggest number of Mayas. Another verse of Yunus is "Ten ölür, canlar ölesi değildir." (the body dies, soul is immortal) The **heart,** in which all of the human values are gathered, in other words, **Soul** is indicated with the word 'can' in mysticism.

1694-1778 VOLTAIRE: The world should be dominated by the intelligence.

1712-1778 J. J. ROUSSEAU: "Humans were good. It was civilization, which made them bad. Greek art remains have passed to Italy after Byzantine has been laid low. That's how the unhappy pedantry of Europeans and their virtueless civilizations have started. **There are 'JUSTICE'** (law) and **'ETHICS'** (tradition) principles in the depths of **PEACE**".

1743-1704 LAVOISIER: "The word 'merge' should be used instead of 'to _be_'; and the word 'separation' should be used instead of to '_die_'. There is a change in the appearance in every change and merge. Essence does not change. _"_

1771-1858 ROBERT OWEN: As the difference of humans come from the neighborhood and coincidences, it is not good to differentiate between humans and have great expectations. The incomes of humans should be same with each other. Humans should be able to survive in equal conditions.

1772-1837 CHARLES FOURIER: "A worker should be the shareholder of enterprise. He should participate in the management and get shares".

1773-1831 GOETHE: "Virtue is the essence of humans. Whatever the devil (Mephisto) do, it will not be able to separate humans from this essence. At the end humans will be elicited, they will manage to preserve their humanity. Happiness of **humans depends on the happiness of community. It's nice to live by participating in the community. The duty of mind is directing this power to the most useful purpose. What makes humans feel happiness and which opens the gates of the heaven on earth, are their feeling it in themselves. Humans understand that by means of the consciousness they reach, despite of the devil (Mephisto), and they can say** 'Stay! Thou art so beautiful!'."

1798-1857 AUGUSTE COMTE: "The first steps of **Social union are prepared in the family.** The purpose of ethics is to prefer humanism instead of animalism. **Individuals are a part of humanism organism.** In the **community, "DUTIES" precede the "RIGHTS". The regime** should be a papacy based on science. **Community should be organized by means of science. VIRTUE is love. Love the humanism like you love a human. PHILOSOPHY** is the **science of sciences, which** unites all of the sciences. There is only one absolute principle. Everything is related with each other."

1821-1881 DOSTOYEVSKY: He admires the European civilization. But he believed that the west was in a spiritual retreatment, they would never understand **Russia. He indicates that "Russia is obliged with the duty of uniting the cultures of West and East. Humanism will find a way to reconciliation in freedom."**

III) THE ONES VERY CLOSE TO THE UNIVERSALSYNTHESIS:

B.C. 570-490 TAO-TSU: "Tao gives birth to the **One.** One creates the **Two.** Two produces the **Three.** Three creates all other beings of the world."

B.C. 554-480 BUDDHA: "Happiness is neither in the Physical life, nor in the Spiritual life. It is in the middle of them."

485 B.C. -411 B.C. PYTHAGORAS: "The letter γ (Epsilon) looks like the parting of two ways in front of a human. The ways are **BADNESS** and **VIRTUE. MIND** is the principle of unity."

1207-1273 MAWLANA CELALETTIN RUMI: "Even if there were a hundred books, they all have only one chapter. They turned towards a single mihrab in these hundred directions. All of these roads lead to a single house. These thousands of wheat spica are made of a single seed". Creature is the appearance of Creator. Music is the essence of similarity of souls of humans. Sema (dance) is the appearance of universal ecstasy. **Religions are one and only. They separate on their way.** Left palm of the whirling dervish is towards the ground, the wealth and **MATTER**; and the right palm is towards up to **ALLAH** (to the meaning). And the whirling dervish symbolizes the **WISE** (Intellectual) and **PERFECT** (Living the knowledge) humans.

1210-1274 SADRETTİN KONEVİ: Creator (Halik) **and Created** (Mahlûk) **Universe is not apart from each other. Plurality (Kesret) is in the appearance. In fact, it is unity (Vahdet).**

1571-1630 KEPLER: *created universe* **consists of an astonishing symphony in thoughts and souls order, in physical beings. Everything is connected with each other in a mutual and indivisible relation; and they constitute a harmonious colored whole. He explained the motions of planets in accordance with the harmony laws of music.**

1770-1831 F. HEGEL: Virtue is a reconciliation between good and bad.

IV) SYMBOLS PREPARING THE UNIVERSALSYNTHESIS:

The symbols, which are deemed indicating the universal unity since the ancient ages, and also which today we can still accept that they indicate the universal unity, are:

4000 B.C. -3000B.C. Sun God Sumerians and Chaldeans.

4000 B.C. -3000B.C. Sun God in Egypt

3500B.C. Sun God in Japan

2000B.C. Sun God in Hittites

2000B.C. Mandala in India (symbol of the **Universe**)

Xth Century B.C. Menhirs in Celts of Europe (dual stones)

VIth Century B.C. Kalachakrain Asia Buddhists (symbol of the **Universe**)
Vth-IIIrd Century B.C. Yin-Yang in China (symbol of the **Universe**)
Vth-IIIrd Century B.C. Pi in China (symbol of the **Universe**)
Ist Century A.D. Yantra in Hinduism of India (**Page 71, Figure 8**) (symbol of the Universe).

YANTRA is a geometrical diagram, which is drawn for religion purposes. It means "Tool" in the Sanskrit language. Being the direct reflection of god by means of numerical signs, Yantra is a magical instrument. "Mahayana" of Buddhism indicates the condensation of thought in Hinduism. "Arthur's Circles" which has been built by Celts for adoring to Sun; **"Roland Disc"**, **"Druid Stones"**; obelisks in other words, **"Menhirs"** which are called as **"Cromlech"** sorted in a circle, in Malta island, North Africa, Palestine, Caucasus and India, can all be included in these symbols. In the city architecture, **"Trelleborg"**, an old port city in Sweden; in the building architecture, "Tholos" type circular buildings in the Northern Mesopotamia "Arpaçay Tepe" from V. thousand B.C. **"Vesta"** Fire Temple in Rome (**X. Century B.C.**); **"Etruscan Tombs of the Kings"** in Tsere near Rome; also again in Delphi-Italy, the buildings made for the **Sun God Apollo** (In Hellenistic times, meaning of the word "Helios" was the sun); "Stupas" in India and Burma as buildings in **VIIIth Century B.C.;** circular tribe tents with an ancient oven in the middle of the house, in Asia, Africa, Austria, can all be given as examples for **these** symbols.

V) DEFINITIONS ABOUT THE UNIVERSE PREPARING THE   UNIVERSAL SYNTHESIS:

COSMOLOGY: It is the philosophy field, which unites the various images of the physical world defined by separate sciences. The word 'Cosmology' was used by C. WOLFF and E. KANT for the first time. According to Kant, as **Cosmology** has been confused with the knowledge of relations (**Theogony**) between Gods in the ancient days, today also universe is being confused with the formation of sky objects **(Cosmogony)**. It is obvious from the ancient art objects that these relations between **"Cosmology"** and **"Religion"** have also been present in the oldest civilizations. There are many examples of this claim in pictures, architecture and urban science. Ruins of king cities, ancient city walls and castles in Iran, India, China and Cambodia, show that the relation between **cosmology** and religion has always been taken as a topic and treated. In all of these artworks, symbols originated from the celestial sphere have been used abundantly. The most common ones are the flat circle shaped ornament which shows the pyramid and tower, celestial sphere and the connection between earth and **sky,** luck and rebirth; egg which shows the universal power of both gods and the kings; and the golden ball. In architecture, four corners of a building showing the four directions of the world, columns showing the axis of universe; dome showing the celestial sphere etc. are cosmological factors. In some of the Chinese temples, the circular shape or foursquare plan of the structure, **sky** or earth have been tried to be revived. In the holy book, there is the same universal relation in the part related with **Suleiman's temple**. In the ancient Egypt, Heliopolis theological opinions have been based on three gods. "Nut" (**Sky** god), "Geb" (**Earth God**), "Shu" (**Air God**). **Babylon tower** (Ziggurat) has been a temple, which has been joining the Earth with the sky. In the older times, mountains have acted in the same way. According to the holy books, **Ecbatana city in Iran** has been composed of 7 nested and circle shaped city walls, which heighten towards the center to point to the sky. It has been assumed that there was a king in the middle of circle. In some of the documents in Paris State Library, there are pictures on the chalices from the age of Sasanians, which show the king on his throne in the center of the celestial sphere. There was a pole star in the center of celestial sphere. It is known that in the Chinese and Indian temples, apart from the pictures representing the universe, there has always been a separate yard named "Garden of Eden" in the traditional Chinese house, and this yard has been pointing to the relation between earth and **sky**. There is a very big ring in the Buddhist temple named **"Borobudur"** in Java, and this temple represents the universe. Similar examples can be seen in Africa, Oceania and America. Holy expression of universe in various handicrafts as big pyramid shaped temples, plans of the villages, columns showing the totems, pictures on the manuscripts can be seen in the ancient Mexico. In the first age, this subject has been treated by Pythagoras, Plato, Aristoteles and Stoic school philosophers and in the middle age by Saint Thomas D'aquin and Dante, and transferred to humanism, Renaissance idea and today's science worlds. In Urfa / Anatolia, Hagia Sophia Church is one of many churches, which can be deemed as an example of the middle age cosmology with its dome as the symbol of 7 layers of sky and golden inlaids as the symbol of bright sky objects. Similar examples can be seen in the Renaissance art. In the **XVII. Century, Modern Nature Science** appeared and these ancient symbols were left gradually. But new space understanding created pieces in baroque style. Today, we can see the universe symbols in the figurative arts, in the newest pictures and sculptures.

COSMOLOGY It is the science, which researches the general laws ruling the universe. Classical philosophy assumes cosmology a part of metaphysics. Scholars, going beyond each other and including each other, expand the knowledge of humans, in time. According to Einstein, the purpose of Cosmology: **"It includes the minimum number of assumptions and axioms (absolute facts), logical deduction and maximum number of experimental facts**. The number of elements which has been assumed to be infinite, was reduced to a hundred something and then to several particles". Einstein collected the universal notions, which are reduced in five attributes as **space, time, gravity, energy and material**, in terms of 'Relativity theory' and **Matter** and **Energy**. He tried to combine **infinite 'bigs' within finite 'smalls' in the same law by means of "Unified Field" theory.**

UNIVERSAL UNITY: The physical integrity of all events and facts in the universe, which appeared from the actions, and changes of materials, is a result of actions, changes and developments of materials in time and space. Humans are the most organized form of these events and facts. **Everything in the universe is in coherence with the physical unity. Therefore, they are bound to each other.**

UNIVERSALISM: It is the common name of doctrines, which are biased to universalism or completion.

PHILOSOPHY: (In Greek: Philo (**Love**) Sophia (**Wisdom**)) **It is the knowledge which tries to explain the living being completely in accordance with Metaphysics and Anthology. According to the science doctrines: it is a doctrine, which studies on the notion, principle and methods of sciences. It is the brainwork, which studies the matter, life and various indications of 'Acun' (Universe, cosmos) like community, soul in terms of principles and targets; and the efficiency of this work.**

DIALECTICAL MATERIALISM: It is the science, which determines the fundamental laws of all kinds of developments. It was suggested by German philosophers K. Marx and F. Engel.

It is composed of two doctrines:

1-Historical materialism.

2-Dialectical materialism.

Both of them complete each other in a dialectical dependency (**Dialectic: The way of finding the truth by evaluating the notions together with their opposites). Dialectical materialism: It completes the Nature, Community and Consciousness facts in a universal understanding; and presents that this completeness developed by the same law of contradiction.** Idealism and Materialism, and also **Metaphysics** and **Dialectic** were overcome with Marx's Dialectical Materialism. Being the doctrine of both 'to know' and 'to do', **Dialectical Materialism** presented the dependency of theory and practice. There is no application without a theory; there is no theory without application.

[**Note**: Here, **Marx** and **Engel's "Dialectical Materialism"** definitions were fundamentally Materialist. According to Einstein's $E=MC^2$ formula; for the complete universe, we get reach to the reality with the opposition of two notions, **Matter** and **Energy** (or Soul), in other words, with **Dialectic**. So, if one would say let's try to find the truth by thinking the SPIRIT and *matter* together, but ultimately become *materialistic*, it would not be true but "Dialectic realism" more accurate. The *balance* that kept the universe alive would not be ignored. Reality is in their universal harmony. And that is symbolized on the front cover of the book with Mawlawi dervish who represent the micro universe, in other words, the humans. All of the sciences associate the development of facts indifferent fields, only with special laws, which are valid in these fields. These laws are called as "unity and struggle of oppositions", "the law of passing from quality to quantity and from quantity to quality" and" the law of negating the negative". Dialectic is positive and necessary in these subjects, but materialism, never.]

HISTORICAL MATERIALISM: Marx' doctrine which proves that communal development is based on matter. Studying by means of historical dialectics methods, K. Marx ascertained that communal development is based on matter. Before Marx, it was attributed to the **Soul**. The humans (community) make the history. Classes and class conflicts lie behind the thoughts (ideas) of humans. Natural and communal phenomenon's both effect and be effected. And since the conscious human is a communal and natural phenomenon, he attended this dialectics and started to build the history himself. It prepares the conditions changing and forming itself, by means of its opposite effect.

History of humans:

1- **Primitive** community (production type is collective and there are no personal properties).

2- **Slaver** community (production depends on the personal properties of slaves).

3- **Feudal** community (production depends on the personal properties on the ground).

4- **Capitalist** community (production depends on the personal properties on the means of production).

5- **Socialist** community (production is shaped by collectivizing of the means of production).

The main characteristics of historical materialism and later dialectical materialism is the integration of **nature and community**. Historical materialism showed that community develops by means of unique laws like nature; and that these laws are also essential like the natural laws.

HARMONY: In Greek, it is the composition, which appears by playing various sounds at the same time, composing an accord, in other words creating a **"Voice order"**. (Accord: In French, it means each of the voice groups used to create the harmony).

SCIENCE PHILOSOPHY AND ITS METHODS: Generally, methods special to the particular disciplines for every science discipline can be assumed as an explanation to the experiments. **There are four principles to go towards this direction:**

1- **Technicality principle** (it is necessary to develop technical means for the experiment)

2- **Reviewing principle.**

3- **Duality principle** (It emphasizes the existence of a **theory**, which should be in a continuous reaction with each other; and a **testing apparatus**).

4- **Cohesion** principle: Every development towards the preciseness of a limited scientist should be filled with the implementation of an increasingly expanding knowledge. **There are 4 steps of these principles:**

1- Encountering a question, problem or a contrarian expression in acknowledge.

2- Arising of an assumption on the way to the solution.

3- Retrying, if the assumption cannot be confirmed by returning all processes to the starting point.

4- If the assumption is confirmed, this stage is reached.

Philosophy of Sciences, thus appears as a mentality between all of the disciplines. This way of thinking tries to establish a consistent discipline by adopting the principle to take all of the disciplines as a test field, and by being prone for new trials.

MODEL: **"Model building"** is building a consistent substructure based on assumptions. The results of this model will be deemed successful if they are in conformity with the observation results.

THEORY: The model should both show these properties and represent the nature at the same time (theory is a composition and it should be in unity like art and all technical compositions). Also as in all of the compositions, it continues its existence autonomously.

GEOMETRICISM: It is the system, which reduces everything in the symbols and methods of geometry.

RADICALISM: It is a behavior and politics era suggested by some of the English philosophers, based on the basic values like liberalism, individualism, believing in mind and moral effectuality. It is the tendency to make radical changes in science, religion and politics.

MATTER: It is the being occupying in the space. It means everything, independent from consciousness. The word comes from the root "MATR" which means "MOTHER" in the Indo-European Languages. (Matter-Materia-Materie etc.) In the Ottoman philosophy, it corresponds to the **"essence"**. In the Greek philosophy, it is called as **"hyle",** meaning the object, which the efforts of humans are directed to. This word has become **"materies"** in Latin, which means **"matter"**. In the first ages, it was assumed in India, China and Asia that the universe has been built from a fundamental matter. In the oldest Chinese consideration, it has been known that the senses (like sweet, bitter, salty, negative) were reflecting from materials and there were **Positive** and **Negative** primitive particles in the universe. Also in Asia, it was believed that the universe was composed of the basic materials like **Fire, Air, Water, Earth. Descartes** defined the matter as an extended (the place it occupies can be measured) essence (unchanging essence of changing universe) in the new age and separated the **material**

and **soul**. According to Ernst Mach; "Objects do not create the senses; senses create the objects". According to dialectic materialism, "Material is a philosophical notion apart from consciousness; but reflected by consciousness and perceived by senses. It mentions all objective reality". Later, not only elements which generated the objects or small particles; all cosmic universe, nebulas, planets, radiations, electromagnetic and nuclear fields have changed infinitely and in an interconverting structure (like transforming of the particles like positrons and electrons to the light quantums, and later their transforming back to positrons and electrons). In this continuous infinite period of change, the dandiest living beings, the **"humans"** emanated as a result of the evolution of creatures in the universe. **Humans are conscious materials.** At last, **Consciousness** constitutes the **"Substantial power"**; and this power constitutes the **"Substantial objects"**, in Theory. These human products, **"Materials"** interact to create new theories and new powers, new substantial products; they are developing in an infinite progress, quantitatively and qualitatively. **Thus the consciousness of humans, not only reflected the substantial world, but started to create it**. But as matter itself, its motion is also infinite and all are dependent to each other. At last, Einstein presented this dependency with $\mathbf{E=MC^2}$ (energy=mass x speed of light squared) formula. (We call it **"Radiation"** or **"Energy"**, if the matter leaves its mass and continue in the speed of light.). On the contrary, we call it **"Matter"** if the energy freezes and takes a different shape. In June 1945, it was proved in Mexico practically (atomic bomb**). But again material was relative according to Einstein. It was suitable for us. It was suitable for our sense organs.** Feuerbach indicates that: "My senses are subjective; but their principles or reasons are objective". This practice is the matter. For Lenin, material is: "a philosophical category which exists independent from Consciousness and which indicates the objective reality reflected in the consciousness". According to modern physics, 7 states of matter are: **Solid, liquid, gas, plasma, electromagnetic field, gravitational field and Nuclear Field**. (7 basic elements of nature and infinite variety of observable universe depend on mutual neutralization and interactions of these 7 basic elements. These 7 basic conditions or elements are as per the senses of humans and as per our knowledge and opportunities, today. Since the material is defined as being transmitted by our senses, infinite, limitless and real; it is open to an infinite and limitless renewal.

SEMANTICS (1883 A.D.): Symbol categories system is composed of opposite words. **French Michel Bras** prepared a **"Semantics differentiation table"** with **50 pairs of adjectives.** Main severely adverbs in the language accompany every opposite adjective; and a negativity accompanies every pair. **Thus, there are 7 possible conditions:**

Extremelystrong+3, verystrong+2, quitestrong+1, neitherstrong-norweak0, quiteweak-1, veryweak-2, extremelyweak-3.

Besides science of semiotic signs is divided into three parts in the symbols theory:
1-Denetatums (The indications which have an existential relation with the designated object)
 (Like noticing the fire by means of its smoke)
2-Demotatums (Pictures which have a similarity relation with Denetatum) (For example, images)
3-The symbols in which the relationship is completely based on cooperation (as the national flag).
Language is also one of the symbol types.
Semiotic in the strict sense; It intends to take the inventory of current sign systems, and to get to a general theory about them.
Semiotic in a broad sense; It is started from the principles which are indicated by all kinds of activities of humans (humans are animals which use symbols). The field of **Semiotic, is the field of science of all humans.** But today we can talk about a semiotic, which studies the science of humans consecutively, more than a single semiotic science headed towards the signs.
SYMBOLISM (1885): (Latin origin word) Expressing something with another thing. It is used in logic, art, mathematics and military.
MATHEMATICS: (In Greek) It is the common name of science, which analyses the properties of quantities by means of numbers using Arithmetics, Algebra and Geometry.
ARITHMETICS: (In Greek) It is a branch of mathematics, which deals with the numbers, properties of numbers and operations. It is the science of numbers.
ALGEBRA (ALGEBRA): (In Arabic**): It is a branch of mathematics, which generalizes the problems on quantity and simplifies by converting to *equations*.** The oldest algebra book known is the one which Mehmet (son of Musa) from Harzem (Horzum) Turks has written in **830A.D.(P-81)**

GEOMETRY: [(In Greek) **Ge**: ground, **Metro**: Measurement] It is a branch of mathematics which takes the **space in terms of line, surface and volume; which examines the properties of figures; and shows how to measure them.**

ALGORITHM: The first algorithm is introduced in Al Harezmi's book Hisab al-algebra and al-mukabala. (S-81) Today, generally in mathematics, computer science, there is an algorithm at the base of programming languages. The path or process steps designed to solve a problem or achieve a specified objective is called an algorithm.

SCIENCE: It is the struggle using a particular method to take a particular subject to a particular purpose. In other words, it mentions the investigations, which intend to find the laws of the events. It is the organization of technique, application and reflective knowledge in a particular system (theories, synthesis). Science is the knowledge, which is obtained by methods and verified by experiences. The main condition of scientific development is the mutual and continuous interaction of **Intellectual Theory** and **Operational Practices**. Science reflects the universe with realities, activities of humans, validated notions, groups and laws. Science gives the knowledge of objective laws to the humans. And humans are in need of this knowledge to be able to realize these actions. It became a tradition to separate science in two parts: **"Science of humans"** (philosophy, history, economy, politics, etc.) and **"science of nature"** (physics, chemistry, etc.)The production needs of community is the propulsive force of science. According to Karl Marx and Engels, someday science of humans and science of nature will combine to form a single science. (For me, this can only be with the theory of everything)

COSMOGONY: It is a branch of philosophy, which combines the various opinions of Physical World, defined by separate sciences. It is the science, which examines the creation of universe. According to Sumerians; universe emanated from the fresh and salt waters of which one was female and the other was male.

DETERMINISM: To get the unknowns using the knowns. (Specific results are get in specific conditions.) There can be no science without determinism. It is doctrine, which proposes that every event is the necessary and inevitable result of another event.

SYSTEM: It is the collection of principles which are bound together to form a **'scientific whole' or a doctrine** (like the astronomy system and philosophy system). In abusive meaning: It is a complement of prejudices appealed for reasoning and classifying the events (system set back the time; and time set back the science). They are the pieces came together to get to a particular result or to constitute the whole (nervous system, solar system, etc.). It is the order of methods to get to a result (education system). The means used to succeed (a system, which can make humans gain wealth). Equipment, system (like the lighting system and brakes). **Knowledge processing system:** All of the units, which constitute a specific type of materials and process methods used together. It is the group of organs, which are composed of various tissues, especially specific tissues in anatomy and physiology; and which are common in all parts of the body (Muscles, nervous system, etc.)

MONOGRAPHY: It is a writing, which represents all of the sciences. (Based on the geometry of the RABIA, an alphabet, corresponding to the golden proportions like the example of thuluth script, or a Scientific World Language like the examples of Sanskrit, Turkish can be investigated)

LOGIC: It is the science of verifying a thought with another thought. Aristoteles calls the rule of right thinking **as "Organon"** which means "Tool". He considers the logical rules equal.

Aristo's formal logic, which did not change up to day, is based on 3 principles:
1- **Identity principle:** Something is the same as itself.
2- **No contradiction principle:** Something cannot be right and wrong at the same time.
3- **'Impossibility of the third condition principle:** Something is either right or wrong. There is not a third probability.

Although "Formal Logic" of Aristoteles verifies the thought with another thought, it is interconnected with reality in the external world. Apart from the reality in the external world, also more than the intellectual relations, "Symbolic logic" is based on the relations between signs constituting the logical formulas. All of the logic currents are symbolically linguist. They assert that philosophy would reduce the language to a logical solution in terms of both "Syntax" and "Semantics".

CYCLUS: It is the time period between the start and end of a biological event. Biological rhythms keep the living creatures alive. "Night-day" sleeping at night, living in the day. While the harmony of these rhythms is a sign of health, their inconsistency is a sign of illness. Like the menstruation of women, heartbeats, respiration rate, insect-killer aerosols being more fatal in some periods of the day and year; the period between rhythms is called as "Biological and Physiological clock". Besides they may also be based on heritable reasons. (Chronobiology) (Biorhythm).

VI) DEFINITIONS ABOUT THE HUMANS PREPARING THE UNIVERSAL SYNTHESIS:

STATE: It is the political community gathered under the management of a state. Socrates has founded Ethics science for the good citizens who would realize a strong state. In "State" of Plato, state is the idea of community. Machiavelli finds all ways permissible to strengthen the state. Rousseau revived the old Greek democracy. According to Hegel, state is rationality. It is the most perfect formation of universal thought that is realized in the humans and community. But Fichte suggested that state would one day disappear since it prevents the development of humans. **Marx and Engel** suggest that humans are not political, but social beings "quintessentially"; and that **state is the organism, which the dominant class has realized for exploitation**. A state giving orders to the **humans will be replaced by a management ruling the nature and production process**. It does not matter if the state is capitalist or socialist; it is always the dictatorship of a small class. In the historical progress, there were no states up to the day when humans were separated into classes and slaver communities appeared. Human groups were communal. It was constituted for protecting and maintaining the dominant class in state formation in a classified community. **State will be exalted as much as it exalts the individuals.**

STATE PHILOSOPHY: It is a part of philosophy, which analyzes the state organization. The records on the best regime are in the history of Heredot. Later, Plato, Aristoteles, Roman Cicero, Seneca; in the middle age Augustinus and Thomas d'Aquino were in the aim of founding a divine Roman Empire as the state of God. John of Salisbury, Marsilius of Padua, Machiavelli, Calvin, Brutus, Bodin, Hobbes, Locke, Spinoza, Sieyes, Robespierre, Burke, etc. argued that the state was an artificial organization. Fichte was the first person to assert that someday the states would be annihilated. Hegel idealized the state as Plato also have done. In the **XIXth-XXth Centuries,** Bentham, and John Stuart Mill have argued for the metaphysical state; Kropotkin, Bakunin and Proudhon have argued for the acephalous state; Bernstein and Kautsky have argued for the opportunist (behaving as per the condition) state; Mussolini, Hitler and Salazar have argued for the fascist state.

FASCISM (1922-1943): It is the only authorized state method, which represents the professions in Italy. On the surface, it is based on the guild system. There is no freedom. Everything is for the state. Their principles are: to believe, to obey and to fight.

GERMAN IDEALISM (XVIII-XIX centuries): It takes Kant as a starting point. Fichte, Schelling, Hegel, Schleimacher and to a great extent Schopenhauer are being remembered with this name. All of them are materialists. They do not criticize Kant. All of them are system philosophers. In other words, they pursued a goal to build a universal system with their philosophies and to leave nothing to say.

SYNCRETISM: It is the doctrine or condition combining of different, often seemingly contradictory sects or beliefs.

VIRTUE: It has the idea to assert the advanced style of individualism, **altruism** (opposite of individualism)**,** against the resistance to individualism. It is the power to defeat itself to overcome itself. [1-**VIRTUE is knowledge (the knowledge of age which the person lives in) 2-VIRTUE is freedom. It is the common name of skills like beneficence, modesty, bravery, honesty, scrupulosity, merit and kindness.]**

ALTRUISM (1830): It is living for the sake of others (Selfless, altruist, devoted). It was suggested by **Auguste Comte**. Making an effort for the benefits of others. It is the **social understanding of Japanese people**. The obligation of humans to take working for the health, sake, happiness of their families, communities and humanism, as a religious duty, besides to be ready to sacrifice their life, properties, freedom and family; are also in the same direction with altruism. **In the VIIth Century B.C., altruism was** one of the principles of Confucius, either.

SOCIOLOGISM: They are the societies, which keep the conscience of communities above the conscience of individuals.

CIVILIZATION: It is the collection of common institutes between communities. It is the living of a nation coming together adapting themselves to the customs. Apart from the ones created by the nature, **civilization** or **culture** are everything, which have been created by humans. All of the results of the efforts humans show for living better collectively, appear as science and culture.

CULTURE: It is the complement of all kinds of living, thinking and art assets in terms of traditions which generates the material and moral (intangible, spiritual, psychic) characteristics of a community; and consensus in perception and mentality.

COSMOPOLITANISM: It is the general name of doctrines which recognizes humanity as the **nation; and universe as the homeland.** When they asked to **Socrates** where he was from, he used to answer: "I am from the **World community**". The religions like Christianity and Islam are also considered cosmopolitan. For the Renaissance people, **Dante** says: "Their homeland is the world". **Bertrand Russel** says: "If we could reach to the science of being a world citizen, instead of making ourselves miserable, we would prefer to be one of the soldiers of the army to take the humanity to civilization". **Tevfik Fikret** says: "**Earth is my homeland. Humans are my people. But humans are humans. I believed that with my mind**".

STATE CORPORATISM: Fascist bourgeois dictatorship (cooperative state that was developed in Italy and Portugal) which assumed that the state (considered as the guild of guilds) would solve the conflicts for the public welfare, after organizing the capitalists (the owners of production means) and labor as a guild. But there is also a "**Communist corporatism**". State plays the role of a referee between the opposite benefits, by representing all classes of the community. Mussolini indicates that: "Fascist state also proved itself in the economy field by means of cooperative, educational, communal institutions which the state, itself has founded. So the State notion would reach to the furthest points; and all of the political, economic and spiritual powers of the nation in the own organizations of the state in an order were functional. Fascist state limited the 'useless' and 'harmful' freedoms; but kept the essentials. On this point, only state can decide, not the individuals". 'Fascists' do not allow free elections. Communism mentality (Corporation socialism) was also developed by English **Penty,** and **S. G. Hobson, A. R. Orage** and **G. H. Cole**. In the age of **Etruscans**, there were coppersmith, shoemaker, carpenter, plasterer, painter, weaver, worker associations, in Rome. They were also checking the prices (**X-III Century B.C.**)

RACE: This term was used by French **Barnier** in 1684 for the first time. And later in the **XVIIIth century,** classifiers and naturalists used this term. But **Kant** declared its modern meaning: "Race is a biological term which is one of the sections of zoological classification. It is a wide grouping in the species. It is the collection of individuals with a determinative and biological character, who have been descended from the same ancestors. It is based on genetics, anatomy, physiology, pathology and related measures. Today's humans 'Homo Sapiens' are divided in 4 race groups and various races as black, yellow, white skinned and primitive; and each race is divided into branches.

The difficulties of exact limitations are as follows:

1- It is difficult to select the criteria.

2- Today there are too little pure races since they all are mixed together.

3-Characters, which define a specific race, are transitional. **Races show continuous evolution and alteration like humans.** Humans get more complicated while moving away from the center in the classification of type, species, race and sub race and kind. Although the nations are mostly composed of various races, there are nations composed of people from the same race; nations composed of people from the same race and speaking various languages; and nations composed of various races and speaking the same language.

RACISM: It is an idea which suggests the superiority of a race among the other races. This idea was founded by the German philosophers like **Nietzsche, Marx, Weber and Werner Sombart**. According to Nietzsche, weak humans should be extinct for the victory of superior humans. According to Weber, emotions lead the nations, not mind or science. According to Sombart, what makes humans superior is the racist life and scientific superiority. Also after examining the fossils of all-time animal species in the world, **Darwin** concluded that strong animals can live on, but the weak ones are extinct. Darwin indicates that: "**Virtue** is a product of evolution which arises from the necessity to comply with the environment and conditions. It arises from the protection instinct. It is a must for the humans to survive". **Adolf Hitler** indicates that: "The races except the superior races are obliged to help for the victory of the best and the strongest, the big will ruling this world; and also obliged to ease the work of it. Superior race has the right to ask for the lower class and weak ones to resign themselves to itself. Aristocratic principles of the nature should be respected". **Thus, racism has been used to hide the communal inequality, imperialism and imperialist requests up to today. (Note: European nations which are saved from the exploitation of Feudal lords, kings and the church, tried to batten on first themselves, and later the other nations of the world. They are innocent. They did what they have seen).**

ANTROPOLOGY (SCIENCE OF HUMANS): In Greek: Anthropos (means **humans**), logos (means **science**). The subjects of anthropology are communities and cultures.

The topics it researches are:
1- Why do not humans and communities look like each other?
2- Why do humans and communities look like each other?
3- Why and how do humans and communities change?

The fundamental sciences of anthropology are: **Biology, Sociology and History**. Being the topic of social and cultural anthropology, "**Culture**" has 164 definitions. Cultura (means crop in Latin). French Voltaire used this word in the meaning of the formation and development of human mind. The word, then, passed to the German language. It was used for representing the '**civilization**'. Later it passed to Spanish, English and Slavic Languages. And then, the word '**Civilization**' was preferred instead of 'culture'. French and Germans preferred the culture. According to German Hegel: **Culture is everything that the humans have created against the creations of the nature.** In scientific writings, the word 'Culture' is used instead of 'civilization' (The word '**Uygar - (Civil)**' means well-adjusted human in Turkish and is attributed to the Uyghurs).

ART: It is the creativity which humans use to stir the excitement and admiration for the beauty; it is the search for good and beautiful.

TECHNIQUE: It is the way of applying the findings obtained by the science via theoretical means, into the daily life. (Technology arises when science tends to satisfy the practical needs) "**technique**" is objective, "**science**" is rational. "**Science**" looks for the reality (truth); "**technique**" looks for the useful (convenient); "**art**" looks for the beautiful and good.

EVOLUTION: It is the progress of oneself to adapt to the changing environment (structures in the environment). It is the continuously maturating development for the creatures from a single cell to the humans (It is a progress of core structures developing from the beginning in a new environment composed by the core structures).

SOCIALISM (COMMUNITARIANISM): The idea of providing the development and, thus, happiness of humans starts with humans' having a piece of land. Nomadic communities were using the knowledge and human power to be protected from nature and to defeat the nature. Equality and freedom were destroyed when some persons and coteries arose with **lands** and **production means**. In the beginning, some utopic communal orders were asserted to fix it. **Plato, Thomas More, Francis Bacon, Tommaso Campanella, Valentinus Andreae, Barclay, Heywood, Winstanley, Harrington, Gabriel Foigny, Morelly, Gabriel Mably, Etienne Cabet** suggested imaginary organizations **based on community of goods and equality of humans.** Besides **J. Jacques Rousseau, Gracchus Babeuf, Charles Fourier, Saint Simon, De Sismondi, Pierre Leroux, Robert Owen, Louis Blanc, Rodbertus, Mikhail Bakunin, Blanqui, Ferdinand Lassalle, Joseph Proudhon also presented works in the same direction.** At last, the communist proposals of **Karl Marx and Engel,** which are asserted to be the most consistent, can be taken into consideration.

CYBERNETICS: Living and non-living beings continuously exchange knowledge with their environments; and they manage themselves as a result of this exchange. Taking this system as an example, self-deciding machines (like a living being) were built based on mutual information exchange. **Prof. Dr. Norbert Wiener** defined the cybernetics as "**mutual knowledge exchange, control and management science in living beings and machines**". **Computers** were produced thanks to the cybernetics.

It covers 3 processes like communication, control and balancing science:
1-Information exchange.
2-Control.
3-Balancing.

Greek philosopher **Plato stated that: "Kubernetes saves not only souls, but bodies and goods, as well, from great dangers". In Greek, "Kubernetes" means** steersman. Thus, Plato suggested the exchange of knowledge, **control and balance** as a "MANAGEMENTSCIENCE".

B) INTERPRETATION OF MAWLAWI DERVISH AND HIS GEOMETRY "TRIANGULAR PYRAMID" FOR HUMAN:

Humans are the most perfect work of the divine power, which creates the universe. Since the divine creator has given an ability to the humans to comprehend their own perfection and the perfection of universe, sooner or later humans will succeed to establish a physical and spiritual order on the world, which suits, to their own perfection and the perfection of universe. Divine creator has created the universe with mathematics. Universe is composed of **"Matter"** and **"Energy"**. Both of them have a higher mathematics, interdependently. ($E=MxC^2$) Throughout the history and today, the reason of wrongs giving pain to the humans, is that the universe is not completely known and the universal truths are not told to the humans as they should have been.

In order to find the answer of the questions "**Right for whom? Right for what?**" which would resolve the chaos today, let us define the Mathematics, which is the alphabet of all sciences, and the universe that we live in: **Mathematics is a science, which tries to solve the problems in the nature, by indicating with "Numbers" and "Figures". When the truths are told as per the geometry, in other words the most comprehensive geometry of the universe, which is the visual form of mathematics; like using a "Magic scepter", perhaps a "Golden age" may be started to end all of the sufferings and evils humans face in the world, to lead us.**

We have told in the previous parts that this geometry was the **"triangular pyramid"** (According to Pythagoras 1,2,3,4, i. e. RABIA). Humans first think, and then attitudinize with words or verbs parallel with their own opinions. If the truths are not told to the **humans, who can find fault with the person doing wrong?**

The only objective of the information given from the beginning of this book to the Theory of Everything and quantums of subatomic physics at the end; was doing a research about what kind of messages would the humans obtain from these information. For this reason, I dared to penetrate into the deepest modern problems of the Physicists. The objective was finding the facts of **Universe and humans. Hence, the universal message, which showed up until now, in my opinion, is the MAWLAWIDERVISH and his geometry TRIANGULAR PYRAMID** (i. e. RABIA).

I) HISTORICAL DEVELOPMENT PREPARING THE INTERPRETATION OF MAWLAWI DERVISH AND HIS GEOMETRY:

History is the numerical equivalent of time. While searching for the universal synthesis in the Part I, as we have started from the History of Universe; let us start from the History of Humanism while searching for the comments of this synthesis for humans, in other words, the universal facts:

It is suggested that the first humans looked like more of a monkey. The fourth humans type which also includes us, "Homo Sapiens" (thinking humans), appeared in Central Asia approximately 50-100000 years ago, in the last ice age. At the beginning, hunting in the caves near the rivers and mountain slopes, population of our ancestors, later increased in Central Asia approximately 20000 years ago, by means of stockbreeding and agriculture. Their first communal life experience could have been started in the tents. In Northern Siberia, humans are still living in such tents covered with reindeer skin. If we assume that Civilization is "the communal life in houses they have built themselves", we can start the civilization from Göbeklitepe, which is the oldest site of the world, in Southeast Anatolia, in the vicinity of Urfa. Göbeklitepe is dated to 10000 B.C. And the Biblical Flood is also dated to 10000 B.C. In other words, It is the date when the last ice age has ended, the ice has melted on the Northern hemisphere and the seas of the world have raised for 80 meters. Previously, Northern hemisphere was covered with ice to Anatolia, the Alps, Himalayas and Central America. In other words, since the climate of Anatolia was colder in that age, it is obvious that people were going in their houses from chimneys in "Çatalhöyük" village, which is dated to the years 6-7000 B.C. in Central Anatolia, in the vicinity of Konya. There were no doors or windows. In the houses, there were sets higher than ground, which were used for sleeping and sitting down. In today's Anatolia, in the villages, there are wooden bedsteads named "Sedir" which are higher than the ground, being used for sleeping and sitting down.

In **Mohenjo-Daro and Harabba** cities, the ruins whereof were found in the Indus Valley, with top layer dated at least to **4-5000 B.C.,** that is approximately 5000 years after the Flood and the bottom layers which cannot be dated because of being under the waters of river, there were intersecting roads, fresh water and waste water channels between two-three storey houses, grain mills, as in the modern cities of today. Thinking that **Sumerians** have come from the region between Ceyhun

(Amu Darya) and Seyhun (Syr Darya) rivers on the East of Caspian Sea in**4-5000 B.C.,** to the region in Mesopotamia between 'Fırat' (Euphrates) and 'Dicle' (Tigris)rivers, and that they have brought an advanced culture, it can be predicted that the researches in this region will shed light on the history of humanism. The communal life experience in the history of **humanism of approximately 20000 years, in other words Civilization**, has formed the **Islam mysticism** from Central Asia to the villages in Asia Minor, to Sumerians, to Egypt; from North to maybe Europe, starting **approximately from 800 B.C.;** to the Greek trade colonies on the seaside of Mediterranean Sea andto Rome, starting **approximately** from **700 A.D.;** and after accepting Islam, based in Baghdad and Al-Andalus.

Universal synthesis of humans which Leonardo Da Vinci has learned from Al-Andalus, and which he has rendered with a human figure in a square and circle(Vitruvius man), and transferred to our age; in which square has represented the "**matter**", and circle has represented the "**soul**"; and that humans were composed of soul and matter. But unfortunately, the human in the middle of the symbol of Da Vinci, could not be the human who should balance the **spirit** and **matter** which mysticism tried to describe; on the contrary, it made a materialist association. This did not conform to the concept "**civility**" of Sufism (**page 61,180-185**). Also humans in the world do not consist of only men. Thus, the third element, in other words the human, who should ensure the **balance**, has been forgotten. To-day, as a result, **Materialism** is dominant in the West, and **Spiritualism** is dominant in the East, and since the third element, the **balance**, that is, the WISE and eugenical is not available, we have the fight for either you or me between the first and second.

The social qualifications, which are the prerequisites of being a Wise and Perfect human; the spiritual and humanitarian values of Islamic mysticism, which Da Vinci called "spirit" and defined as energy by the west today; such as respect, love, devotion, tolerance, morality, decency, virtue, con-science, modesty, charity, credibility, generosity, mercy etc. have not been fully transferred to the West to some extent to the East from Andalusia. **On the other hand, the knowledge about material inherited from Al-Andalus to Europe have started first the Renaissance, and then today's age of knowledge. It would be injustice not to appreciate that, but Soul was only left to the church. However, Soul and Material which Da Vinci has learned from Al-Andalus, was not only an Is-lamic orin other words religious knowledge, it was the synthesis of Islam religion, and Islam Mysticism which tried to synthesize science of the age. There was a third element, in other words wise (intellectual) and perfect (eugenical, living his knowledge) humans who has to balance the matter and soul completing this dual synthesis (duality); and Da Vinci misquoted it.**

II) COMMENTS OF MAWLAWI DERVISH ON THE INDIVIDUALS:

In the most accurate interpretation of Aristo's "hierarchy of pleasures and happiness" **ac-cording to Mawlawi dervish, it can be said that the dervish holds "Physical happiness" on his left palm, "Happiness with Knowledge" on his head (Given as the "most supreme Rabia" on the head of the dervish as shown on the front cover of the book) and "**Happiness with Morality (Ci-vility)" **on his right palm. The synthesis of** RABIA, which is the algorithm of the survival of humanity**; may describe the "happy human" with the Mawlawi dervish, having "morality (civil-ity)" on his right palm and "matter (body) under his left palm.** Both of them are sine quo non. **If we cannot interpret the knowledge synthesis of the 21st century in this way as a whole with Mysticism and mawlawi dervish and explain it to the world, neither faults nor lessons come to**

previous universe; or at least "**Big Bang**", as can be seen in the "Quantums of the **universes**" table in **Annex-Ia**.

'**Respect to the God**' brings the 'respect to the elders' in the community; elders in the humanism history, today's living scientists, statesmen, at last the older humans, grandfather, grandmother, father, mother, elder brother, elder sister. It is the first condition of being human to "Respect" to the elders (upwards) and "love" to the youngers (downwards) hierarchically. The respect of youngers to the elders is in a sense their respect to themselves. Because, 30-40 years later, they will become elders, too. Today, the youngsters in America, put their shoes on the table, towards the nose of their fathers, cross their arms, and accept what their fathers say if they are convinced. Also imitating to them, European and World youngsters dare to exhibit similar behaviours with justifications like **rights, law, liberty, freedom and 'wisdom does not come with age'**. These behaviours can be very well received to some extent, provided that the father gives consent. An Anatolian proverb says: "**Speech for adults, silence for youngers**". Younger will of course do decently provided that he is convinced, with respect to obedience and doing what is said. The younger should have no rights for objection with respect to obedience. Because the brain of some of the humans are able to complete their development medically up to approximately 22 years. But in some people, it can be longer or sooner. A young person is like a green fruit. It may not be enough to understand what the elders say, as much as the elders try to tell. They tend to see the reality through a tulle curtain. Perhaps he will understand what is said, ten or fifteen years later. The elder's sayings include everything from humanism history, living environment, community, what their elders have transferred to them, to the manners they learned in their life. He gets closer to the end of life. He is striving to tell what he knows for the successors not to face the difficulties he has once faced. His brain, nerves and body are not as strong as it was in his youth. He should feel relaxed to be able to tell properly what he wants. This is a relay race, which should be relayed from generation to generation. Let us assume that a very wrong old man is trying to tell very wrong things; and his listener young man is a young professor among the most intellectual humans in the world. If the young one could evade without showing a lack of respect for the elder, he is deemed to pass the exam and to glorify himself respectively. The words of a right old man is a bigger achievement than he would get from the right knowledge. He has succeeded in education. Humans cannot implement all they have learned. In case the humans can succeed in that, they will advance in a trial and error method.

2- LOVE TO THE YOUNGERS: Love to the youngers is the duty of elders towards the supreme, the God. Because love is like the sun. It heats, brightens, regenerates. Here we need to touch a sore point. Everybody, elder or younger, are entrusted to each other. But humans are also the entrust of Allah. It is the prior duty and responsibility of humans, elder or younger, not to give harm to the others, as well as not to allow somebody to give harm to them. As a result, we can see that the solution is again in Plato's "golden point", in other words in the kindness of "**Balance**". In case the elder who is treated badly, behaves in the same way to the younger, he is no longer an elder. They should tell the truth to the younger tolerantly, provided that the younger is willing to listen.

3- DEVOTION: It is the second basic condition of being human. In the Anatolia traditions, there is a principle like "**You first**". If everyone say "**Me**", then the life would turn into a struggle for bones. To offer to the guests first, while having meal; to listen much more than talking, to give way to the others, to welcome. These are underestimated behaviours, which would sublimate the humans. Real nobility is hidden in such underestimated details.

Nowadays, there is a "**Me first**" principle among humans from various nations, various ethnic groups, with different languages and religions, workers, employers; and among the individuals belonging to the same ethnic group, and also the same family. Motor vehicle traffic is the most striking example. The one taking the wheel walks on air. So to say, they are making a film. They are almost going for a fight. Everybody is trying to pass, almost trying to run over and punish each other. But this view of cars, is a view of herds. Drivers should get rid of appearing like that. People giving way to the others will earn respect among the humans. This should be the main thing to learn, the main behavior. But, giving way to the emergency situations is a matter of courtesy.

Also cheers for success in sports, cheers for violence in music, can be assumed as harmless fun style. Humans may need to have fun unconditionally, to freak out with today's words, to stop and smell the roses, to some extent. This should be done without giving harm to other people. **Ethicists'** philosophic definition is: "**Bearing the responsibility to obey the rules**". The answer of question "**king or rules?**" is not being told to the humans. But the forest residents try to be the king sometimes; and they are innocent. **Kindness is what sublimates the humans;** in other words obeying the rules. The scientists who are described in this "**The synthesis of universe and humans**" chapter as mis-thinking about humans, unfortunately played a part in today's bloodthirsty world mentality by committing blunders in the last centuries. English Nature Scientist **Darwin (1809-1882)** indicated after researching the animal fossils from the ancient times that: "living nature (including the humans) emanated by means of evolution. The propellant power of evolution is the fight for life; and the result is the natural selection. In other words, strong animals survive, weak ones die out and they are annihilated by the strong ones". At last, **Nietzsche** (1844-1890) told that: "Weak humans should be extinct for the victory of superior humans". These ideas were associated with the faith in Messiah who has been waited for the ancient times. It was concluded that the struggle between humans is a law of nature; and that a superior rage would annihilate the other humans to save the world. And the world lived a **Hitler** tragedy. Believing in the scientists, maybe he also has started out to correct some things on the world. Unless science presents the real universal truths, now both the strong and weak communities worldwide live the wars and we are losing millions of humans.

In fact, there is no evil; wrong ones create the badness. For the elimination of the evil, first of all, *the truth should be set forth incontestably clear, scientifically and universally not to be objected to make everybody understand.* Darwin's theory can be true for the animals. It can be true for the "**Homo Sapiens**" (*thinking human*) type humans (today we are a member of it), previously for the "**Neanderthals**", and also for the most previous ones. At last, we cannot deny the wars and struggles of humans throughout the history and also today. But it is another thing to accept that this struggle between humans is an inevitable law of nature. This would be an approval of animalism and primitivity, and it will be beneath the humans.

Today, the existence of nuclear weapons and biological mass destruction weapons which could be enough to annihilate the world hundreds of times; is the warning indicating that it is time for the humans to recover by means of mind and science.

There is no dandier living being than the humans in the known universe. And in our century, the world has financially become a heaven in terms of life comfort, which has not been seen throughout the long humanism history, at least for the humans in the middle-income group in the developed countries. Despite the last economic crisis in the West, it still continues to some extent. We are in the knowledge era and everything changes very quickly. It would be a shame to spoil these quickly obtained values because of policies which are not based on science and which are based on misinformation.

Struggle and competition should of course be present in sports, production, trade, democracy, art, music, etc. What is incumbent on the humans, is to make the struggle conditions and process, gain the footing suiting the kindness of humans. And this can be achieved by balancing between **Matter** and **Spiritual** values, without capitulating to any of them; in a sense by means of "**Kindness**".

Communities, which find themselves superior in terms of race superiority, have no difference from a person who find himself superior. The real superior humans have no right to voice an opinion in this subject. Only the other humans can decide their superiority. World needs the persons and communities, which deserve love and respect. Allah and His slaves would bless them. Superiority's being based on force by means of guns, power, money and hidden or obvious financial capabilities should become unfashioned. Otherwise, it would not be different from the supremacy of forest residents between each other. We must not forget that, we are no longer in a sword & shield age; though the term gallantry was appraised at these times, but now there are nuclear weapons. In Australia, in the twentieth century, they found a village unaware of the civilized world, and they asked; "In case you are in a forest, and you see another person coming towards you. What would you do?". And the villager answers: "First of all, I ask if this person is one of my relatives, or not. If not, I kill him, before he kills me". This is a particular example for the 21th century's sensible, civilized humans.

Humans may say "**Me, first**" for a thousand times. And in all of them he can run over, steal, grab and win. But in a world where everyone says "**You first**", one day, billions put every person on pedestal. Its income is above what may be conceived. The ones who are in an effort to sublimate the others will be sublimated. The ones who show respect will have the right to be respected. In the Anatolian life style, people see themselves as the others do. Before television standardizes the people of the world, the behaviours and way of thinking of all world communities, including Anatolia, should be scientifically examined.

4-TOLERANCE: is the forth and the last precondition for being a human. When people encounter with a faulty and annoying occurrence, they usually give reaction to the "Occurrence". For this reason, some occurrences, which had ended up in hospitals, prisons and graveyards throughout the history and in every single minute, we witnessed. However, the first reaction and emphasis should be directed towards the "Person" rather than to the occurrence as a reflex. The person who should be excused is the excusable one. He/she may misbehave by reason of unawareness, tiredness, drunkenness, intelligence level, or mental health. If you do not take attention to these factors previously and consider the addressee as you are own, the necessity of compensating the situation with a counter attempt arises and eventually the images we get accustomed to see in animals take place. However, when we rapidly take attention the "Person" to understand the reason of behavior of this person, his/her state of excuse is ninety percent comprehended, while it may not be comprehended with ten percent possibility. Then you should consider this hundred percent, there is absolutely something significant. Saying that "You are right my friend" and apologizing kindly would not lead the person to lose anything. Of course, you may actually be faulty. Then, you should also know to apologize. If the problem were solved with mutual comprehension and politeness, both parties would have gained considerably in the society. According to a German proverb "Ende gut, alles gut", if the end is good, then everything is good.

III) INTERPRETATION OF THE MAWLAWİ DERVISH FOR THE SO-CIETY (Annex-IVa):

In the 21st Century, people should establish the social, mathematical and systematic structure that is suitable both to their own qualifications and structures and to the divine harmony and order of the Universe they live in and they become a part and then timely prove that they deserve to be the most developed being of the universe by becoming integrated with each other and the World and the Universe.

It is understood that matter beings of the Universe which grow infinitely beyond the galaxies and become infinitely smaller towards the basic particles of atom pass through the stages of birth, life and death within the infinite time and space dimensions in the Astral and Atomic scale, and while they experience these stages, everything displays a divine integrity with infinite perfectness composed of nested infinite systems in a Rhythm, Accord and harmony with infinite qualifications, mutual actions and reactions which may be expressed with mathematics. In Annex-IIf and Annex-Ia, the Schedule of Universes' quantum, it is seen that the life adventure, namely destiny of each universe should exist inside the only energy status (**NUR-U KADİM**) just like a computer program as the Time and Space in the Astral and Atomic scale. Here the Science and Religions unite. This unification is more clearly specified in the last page of the New theory of everything as the "**Bose-Einstein quantum concentration**" (**P. 133,156**).

People have oriented towards the power, which created them throughout the history from the era of cave pictures to monotheistic religions to an extent they are able to understand. It is worthy of commendation that this brings along the respect for ancestors and elderly people in many cultures. In the current era we live in, since Disbelief gives rise to the thought that "Any good and bad action held by a person will get away with him/her, there is no one who would call them to account", the life in the world is becoming a wild fight and the world we live in is turning into a unlivable place and creating moral and matter dirtiness.

Three main factors have been based in all works of people beginning from stone axes to current spacecrafts; "**Matter**", "**Human**", "**Knowledge**" which may be called as three main factors of the dervish, because both Knowledge and SPIRIT are abstract terms. This is because, Spirit, Reason and

Knowledge are all abstract concepts. Human gives shape to the matter with his/her knowledge. Mawlawi dervish is a model for an individual, just like it is a model for the society. Human is substituted by the "State", Sense is substituted by "Knowledgeable" (universities) and the Matter is substituted by "**Producers**" (makers or professional organizations). The universal order, harmony, rhythm to be brought by this common model in people for individuals and society will destroy the chaos, negative things and all problems rising from deficient and wrong information which have been the source of disaster, evil and misery for people throughout the history for individuals and society, and will cause people to meet a totally one single happy and healthy organism structure. The application of dervish in a society is given in Annex-IVa in the "Table of Classification of Sciences" with all its details. **PLATO** says, "The state is an enlarged symbol of the functions of individuals".

Only the "**Double Rabia**" (P-175) can make it possible for the orientations towards the supreme power, order, perfection, the truth, which is the source of all things, the creator, to lead us to the "Golden Age", described at the end of the book, either individually or in communities through religion, morality, intuition and healthy physical and spiritual training until today in the history of humanity. Everyone should accept the Unity in Plurality over the head of the Dervish, representing both the state and the individual, that is, everything in our universe gather in the unity of God (Great is His Majesty). This truth may be attributed to whole of the humans in the world by being renewed, updated by the Cosmology Faculties of the Universities in the light of contemporary synthesis of Cosmology and constant developments. Such syntheses will not only assure that the humanity on the earth to maintain their existence, but also the harmony, concordance, happiness it will ensure to the humans, other living creatures and even to the non-living things, like the divine balance in the Matter and Spiritual existence of the humans on the Earth, in the Matter and Energy of the Universe as well as in the causes and effects on the individuals and societies as thoughts and behaviours, will be inexpressibly magnificent.

C) HISTORICAL ANALYSIS OF MATERIALISM DOMINATING IN THE WORLD TODAY AND A UNIVERSAL SOLUTION FOR THE FUTURE:

Before monotheistic religions, polytheistic religions had a serious binding and frightening pressure on the people of Asia and Asia Minor. Although the Greeks, who acquired the information gathered by the Egyptian temple priests in 7 branches of science through the studies for 3-4 thousand years, adopted polytheistic religions by giving Greek names to the same gods, these gods became heroes in tales. They lost their fearsomeness. Contributions of the Greek culture and science world to today's world cannot be ignored. However, storytelling over religion, gods resembling human, stealing, killing, and envying each other removed the concept of "Sin" in the Greek. In Greek colonies, life became more matter with the wealth brought by trade between the east and the west. For example, if a thief was not caught, he did not get the blame.

Our Greek fellows should not take offence; the transfer of the information and **freedom** that was transferred from the Greeks to Muslims first, and then to Rome, to Medieval Europe, to the Renaissance and thereafter to Europe is okay, but **Materialism**, starting with the **craftiness**, **selfishness**, **individualism**, **egoism**, and then extending to and still continuing **racism** in Europe caused the world to suffer of two world wars. Eventually, it is notable that the economic crisis in Europe and the United States affected Greece extremely severe. **The societies where everyone counts for self cannot survive.**

Greek civilization could not survive; Rome emulated them and collapsed as well. The aristocracy of the lords, kings, emperors and the church was the continuation of selfishness and they became a thing of the history. But while each of these Aristocrats developed their own musical style in Europe, the oppressed European peoples developed the European Folk Music as a means of struggle. In the lyrics of the old European folk music before the feudals and survived up to present; there is the **longing for leaders living with the people like the people.** Such leaders might be **Attila**, who ate from a wooden plate; **Odin (P.181),** who was the Norwegian King and then German God and might have emigrated from Asia, the lands of the Turks; and even the leaders of the **Alpines** who traveled from Asia to Europe in 9. 000 B.C. and the **Etruscans** in the 3rd millennium.

The Greek lifestyle affected the Romans. Examples are having people killed and torn by others or animals in the arena for the sake of joy of Romans and ruthless treatments to the societies in the countries they invaded. In a sense the tricky rich aristocracy of the Greek turned into a cruel aristocracy in Rome.

Christianity was founded by Jesus [Jesus Christus (Savior Jesus)] in Jerusalem as the savior of the poor and the oppressed. After Christianity was adopted, Europe lived through a dark Middle Age with ruthless practices against science of the church, as the sole ruler in Europe above all kings. 800.000 women were burnt down in Europe where ignorance, poverty, wars and diseases prevailed. While disappearance of the concept of sin **in Greece** and **Rome** created certain problems, the perception of the sin by the church **in Europe in the Middle Age**, which tried, explain anything only basing on the Bible with an ignorant point of view also created others. The church insisted on suggesting that the world has a flat shape. In the Middle Ages, Europe struggled with misery, ignorance and plague.

During and after Renaissance, Europe transformed the science inherited from Andalusia and Islamic works into technology and industry and with the military and economic opportunities and in response to the pressure of the church in medieval times, and the Roman mentality, having no sense of sin turned into the **racial supremacy in Europe, and then to the aristocracy, colonialism, imperialism of the rich nations in the world,** the armament race. Finally, with the French Revolution, the national states in Europe and then the idea of supremacy took on the same selfishness. They did the same to the less developed countries by colonizing them what the kings and aristocrats had done to them in the past. Ultimately, the level of prosperity has risen along with the industrial revolution and information age in Europe, including America and Canada, as the wages rise, the last economic crisis is pushing Europe and America, while the companies tend to produce in China and other countries where labor is cheap. If that continues, even the formula of Karl Marx "**Let's cut the bosses**" will not end the crisis.

After XVII. century, in the Ottoman Empire, science was abandoned despite warnings of Katip Çelebi and other scholars, everything began to be regulated based on false interpretation of verses of the Quran (Sharia). Despite contrary statements in the Quran, engaging in science was perceived as interfering with the work of Allah and creating perfect artifacts as being disrespectful to Allah. Thus, the Ottoman made the same mistake with Europe in the Middle Age and started to colonize itself even it was relatively strong first with the Capitulations. Then at the beginning of the 20th century, it had to leave a certain part of its lands to Europeans to be colonized after the World War I. Turkish Republic was founded in Anatolia along with 30-35 states on the former Ottoman lands.

It can be concluded that "**Capitalist materialism and aristocracy**" became prevalent first in the English colony North America, then after independence as **USA**, with Canada and in Europe and North America at least after its economic and cultural integration; and "**Communist materialism**" to some extend in Russia and China. Previously, Spain, Portugal, England, France and Italy had colonies in the world. "**Racism**" departing from **Darwin** and developed by certain other scientists, can be considered as the request for "**Roman privilege**" of European countries aspiring to Romans. But, due to the fact that Christian Europe was not far away from the concept of sin, their ambition to legitimize themselves and appealing to science by thinking "**if struggle is a law of nature scientifically, everybody shall arm himself and let's see who is more Roman in the arena**" caused two world wars.

Torture imposed by the Soviets, before its disintegration, to its own people and people of its dependent republics can be considered within the view of "Roman wannabe" because communism is against religion. In the end, after demolition of communism and independence of former Soviet states, the world has become largely dependent on capitalism.

After the dissolution of the Soviets, the world became somewhat dependent on America and Capitalism. The second world war ended upon America's use of atomic bombs. Today, not only the five biggest, even Pakistan and India can produce atomic bombs. The understanding of the West, "**What can we do? Life is a struggle, sooner or later the superior will win**", can make the world uninhabitable. **In 1960**, if all the atomic bombs that the countries held, would be exploded at once, they might destroy the world not only once but 250 times. We do not know how strong they are today. Actually no need to know it anymore. Let the war aside, even if an accident would occur by human fault, and it would trigger the other bombs or atomic power plants, would that not be a waste? What will the responsible people do? What will humanity do? The truth that strong animals survive as declared by Darwin may be true for the animals. Applying that to human is disrespecting firstly the human. This mentality has already surpassed the goal of obtaining colonies. It is clear that it has reached a very dangerous stage.

It would be wrong to assert that materialism is dominant in Europe, where the rate of Christians has decreased to 15%, as in the rest of the world irrespective of dominance of the concept of sin and racism. How can a more balanced, smart and fair world be founded by getting over the materialism dominant in daily lives of world people now? We can certainly indicate under the light of what is being told in this book that the solution is Geometry.

KEY TO SOLUTION: As seen on the front cover of the book, the formula $E=MC^2$ lies in the foundation of the Triangular Pyramid, which is the geometry of the "**Mawlawi Dervish**" representing both the **individual** and the **society**. In a more meaningful expression, it is **RABIA**, i. e. **1,2,3,4.** These 4 numbers are also seen at the top of the steps of the Pythagorean triangular pyramids on **page 137**, the mathematical code of the universe.

After drawing a roughly comprehensive framework of all modern and universal information provided from the beginning to the end of this book and archived in the libraries of the world, it has been concluded that every being in the universe we exist is made up of **energy** and **substance** and continuity of existence of every being depends on Balance, the third factor between these two. Achieving this balance in our world is up to the **wise** and **eugenical** man, representing the **micro cosmos**.

After this conclusion, if individual and collective materialism, hostility and egoism continue to prevail on attitudes on humans, how possible is it for the life and existence of humans to continue on earth? We have to look into this now in the 21st century.

We need to begin with bringing a balance of a Sufi and a dervish between irresponsibility and disorder taking refuge in the concepts of "**tolerance**" and "**repentance**" in Islam and the lifestyle which gives prominence to matter value rather than humanity in the west. In other words, the **balance** between these two elements in the drawing of **Leonardo da Vinci** representing that human is made up of **soul** and **substance** can release people from Materialism. First a call is needed for everyone to realize such balance socially. As **in the XIII. century**, there is no thinking other than Sufism and especially Mawlawi Sufism which makes such a call better.

Let's begin with repeating the call of Mawlana. It is written on the door of his tomb in Konya: **"Come, come, whoever you are, wanderer, worshipper, lover of leaving, Ours is not a caravan of despair. Even if you have broken your vows a thousand times, it does not matter, come, come yet again come"**. Mawlana also says **"Religions are the same, what is different is the route. Even there are one hundred books; they all have one chapter; all revolve around a single mihrab; all roads lead to one home. Thousands of spica originate from a single seed"**. It would not be wrong to say that the **Big Bang**, which we explained under the title of "**Theory of Everything: Rabia**" **(P.168, 178)** at the end of this book, is the only seed in which thousands of heads occurred.

If we would update the first expression above: **"Come, come. Wherever you are, come again. Even if you do not believe in God, if you are an atheist, a communist, a capitalist, a satanist, a materialist, a westerner or an easterner. Come, it does not matter which color you have, which language you speak, which race you are from, which religion or thinking you believe in, our lodge is not for despair. Come if you have changed your mind and broken your vows a thousand times"**. This call can lead us to a more livable world without having on conscience. First Iraq and Afghanistan moves of USA, the missile shield project in Poland, then Georgia incident and nuclear activities of Israel and Iran, and finally the **Arab Spring, and more recently the events in Syria, Iraq, Iran even North Korea;** all indicate that the setting of conflict has been escalated. Without any further despair, what can be done for solution? What is required by the common sense? Let's try to find an answer: A composer makes music with pleasant **sounds** (notes), a poet brings **words** and **sentences** together to write poems, painters use **lines** and **colors** and sculptors make three dimension **sculptures**. Now **let us strive to lay a World and a World Society, which suit the excellence of the universe and the humans** on the bases of the knowledge in the New Theory of Everything **(P-167)** in the fourth chapter of this book. This is a responsibility before the creator and owner of our universe, only by doing so we can come close and be exalted. As the best creation of our Creator, it is a must for us to show that we deserve the perfection bestowed upon us. We can approach Him only through that way and be exalted. In the 21st century, we can achieve the purpose of **approaching Allah** instead of whirling mystically as the Mawlawi Dervish did, **through science**.

The physicists put out three theories of everything, but they couldn't prove them. Without waiting for the new theory of everything they can prove, we can start with **RABIA** to avoid the Third World War.

D- TWO REASONS THAT GRANT US THE OPPORTUNITY, WHICH HAS NEVER BEEN OFFERED IN THE HISTORY UNTIL OUR ERA:

Young people in the world need to know they have an opportunity especially in the 21st century that has never been seen in the history of humanity.

THE FIRST REASON: It can be asserted that the "Information Age" began in **1990s** in USA. The history of humanity has never seen such an era. Not only young people, but everybody has this opportunity. While computers were the size of a double-decker, we can carry them now in our pockets. It is just like the Aladdin's magic lamp, everything is possible.

THE SECOND REASON: Again in **1990s**, the "**Union of Soviet Socialist Republics**" was disintegrated. Similar to occurrence of 30-35 republics following the collapse of the Ottoman Empire, many states declared independence after the Soviet Union. Before **1990s**, the north hemisphere was under the rule of **Russia** and **China**, the north was "**Communist**", the south was "**Capitalist**" but unfortunately, **communism** in theory and **capitalism** in practice, they were both **materialist** countries. The border in between was called the **Iron Curtain**. People trying to cross the border were shot. Germans also shot Germans at the border. This was removed in **1990s**, then the Berlin Wall was demolished in **1991**. **All doors in the world were open to people in the beginning of 1990s** and and a magic box, **the computer was presented to humans like the magic lamp of Aladin**. Thus travelling to all countries worldwide has become possible **to some extent**. That is, **the chance is achieved**.

So, what the young people should know; **getting the chance is something, and benefiting thereof is something else.** Luck flies like a bird, if you throw it, it falls. You cannot say "That bird comes again, then I will try my chance again." Because when you are young, it feels like that life will never end. However, it passes as quickly as blinking.

I) FOUR CONDITIONS TO USE THE OPPORTUNITY OFFERED BY THIS ERA:

1) LEARNING ENGLISH: You need to learn English at the youngest age possible, because both French and German are hard to learn and not used in every location of the world. English is easy to learn and applies to all. We tried to learn English from books in the past. But we used to read the words in English as if they were written in Turkish, I mean; we used to speak English like Tarzan did. You can upload many lessons to your mobile phone in the evening and you can listen and try to repeat these lessons the next day when you are on a bus or on a boat. You can complete the entire book in a year and begin to speak like a native American or English and learn English in the half of the daily life wasted away.

2) FOLLOWING PERIODICAL PUBLICATIONS: Follow an English periodical publication, which is published on weekly or monthly basis but gives any recent updates in that profession.

3) READING BEST SELLERS: Books. Follow English books of the world champions of the profession in that era, acknowledged to be specialists in that area by the entire world. New champion books, which update former information under the light of new studies, can only be published after a week or a year.

4) FOLLOW THE INTERNET: The Internet provides developments even in the last half hour in the world.

Everybody who follows these periodical and non-periodical publications and the Internet in their own profession for 5 years, 10 years and 20 years can use the chance offered by our era to become one of the bests in his/her profession.

SOME PICTURES AND DRAWINGS IN ADDITION TO PART THREE

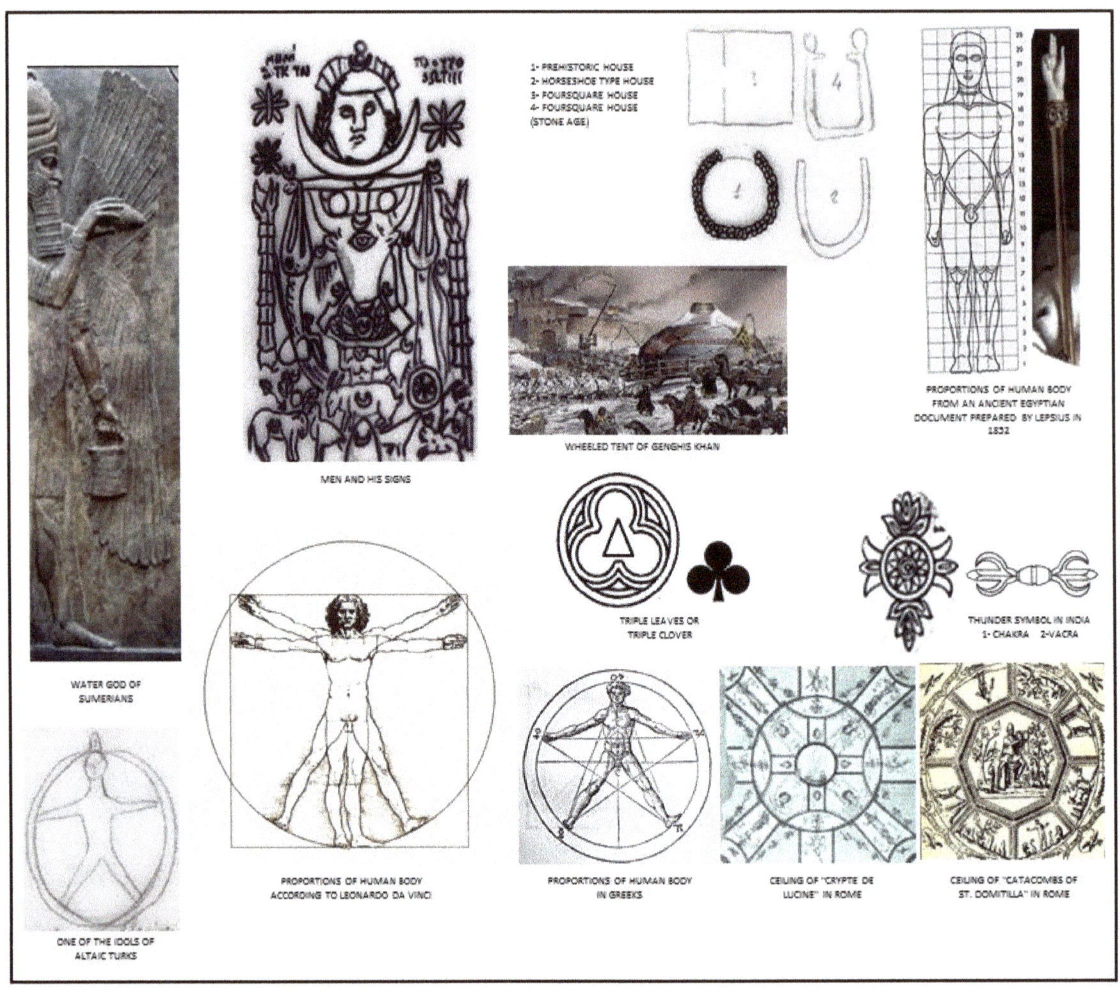

PROPORTIONS OF HUMAN BODY
FROM AN ANCIENT EGYPTIAN
DOCUMENT PREPARED BY LEPSIUS IN
1832

WHEELED TENT OF GENGHIS KHAN

MEN AND HIS SIGNS

WATER GOD OF
SUMERIANS

TRIPLE LEAVES OR
TRIPLE CLOVER

THUNDER SYMBOL IN INDIA
1- CHAKRA   2- VACRA

PROPORTIONS OF HUMAN BODY
ACCORDING TO LEONARDO DA VINCI

PROPORTIONS OF HUMAN BODY
IN GREEKS

CEILING OF "CRYPTE DE
LUCINE" IN ROME

CEILING OF "CATACOMBS OF
ST. DOMITILLA" IN ROME

ONE OF THE IDOLS OF
ALTAIC TURKS

B IN PART THREE
ADDITIONAL FIGURES AND DRAWINGS ON THE REASONS OF WHY THE UNIVERSAL SYNTHESIS IS
"MEVLEVI DERVISH" AND ITS GEOMETRY IS "TRIANGULAR PYRAMID":
OTHER EXAMPLES ON THE HUMAN BEINGS IN HISTORY:

ADDITIONAL PICTURES AND DRAWINGS
OTHER EXAMPLES ON HUMANS IN HISTORY:

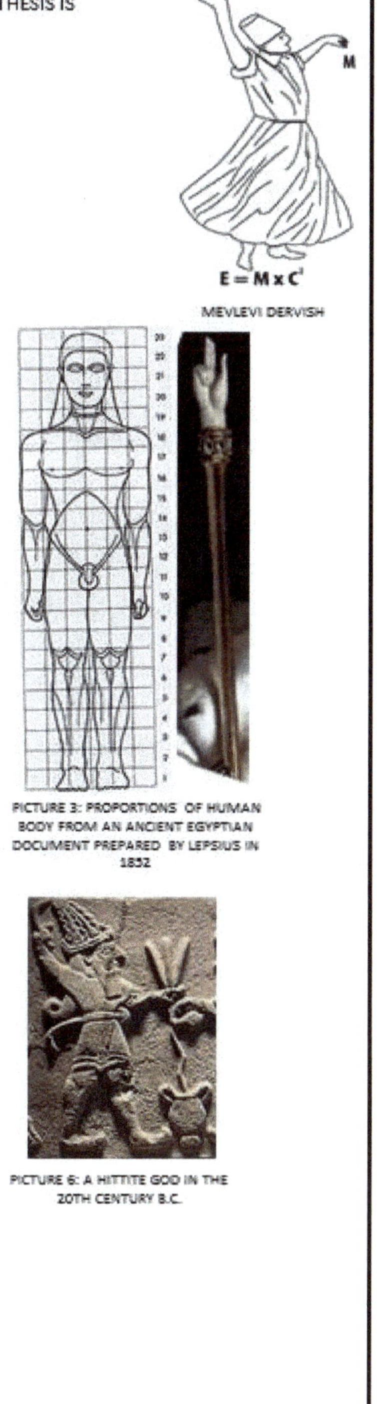

MEVLEVI DERVISH

PICTURE 2: ONE OF THE
IDOLS OF ALTAIC TURKS

PICTURE 4: PROPORTIONS OF
HUMAN BODY ACCORDING TO
LEONARDO DA VINCI

PICTURE 3: PROPORTIONS OF HUMAN
BODY FROM AN ANCIENT EGYPTIAN
DOCUMENT PREPARED BY LEPSIUS IN
1852

PICTURE 5: PROPORTIONS OF
HUMAN BODY IN GREEKS

PICTURE 6: A HITTITE GOD IN THE
20TH CENTURY B.C.

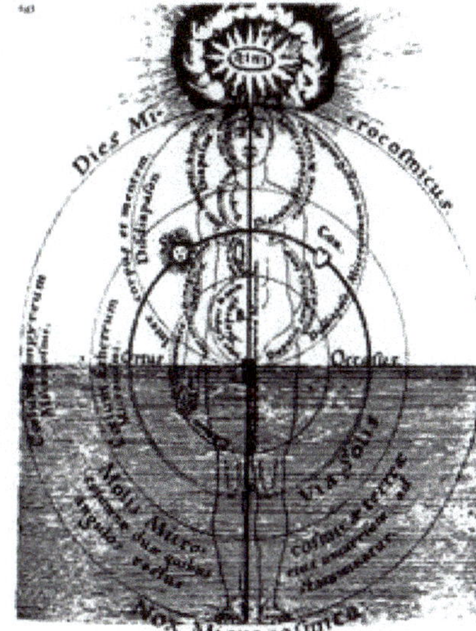

PICTURE 7: A DRAWING ABOUT "MICRO
COSMOS HUMAN" ON 1619 IN
OPPENHEIM/EUROPE

BAAL: THE LEADER OF GODS WHICH
SOME OF THE SAMI PEOPLE ADORE IN
THE NEAR EAST AND NORTH AFRICA

PART FOUR "THEORY OF EVERYTHING: RABIA"

A) NECESSITY OF SUCH A THEORY

While looking for the universal facts to explain everything in the universe and to be valid for everything (as otherwise, this world will not be in order), we should add some of the new thesis to the knowledge present in the modern literature. Because today science still could not explain neither the **Quantum Mass Gravity Theory,** which Einstein could not prove for 30 years; nor **Unified Field** and **Super Mass Gravity Theories**. Getting to the **Universal Synthesis** can only happen by new theses, which are open to anti-theses.

B) PREVIOUS "THEORIES OF EVERYTHING":

I) FIRST THEORY:

Several Super String types which are looped in 10 or more dimensions, and are composed of theories combining the following "**Electromagnetic Wave**" (Photon) and **Matter**[(Particle), Graviton] nodes, huge energies in the "**Quantum Universe**", "**Quantum Mechanics**", "**Relativity Theory**" and other universal geometries, were mapped(see **page 130,** String theory)

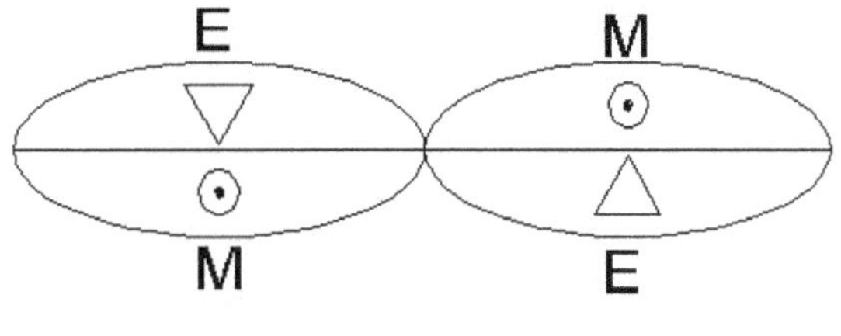

LIGHT QUANTUMS
(PHOTONS)

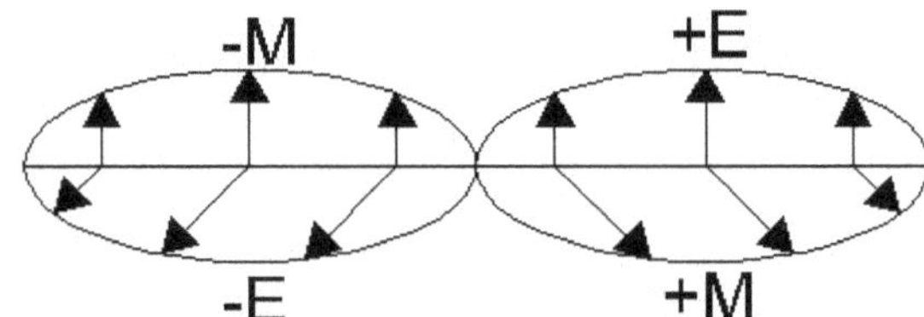

ELECTROMAGNETIC WAVE QUANTUMS
E=ELECTRIC FIELD: THE FIELD, WHICH IS
CREATED BY ELECTROMAGNETIC FORCE
AND PHOTON QUANTUMS OF ELEC-
TRONS
M= MAGNETIC FIELD: IT IS CREATED
BY THE GRAVITATIONAL FORCE AND
GRAVITON QUANTUMS

II) SECOND THEORY:

They created structures having super symmetries in 11 dimensions connecting subatomic particles together after combining some of the quantum models and mutual interactions with some images and events, and decided that we could be in one of the infinite number of universes.

III)THIRD THEORY:

According to the last appearance as can be seen below: An "**M theory**" was presented having weaves composed of *Superstrings* and *Super Earth Gravity*. According to this theory, "Big- Bang" (Big Collision) was created from the colliding thin membranes. And it was indicated that our universe was where these mem-

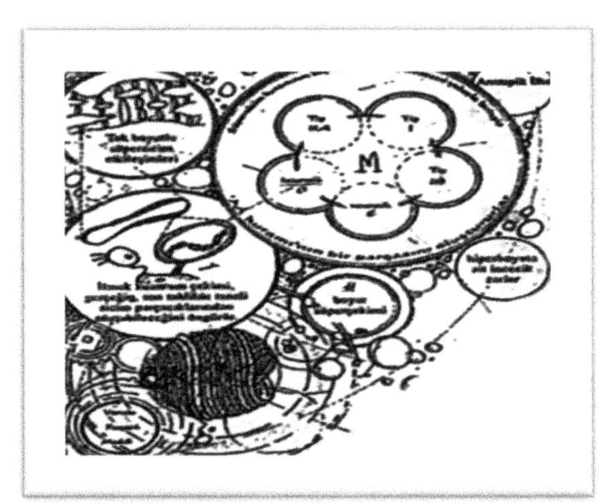

branes intersect. Here, before starting our Theory of Everything, let's sort some knowledge to constitute its fundamentals:

C) PRELIMINARY INFORMATION ON THE UNIVERS PREPARING THE THEORY OF EVERYTHING

For the "**New Theory of Everything**" which we will present here, **"Quantum Table of the Universe"** was given in **Annex-Ia,** including the **Light and Electromagnetic Wave Quantums**. According to this table, we can see the **Quantums of Universes** which have expanded in Big Bang by means of kinetic energy of photons, and which will start to be gathered after **Big-Crunch** by means of the potential energy of gravitons, and which will explode again. **Quantums of Universes** in this table are possibly the quantums of the light beam in a gigantic Universe. In **Annex-Ia**, we will be able to see how much the outer frame in the expanding quantum structure, and the theory which we will explain based on the other tables in the annexes fit in the mathematical definitions given in the previous page.

I) GENERAL DEVELOPMENT OF UNIVERSE AND ITSSTRUCTURE:

According to the Quantum Table of Universes in the Annex-Ia: "Photon Quantum" which represents the Kinetic Electrical Field Energy Quantum and Potential Energy Magnetic Field Quantum of Matter and Anti-Matter Universes in the Astral scale, and the Electro-Magnetic Wave representing the kinetic energy in the atomic scale, makes a chain with "Graviton Quantum" which represents the potential energy. In other words, "Graviton" and "Photon" constitute the universes, which explode in astral and atomic scales alternately, one by one and periodically, and then expand, later gather and explode again. Thus, the evolution of universes are managed by the balance between Kinetic Energy and (K_E) and potential energy (P_E) (Universal duality). Since K_E will be dominant in the universe after Big Bang; anti-matter, which reaches to the maximum density before Big Bang, starts to increase as a matter while the universe enlarges after Big Bang. K_E, which will reach to a maximum during Big Bang, will start to decrease and when it is zero before Big-Crunch, since this time Dark matter is maximum in the environment of universe, and since P_E is maximum, transforming of dark matter in the environment into Anti-Matter, in a sense in curveting process of dark matter (Big-Crunch) starts.

In all phases of the universes, **Matter and Anti-Material Particles,** which are formed in the **Atomic and Astral scales** like the "**Granular plurality**" in Mathematics, are in a continuous motion and alteration, in a **divine mathematical rhythm, accord and harmony**, both around themselves, and around each other, in nearly equal time periods, in other words **periodically**. Maybe in the future, all this universe will be able to be calculated and viewed in computers as geometrical **simulation models** by means of **Crystallographic Geometry**, Calculus equations of universes (to be told below) and **Automation software**.

"**Numbers and Our Universe,** and **General Calculus Equation**" in Annex-Ib,

"**Calculus Equations as per the values between the Sub-atomic particles of our Universe, Ist and IInd Big Bang**" in Annex-Ic,

"**Calculus Equations as per the values of Sub-atomic particles of our Universe today, and the time of today's symmetry**" in Annex-Id,

Calculus Tables above (Annex-Ib, c, d) and the tables in **Annex-IIa, b, f** about **BBS** and **Big Bang Particles** were the last tables which have been prepared to make the older tables include the newest modern knowledge.

As Calculus Equations in Annex-Ib, c and d still could not be calculated; the existence of **Super partners in Annex-IIa** and **f could not be proved, either.** In the future, all tables and knowledge, which are given to find out the scientific facts, and all of the equations, which have been proved and theorized up to today, should be combined with "the **Calculus Equations of Universes**". When this is succeeded, all of the Universal facts, wrongs, missing things, remainings will appear more clearly. "**Universal Facts**" which humanism is waiting for since thousands of years, and which explains everything about the universe and humans, will be presented by reducing all knowns and unknowns of our universe which we quantitate and limit between two Big Bangs, as a whole into a **Calculus Equation**. And we will be able to achieve this goal, only by means of **Computers** and **Automation software**. According to **Determinism**, everything is the result of other events. According to **Buddha**: 'Everything in the universe is a chain of events. One event follows the other and the previous one causes the latter one to exist. Existing-creation process is such a 'wheel of existence' (**Kalachakra**) (In Turkish it means '**permanent wheel**')'. Suitability of '**Quantums of Universes' Table** given in

Annex-Ia, and **'Calculus Equations of Universes' Tables** given in **Annex-Ib, c, d**, to Determinism and Buddha is remarkable. Here, the chain of consequent events become the chain of quantums.

In the first half of our universe; there is an expanding positive (+) charged **Matter** spending kinetic energy (K_E) and a neutral "**Dark matter**" whose potential mass gravitational energy (P_E) continuously increase, in our universe in the astral scale, from Big Bang on the lower left to Big-crunch on the upper center given in **Annex-Ib, c and d.** In this first half, after the "**Radiation**" belonging to the continuous kinetic energy in the universe is transformed (**Annex-Ia, b, c, d**) to the matter, energy for the expanding will be provided by burning the stars in the center of galaxies. Neutral particles which are the remainings of this burning will be gathered out of the universe, as a continuous neutral mass, as "**Dark matter**". The reason of continuous acceleration and expanding of the universe is the mass gravity of Dark matter which increases continuously in the external environment.

The second half of our universe: It is the part of tables given in **Annex-Ib, c**, from Big Crunch on the upper center, to the second Big Bang on the lower left. As at the end of the other expanding part of the universe, massive stars transform first into **Neutron star** and then into **White dwarf**, and later into **Dark-Holes** in our still expanding universe since the Dark Matter out of the universe reaches to the maximum mass and gravitational force and since there are no radiation left and thus since it gets colder in the astral scale; in the second half the universe starts to collapse from outside to inside with **Big-Crunch**; and at the same time **uncharged Dark Matter,** which reaches to the maximum mass out of the universe in the first half, starts slowly to transform into **Antimatters** in the universe.

While the disappearing speed of **Anti-Reifying** as a **Black hole** in the universe reaches to the speed of light before the next, **Anti-Materials** which reach to the maximum heat and density in the astral scale and which is transformed to radiation, get up with the **Anti-Proton and Anti – Neutrons** in the atom nuclei in the atomic scale of universe (**Annex-Ib, c, d**). At last, negative (-) charged **Anti-Material Universe** becomes pure energy when the last Anti- Gravitons of Antimatters exceed the speed of light and disappear (**Annex-IIg**). Thus, since the gravitational force which is the reason of gathering of the universe disappears, negative **(-) charged Antimatter Universe** which was anti-materialized completely and later became radiation in the astral scale, will start to transform into a **positive (+) charged Matter Universe** like our universe, maybe because of intense negative loads' (by means of **Casimir Effect**) producing a repulsion instead of gravitation, by Big Bangs on all sides at the same time. First, in the atomic scale, the **Subatomic Single Particles** on the right of **Annex-IIe** from up to down in sequence, later "**dual particles**" in **Annex-IIh**, later forming the "**triple subatomic particles**" in **Annex-IIi**, 300 or 700.000 years and later, "**Atoms**" started to come into exist-

	DURING BIG-CRUNCH		DURING BİG BANG		
CK		**MINIMUM**	**MINIMUM**		**BP**
1	o	KINETIC ENERGY (K_E)	POTENTIAL ENERGY (P_E)		1
2		SPEED	MASS	10^{-5} g	2
3		HEAT	VOLUME		3
4		PHOTON (γ) NUMBER (LESS ENERGY)	SIZE	10^{-33} cm	4
5		RADIATION	GRAVITON ( ) NUMBER		5
6		DENSITY	TIME	10^{-43} sec	6
CP		**MAXIMUM**	**MAXIMUM**		**BK**
1		POTENTIAL ENERGY (P_E)	KINETIC ENERGY (K_E)	10^{14} GeV	1
2	10^{19}	MASS	SPEED	Speed of Light	2
3		VOLUME	HEAT	10^{32} K°	3
4		SIZE	PHOTON (γ) NUMBER (MORE ENERGY)	10^{42}	4
5	10^{42}	GRAVITON ($\odot$) NUMBER	RADIATION		5
6	10^{40} years	TIME	DENSITY	10^{88} tons / cm^3	6

ence starting from Hydrogen. Now forming new elements in the stars, a continuously accelerating and expanding Matter Universe will develop until Big-Crunch.

In accordance with application of the table in **Annex-Ia** in the Calculus Equation; 6 variables (**BP1.... 6**) of potential energy which is minimum, and 6 variables (**CK1.... 6**) of kinetic energy which is maximum **during Big Bang**; and 6 variables (**CK1…. 6**) of kinetic energy which is zero, and 6 variables (**CP1…. 6**) of potential energy which is maximum **during Big-Crunch** will be defined as the functions of calculus equation. As four groups of variables cross-change during Big Bang and Big-Crunch, 4 groups of variables (functions) in the astral and atomic scale belonging to the 4 forces of atom which can be seen on upper right and upper left of tables in **Annex-Ib, c and d** cross-change similarly as shown in the above table.

While 6 functions of kinetic energy (K_E) (**BK1….. 6**) given in **Annex I-b, c and d** and above table are always maximum (**+1**) in the first 10^{-43} seconds of Big Bang (**Planck time**), they become always zero or minimum in Big-Crunch; on the other hand while 6 functions of potential energy (P_E) (**Bp1…. 6**) are minimum in zero or sub-zero values in Big Bang (**Planck time**), they become maximum in Big-Crunch. In the second Big Bang, the values of all functions return back to the values in the first Big Bang.

The basis of previous objections to the here defended **oscillation theory** was that: since "**Entropy**" would continuously increase in every explosion and expanding, in other words, since the photons would increase; the increasing photons would not allow the future explosion, expanding and gatherings. In **Annex-Ib, c, and d**, while the high-energy photons (radiation) which has been maximum during Big Bang, turn to Nucleons; the energy of photons decrease. Since while the energies of low energy photons start to increase during Big-Crunch, the numbers and energy of Dark Matter Gravitons which have reached to the maximum mass and number, will be decreased; and since the number of photons and energy will be zero before the second Big Bang, it is obvious that there will be no Entropy increase problem as per these tables:

Above, some **Entropy, Mass and Radiation power** ratios of the universe are given in the **Annex-I a, b, c, d**, tables of the 'changing stages of universes'. In the following table, we can see that as the distance between the quarks in the atom nuclei of our universe which is expanding and which will be gathered, the force of their pulling each other has the specialty to grow infinitely. **Such a property** does not exist in no other particles in the Astral and Atomic scale. In other words, the power of traction power on each other increases as they become distant. I wonder whether the maximum traction power of the quarks that became distant at the maximum have a role in the re-gathering of the universe that will expand until the Big-crunch?

	ENTROPY	M☉/L= MASS OF UNIVERSE/ RADIATION POWER
IN BİG BANG	⌂/⊕ =1/1	M☉ /L=0/MAX
TODAY	⌂/⊕ =10/1	~M☉/L=300
IN BIG-CRUNCH	⌂/⊕ =0/0	~M☉ /L=1500
TODAY THE PROTON, NEUTRON RATIO OF THE UNIVERSE	⊕/⊗ =80/120 =2/3	

	DISTANCE BETWEEN THE QUARKS AND ANTI-QUARKS		DISTANCE BETWEEN THE ASTRAL MATERIALS, ATOMS, BARYONS AND ELECTRONS	
	IF DE-CREASES	IF IN-CREASES	IF DE-CREASES	IF IN-CREASES
PULLING EACH OTHER	WILL DE-CREASE	WILL IN-CREASE	WILL IN-CREASE	WILL DE-CREASE

As indicated in the table of history of our universe in **Annex-Ie**; before BBS which appeared by means of the single particles obtained in the first 10^{-11} seconds of Big Bang, there were protons as much as photons, together with the **WIMPs** appeared in 10^{-35} **seconds** (**Entropy=0**). Protons' being created before the single particles appeared in the first 10^{-11} **seconds**, is a strange phenomenon which shows the magnitude of gravity force between three quarks composing the protons. We know that the resource of gravity force between these quarks forming the baryons, are Color Quantums (**Gluons**) which are interchanged between the quarks as seen in **Annex-IIa, b, e**. While the **Photons** are not electrically charged, **Gluons** are charged with colors. So, I wonder if **the "Matter/Anti-Material effect" was the biggest force of the universe at the Big Bang moment? Or was it this "gravity force between the quarks"?** Or are the gravity force between the quarks and the **Matter/Anti-Material effect** same? If they were the same, then wouldn't it mean that there is a **Fifth Force** apart from the known four forces within the atom, in other words the gluon force between the **quarks**?

As can be seen in the following table, one of two quarks of Mesons which rotate around the baryons in the nucleus of atoms and which prevent the baryons from moving away and coming too close to each other, is **Anti-Quark** (Anti-Matter) and the other one is **Quark** (matter). What prevents these two quarks from moving away and coming too close, are **Anti-Gluon** (Purple) and **Gluons** (Red) as can be seen in the following atom model, which rotates around the **quarks**.

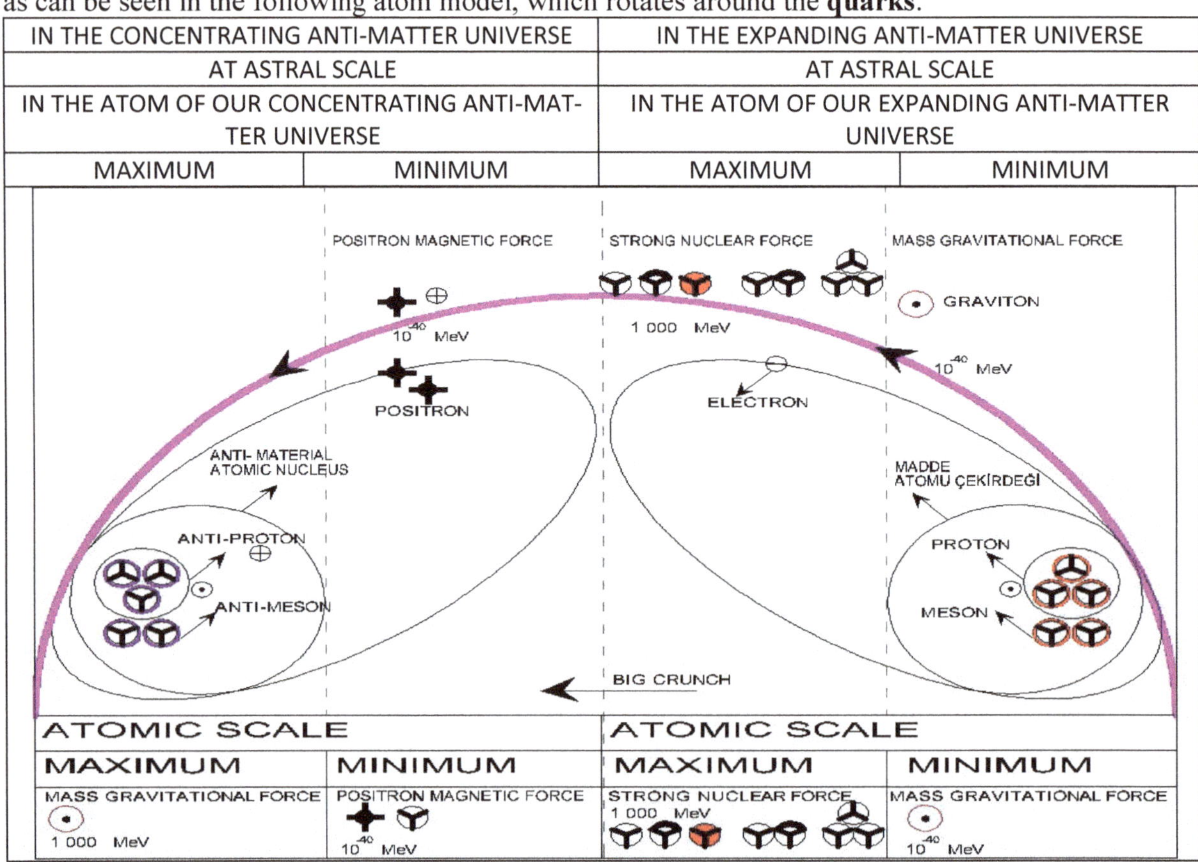

In Annex-Ib, c, d and Annex-IIe, f, Mass Gravity and **Strong Nucleus force from four forces within the atom** are maximum **1000 MeV** as can be seen above. But as can be seen in **Annex-IIe**, the energy of ± 2/3 and ± 1/3 charged **Quark and Antiquarks is between 169, 1-173, 3 MeV**. The energy of some of hyperons in **Annex-Iii** like "**Xi/b Hyperon**" can reach up to**5624 MeV**. At the moment of **Big Bang**, the total kinetic energy (K_E) in the universe can be seen as 10^{19} **GeV** in **Annex-Ib**. On the top of table in the previous page, as the Quark and Antiquarks become distant from each other, this **Matter-Antimatter force** which we can understand that it is stronger than the known 4 forces within the atom; or 5th Force within the **Atom is** originated from the **Gluons** which rotate around Quarks and which are similar to Electromagnetic Force, but carrying color charges. It cannot be a coincidence that the color charges of Gluons have three denominator values as the electrical charges of Quarks (**Annex-IIe**). Unfortunately, here the color force in question, which is the 5th force within the atom, was not shown in the attached table. But it would not be wrong to say that the development of history of humanism knowledge based on **QUARKS** and **GLUONS (Annex-Ib, c, d, IIa, b, e)** in our age. In the future studies, works to understand the **Universal Fact** should be completed by including this fifth force into the Calculus Equations. Considering that we are not working with nearly a hundred accelerators which are present in Europe, **Cern,** and the other countries of the world, where such subjects are tried to be shed light on; theoretically, in other words with **Mind and Knowledge**, to be able to get to the unknowns a little bit, first of all, **let us focus on the happenings (Annex-IIf) in the first second after Big Bang and the previous last seconds,** like the workers in accelerators:

Our universe whose Calculus equations are given in Annex-Id, c, b: started to be formed with **Big Bang** after the universe became **pure energy** by sweeping all of the antimatters away, while the universe was gathering before Big Bang. This formation started not only in a specific area, but every parts of the universe at the same time. In the 10^{-43}**th second, before the Planck Time**; in smaller scales than **Plank scale ($<10^{-35}$ m)** there were "**Quantum Bubble**" and "**Superstrings**" which were composed with **Planck mass (10^{19} Proton) Black holes.** Since the "**Quantum Mass Gravity Theory**" was not yet known, scales smaller than 10^{-35}**m** are also unknown. But since the temperature was atleast10^{23} **K°,** we know that matter can be composed of Quarks and Gluons. Besides, we also know that there would not be any differences between the weak and strong particles like **Quark** and **Neutrino,** which affect each other, because of high temperature.

Now let us list what we know about the Quarks and Gluons:

QUARKS: Quarks tend to be surrounded by **Virtual Quark and Gluons** of the same color. Passing through a high-energy environment, Quark goes deeper, sees fewer colors, feels a weaker strength. While a quark pair moves apart from each other, the gravitation force of **three denominator Gluon color charges around the Quarks, tends to enlarge infinitely.** While the distance between the quarks decrease, the interaction between them becomes zero. Therefore, we can understand that quarks behave as free particles in the hadrons.

GLUONS: They are the color quantums which hold the hadrons together and which are being interchanged between the quarks. In the first 10^{-45}seconds of Big Bang, in the Planck time, the mass of particle with a temperature of 10^{23} **K°,** energy 10^{19} **GeV,** the smallest scale(10^{-35} m) and the highest density (10^{88} **ton/cm3**), was 10^{19} **Proton (GeV). (Annex-Ie)**In this stage, heavy particles (**Baryons, Mesons**) and light particles (**Electrons, Photons**) tend to have the same behavior because of hightemperatureand energy, up to the first 10^{-35}seconds. Also there is a symmetry between **Strong nucleus, Weak Nucleus and Electromagnetic Forces.** In other words, these three forces were combined. In the 10^{-35}**th second in the Big Integration Period,** when the temperature fell to 10^{10} **K°,** energy level to 10^{15} **GeV,** and the mass to 10^{15} **Protons,** since the Strong nucleus, Weak Nucleus and Electromagnetic Forces are combined, "**Wimps**" (Unstable weak interaction, big mass, single particles) and the number of Photons are as much as the number of Baryons in the universe (Entropy=0).

In the 10^{-11}th second, since the energy level of universe fell below10^{15} **GeV,** symmetry was broken in the universe, in other words, three forces within the combined atom were separated from each other; and while **W$^+$, W$^-$, Z° Bosons** were created, there were "**Weaves**" (assumed to be spherical) (For me, these maybe the BBS given in**Annex-IIa, b**) and "**Strings**" in the universe. (Unidimensional high energy strings of width 10^{-35} **cm, length infinite**) (for me, these should be knotty like the light quantums). There are "**Higgs Areas**" with potential energy density in the spaces where there are no weaves and strings.

In the 10^{-10}th second, four forces within the atom were separated from each other. As **Photon** combines the Electric and Magnetism; **Particles** connect the Electromagnetic and Weak Nucleus forces as can be seen below. Since in this stage, energy level falls below **200 GeV**, these particles were disintegrated and probably160 particles were created which then formed **BBS.**

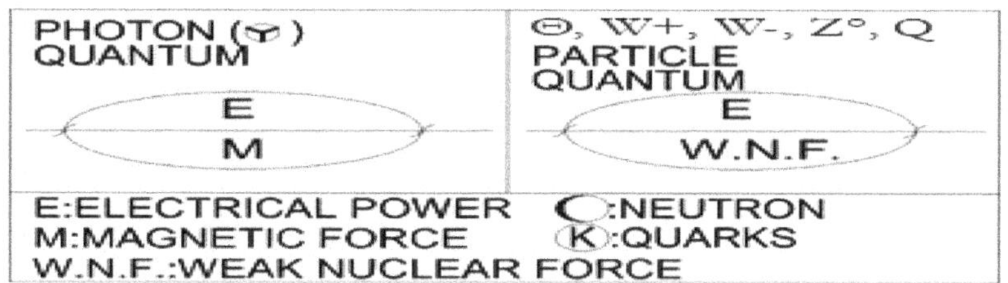

In the 10^{-5}th second, two Quark Meson and Ant mesons were created.

In the 10^{-4}th second, three Quark Baryon and Antibaryons (Nucleons) were created.

In the 1. Second, energy level fell below **1 MeV,** temperature level fell below10^{10} **K°.** Previously created Electron and Positrons disappeared. The number of Neutron and Neutrinos were fixed and 1 Neutron was left for every 10 Protons in the universe. (Today this ratio is, **2 Protons/3 Neutrons**). **(Page 136)**

In the 1stMinute, particle creation ended.

The most significant information among the ones given above is that; Maximum values started to turn to minimum; minimum values started to turn to maximum in the first 10^{-45} **second** of Big Bang, while our universe was taking shape in 10^{-35}**m** scale, 10^{23} **K°** temperature, and with 10^{19} **GeV** energy.

The main thing to present the unknowns of our universe would be the **Calculus Equations and Tables** given in **Annex-Ib, c, d. Therefore let us continue explaining these tables:**

Let us answer the question why the new terms like "**Mass Time**" and "**Density Time**" were included in this table in **Annex-Ib**: We know that since the mass of Sun is greater than the world, while 1 second pass on the **Sun, 11, 5 days (1. 000.002 seconds) pass on the world**. Also at the time of **Big Bang,** as can be seen in **Annex-Ib**, maximum density, minimum mass; at the time of **Big-Crunch** maximum mass and minimum density. Therefore, with the idea that **time and material could be relative to the mass and density,** new "**Mass and Density Matter**" terms were suggested to the science world like "**Mass and Density Time**", unprofessional, but worth to be tested. Although the "**Wimps**" which were formed in the first 10^{-35} **second** of Big Bang and "**Machos**" which were formed 100 000 years later (**Annex-Ie**) were big massive particles; since their function in the universe expired, they took their place around the universe as the first tangible assets of dark matter, and today, they probably constitute the outermost layer of Dark Matter.

It is possible that Antimatterization has started as the Neutrons in the innermost layers of dark matter have become Anti Neutrons, in the universe to be gathered after the **Big-Crunch**. We can assume that the universe, which has been completed in the -1/2 phase of the gathered universe (**Annex-Ib, c, d**), and which has become a black hole until the second Big Bang, has destroyed the current anti-materials. In the last stage, universe whose gravity force was set to zero by the potential Mass Gravity Energy, will enter into an expanding phase, as in the second Big Bang and after the previous Big Bang.

On the table in **Annex-Ib**: "**+ maximum**" is indicated with +1and"**Eksi maximum**" is indicated with -1, symbolically. **Zero (0)** again represents **the minimum of positive and negative values** and also below zero values, symbolically. 2 and 3 denominator values in the middle have been located in the tables between **-1 and +1**. Finally, when we come to the **Nonproportional** (+, -√2) and **Real**

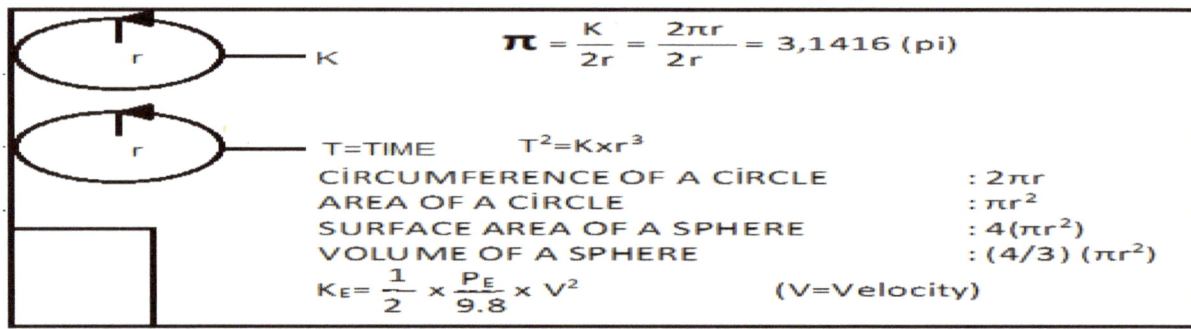

numbers (± **π**)which I didn't put their equivalents in physics; it was told previously that the universes emanated by means of particles rotating around each other and rotating around themselves in Astral and Atomic scales, in specific time periods and speeds.

It would be nonlogical to dissociate the fundamental variables of universe in the Astral and Atomic scales (Functions), from the simple formulas I have given above and the equations of all proven theories, generally in the **calculus equations in Annex-Ib, c, d**. Together with all these equations, it cannot be denied that ±√2 ± **π**numbers on the lower left and right of the table in **Annex-Ib, c, d** and above the **Annex-Ib**, are the essentials of the whole. I will leave the verification and general corrections, and then getting the unknowns from the known function values in Calculus Equation, to the Mathematicians and Physicists.

II) MAGNIFICENCE OF OUR UNIVERSE

As of January 2012, the latest state of physics was roughly as it appears in the graph on the right.

1) At the very bottom of the diagram aside, Einstein completed and proved the **gravitational force (GF)** discovered by Newton within the atom as **"quantum gravitational force" (QGF)**.

2) Upper, **Maxwell** combined **electric energy** in the atom with the magnetic energy and described it as **"electro / magnetic force ((EM)**.

3) **Weinberg and Salam** combined the **EM energy** of the atom with the **"weak nuclear force (WNF) energy"** in the nucleus and described as **"electro weak force (EWF) energy"**.

4) **Einstein** tried to combine the **EM** and **WNA** force energies with strong nuclear force **(SNF)** energy, the so-called **"unified field theory"** for 30 years long, but failed.

5) Finally atom **EM, WNA, SNF** force energies, within the gravitational energy **(GF)**, **"super-mass gravitational energy" (SMG)** has come to the stage of combining. But at this last stage, it has not yet been proven. The ones shown in the graph on the right are indicated with a straight line and the ones that have not yet been proved with a dashed line. In addition to the efforts to integrate the existing knowledge of the universe, let's include the information given in this book and its annexes and try to understand the integrity and magnificence of the universe in which we live:

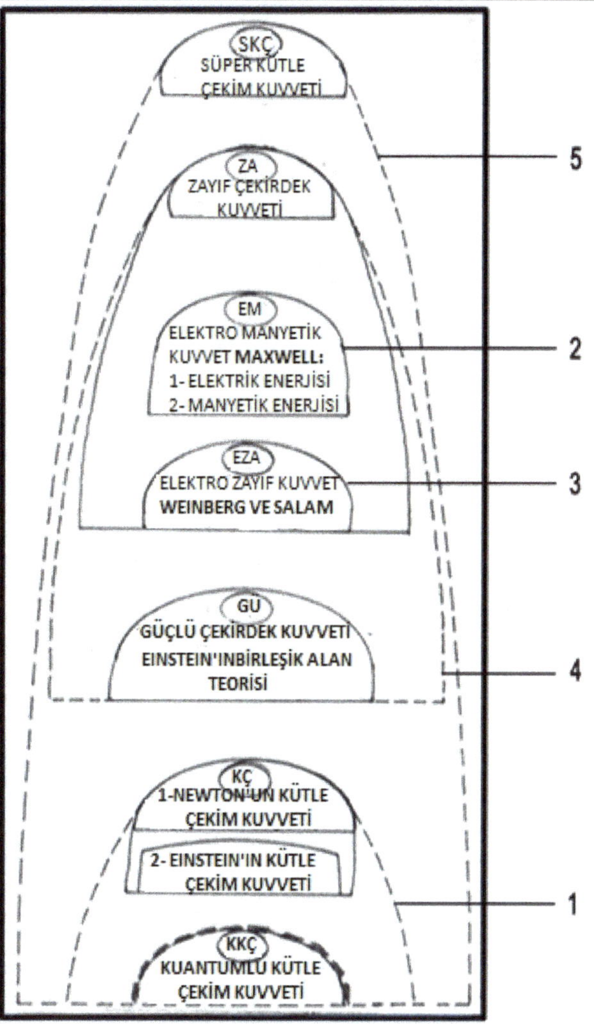

Basic structure of the universe consists of **particles** (spheres, quantum) from sub-atomic scale to astral scale.

As can be seen in **Annex-IIe**; the subatomic particles of the universe, formed in the first second of the Big Bang; First, the **bosons transmitting and influencing the 4 forces of the atom**, then the fermions, which were affected by the 4 forces of the atom, were formed. **Annex IIf** shows the general scheme of the **affected bosons** and the **affected fermions**. These **subatomic particles** formed the building blocks of the universe that we see on the astral scale today.

a) TYPES OF SYMMETRY:

* SYMMETRY IN GEOMETRY

*** SYMMETRY IN SPACE-TIME: (Super Symmetry)** The additional dimensions of the super space are not like our dimensions, they have no dimensions at all, as in **String Theory**, they do not resemble extremely small dimensions. If space were not symmetrical, it would be filled with **massless** and **spinless particles**.

The **S particles in Annex-IIe**, if present, are the products of *super-symmetry*.

Super-symmetry and the Higgs Mechanism match the possibility that the black matter of the universe begins to form with the **LSM (the lightest super-partner)**. So the Higgs Mechanism has already begun mass gaining the particles.

***SPACE-TIME SYMMETRY**: The aim of Einstein's **"Unified Field Theory"** was to **integrate the dimensions of the four forces within the atom according to their geometric properties, including the time dimension.**

b) STRING THEORY:

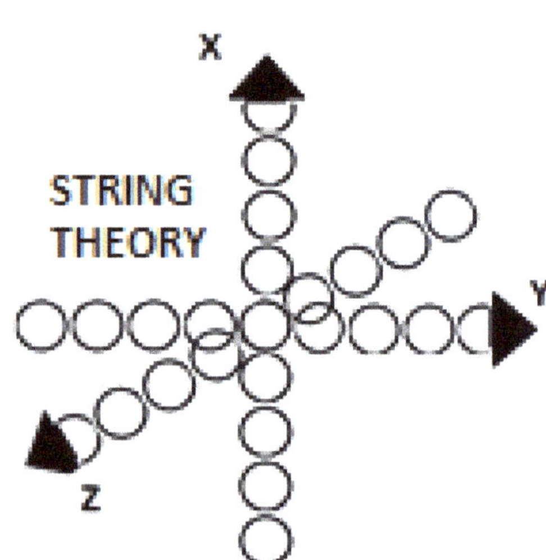

It was introduced in 1980, according to this theory: **3 dimensions (x, y, z) +6 tiny dimensions + time = 10 dimensional gravitational (KW) world** The thickness of the strings was 10^{-35} **cm**, the length was infinite. Empty space is actually filled with **Higgs particles**. The **gravitatons** and **Higgs particles** cause the gravitational field to be emitted by the universe, which is called the "**cosmological binding constants**". **Higgs particles undergo Bose-Einstein quantum condensation.** This empty space has many features that resemble a *superconductor*. **Higgs particles are quantum of the Higgs field of the universe.** When the Higgs field is non-zero, it is in the lowest energy state. **The string is described as a topology from the Higgs field.**

In the first 10^{-11} seconds of Big Bang, **Higgs Fields** with potential energy (P_E) were formed during the phase without strings. So, the **string** was the product of kinetic energy (K_E). The **Higgs fields**, if any, may be from the previous universe. All particles, tuning bosons, fermions, and Higgs Fields derive their mass from their relationship with **Higgs**.

The subatomic particles between the two Big Bangs appearing in **Annex-Ia** are spins in the middle of the calculus table in **Annex-Ic**. **Spins are important** in all phases of expansion and accumulation of the universe, **because the following elements affect the spin**:

1- **Electric charge.**
2- **Space.**
3- **Volume.**
4- **Mass-time.**
5- **Einstein's special relativity and quantum theories**
   (in common "**relative quantum theory**").
6- **Super symmetry.**
7- **Weak force (WNA) of the 4 forces inside the atom.**

When the information on all these issues is explored, not only the importance of spins, but also the grandeur of the universe. In **Annex-Ia, b, c, d** tables, **proton** and **neutron ratios** in atoms of the expanding and collected universe at each stage are given. I think the main factor that creates these ratios is the **spin** of the particles. I believe that future research can reveal this chain of interaction, especially with **calculus equations**. Deeper in this chain of interaction, according to our knowledge today, the possibility of **gluons** appear to be strong. Let us make a few notes about the **weak force (WNA): WNA gives atoms a spiral structure that can distinguish between right and left. A left helicidal particle** can convert its energy from negative to positive, and this particle disappears, instead the right helicidal particle is created. In other words, the **Eta meson** with two quarks, which owes its mass to the instant one, behaves as in "**QCD (Q**: Quantum, **C**: Anti particle, **D**: Go back in time).

INSTANTON: (No-load current) consists of the exchange between **Higgs' gravitational** (KW) and **Z^0 particle, weak force (WNA).** Thus, **Instanton, which is the interaction of no-load current, adds mass (like HIGGS boson) to substances in the expanding universe.** On **page 126;** around the atoms of the expanding matter, the red gluons of the quarks in the protons in their nuclei and the quarks inside the anti-protons, the purple gluons, the gluons that are the opposite of the red gluons. In addition, 120 types of **fermi gluon** are given in **Annex-IIe** with their color and contrast colors.

Gravitational force between **color and contrast gluons**; the heavy particles in one of the **resonances** that make an infinite series, the more energized ones, have larger **spins (EK-Ic)**. So they spin faster around their own axes. This is the case with the **massive graviton** in the Big Bang. it is as

sumed that **resonances** consist of something similar to quarks. The resonances in the tables given in this book are unfortunately absent, whereas resonances in the nucleus of atoms are similar to the sub-atomic particles described as the "**standard model**" that we have tried to give broad information about, but not yet proven. The masses of **resonances** are larger than the particles we know and have more spin. Resonances are assumed to be single, **double (mesonic), triple (barionic)**. Resonances were detected in particle collisions in the accelerators. As the building blocks of **baryons** and **mesons** are called **quarks**, the building blocks of **resonances** are called "**Parton**,, but they are thought to be quarks.

The "affecting bosons" and affected fermions in **Annex IIf**; have created everything in the universe by transmitting the four forces of atom, neither of them would mean anything by itself. The intensity of the 4 forces of the atom varies at different scales, approaching the same intensity around the Planck scale.

According to Einstein;
Non-radiation dark matter + gravity [potential energy (P_E)] and
Luminous matter + kinetic energy [electrical energy (K_E)] are are the
main universal dilemma Duality).

c) MAXIMUM VALUES IN OUR UNIVERSE EXPANDING WITH BIG BANG:

In our expanding universe, **the maximum density in the Big Bang (10^{88} tons / cm^3); radiation, ie photon number 10^{42}; heat 10^{32} K^0; functions such as velocity, light velocity, kinetic energy (K_E) 10^{19} GeV decrease, while the mass, graviton number, potential energy (P_E), volume, elongation and time expands and the rate of passage of time slows down due to the continuous increase of its mass. (Annex-Ia, b, c, d). Prior to the Big-Crunch,** it was also the largest, cooled, 10^{40}-year-old universe that "**the pace of time must have stopped.**"

The maximum of the universe to be collected at the time of Big-Crunch: Mass (**10^{19} protons**), volume, space, number of gravitons (**10^{42}**), time (**10^{40}** years), functions such as decreasing, the universe first becomes anti-matter, then anti-matter gradually accelerating and strengthening black hole goes into the destruction stage. Since the rate of passing of time increases due to the continuous decrease and mass of the mass, the Big rate of passing of time "**will be at the speed of light before the second Big Bang, and wormholes**" (uniqueness) will be formed, which is bet on the existence of **Annex-Ie**. In this uniqueness, also called **quantum foam**, the 4 Force of the atom merges. Quantum foam consists of black holes with **Planck mass (10^{19} protons)** in volumes less than **10^{-31}** meters. The mass of the sun is $\sim$ 330 000 times more than the mass of the earth. **The sun passes 1 second and the earth passes 1 000002 seconds (11.5 days).**

IN THE EXPANDING UNIVERSE:

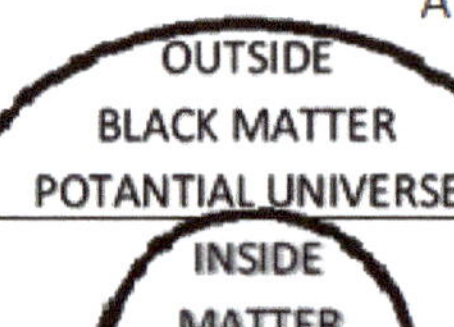

SINCE THE BLACK ATTER MASS CONTINUOUSLY INCREASES, THE PASSING RATE OF TIME SLOWS DOWN CONSISTENTLY

INSIDE, THE MATTER MASS OF THE KINETIC UNIVERSE CONTINUOUSLY SPEEDS UP AND THE PASSING RATE OF TIME SLOWS DOWN BUT MAYBE STILL AT LIGHT-SPEED FOR MASSLESS PARTICLES

AT BIG CRUCNH: TIME STOPS AS PER THE UNIVERSE, BUT FOR US, THE UNIVERSE IS 10^{40} YEARS OLD.

AT BIG BANG: THE PASSING RATE OF TIME AS PER UNIVERSE IS MAYBE AT LIGHT SPEED, BUT FOR US IT IS ZERO

IN THE CONCENTRATING UNIVERSE:

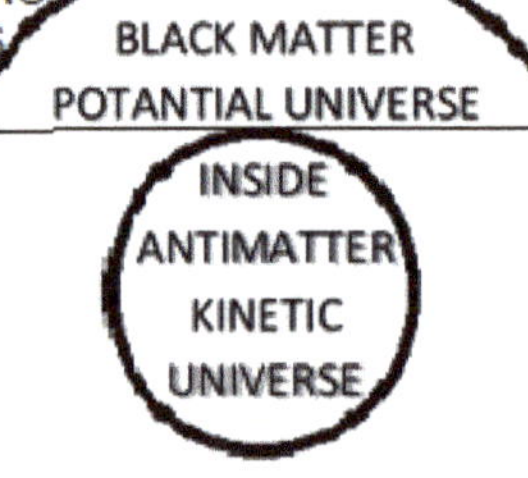

OUTSIDE, SINCE THE ANTIMATTER MASS CONTINUOUSLY DECREASES BY SPEEDING UP, THE PASSING RATE OF TIME SPEEDS UP OUTSIDE POTENTIAL UNIVERSE

INSIDE, THE MASS OF ANTIMATTER DECREASES CONTINOUSLY BY SPEEDING UP AND THE PASSING RATE OF TIME INCREASES. BEFORE THE NEXT BIG BANG THE MOVEMENT IS CLOSE TO LIGHT SPEED, THE PAST AND FUTURE WILL MERGE (SINGULARITY = WORM HOLES

IN THE NEXT BIG BANG: TIME ROTATES AS PER THE UNIVERSE, FOR US THE UNIVERSE WILL BE ~10^{80} YEARS OLD

SPEED = SPACE X TIME

TODAY WITHIN THE UNIVERSE:

$$\frac{\text{GRAVITATIONAL FORCE OF ATOM [POTENTIAL ENERGY } (P_E)]}{\text{ELECTROMAGNETIC FORCE OF ATOM [KINETIC ENERGY } (K_E)]} = \frac{1}{40} = X$$

X = SUPER SYMMETRY EXPLANATION OF THE HIGGS MECHANISM OF THE STANDARD MODEL

OUTSIDE OF THE UNIVERSE AT BIG-CRUNCH:

$$\frac{\text{GRAVITATIONAL FORCE OF ATOM [POTENTIAL ENERGY } (P_E)]}{\text{ELECTROMAGNETIC FORCE OF ATOM [KINETIC ENERGY } (K_E)]} = \frac{40}{1} = X$$

$$GF = P_E \times MASS (M) \times C^2 \text{ (SQUARE OF LIGHT SPEED)} \qquad \frac{P_E}{K_E} = \frac{M}{t} \times C^2 = K \left(\begin{array}{c}\text{CRITICAL}\\\text{DENSITY}\end{array}\right)$$

This shows that gravity slows down as the mass grows due to the effect of gravity on the passing speed of time. So, we measure time according to the mass of the world we live on. So, we measure time according to the mass of the world we live on. **We know that the mass of the center of our galaxy, the Milky Way, is 160 billion times the solar mass. The largest mass is in the black matter of the universe.** We know that most of the graviton, gluons and photons, and even neutrinos, are massless. The perception of time in humans is according to the mass of the world. However, gravity slows down time. Is the speed at which time passes on these particles, is it infinite? According to Einstein; all space measurements are time measurements, that is, time varies according to distance. That's why we see galaxies in space as they were 1 billion years ago compared to the mass of the earth.

Another information about the relation of the mass and speed is that **the mass of an object that travels at the speed of light reaches infinity or the object disappears**. The fact that **1 second on the sun** is equal to **11.5 days in the world**, was the conclusion that Einstein reached with the help of **"tensor calculus"** according to the general relativity theory. According to this fact, the situation that should be in the expanding universe to be collected is seen in the table above. The result is; For each entity on the spheres of subatomic and astral scales, the time varies according to the mass of the sphere on which it is located. In this way, the understanding of time is based on the mass of the world.

In the 1930s, **Sir Arthur Eddington and Nobel Prize-winning physicist Paul Dirac believed that the number 10^{40}** was the key to a comprehensive relationship connecting the micro world of the atom to the cosmic structure. For this purpose, Eddington put forward a wide range of schemes. Here is some more information, supporting the same prediction:

The total number of particles in the universe is about $10^{40} + 10^{40} = 10^{80}$. That is to say the ratio of the electromagnetic power (EM) of the atom to the gravitational force (KW). **(Annex-Ib, c, d) The KW power of an atom is 10^{-40} MeV and one of the EM power of 10^{40}.** In the **Creation monochord** on **page 156**, the outermost range of the universe is given as 10^{40}.

Again, according to the "**entanglement theory**"; It is now accepted that the subatomic research results that substances are connected to each other and are entangled as a single unified network like a universal whole. The **Tao, the Brahmans, and the Islamic Sufism** tried to explain the same thing.

According to Einstein's "**relativity theory**"; it is said that everything is composed of energy particles that are constantly in motion and affect each other. The concentration of **boson and fermion quantums** at the lowest energy is known as "**Bose-Einstein quantum condensation**" or "**vacuum**".. It was a Bose-Einstein quantum condensation da in the vacuum of the energy created by the previous universe becoming a black hole and destroying the antimatter. **Annex Ic**; On the subatomic scale before the Big Bang, **positro-magnetic energy (PM),** which is the lowest energy is 10^{-40} **MeV**, and after the Big Bang, the lowest energy on the atomic scale is the **mass gravitational energy (MGE) 10^{-40} MeV.**

We know that the boson-fermion quantum in each person's own brain, with the Bose-Einstein concentration, reaches its own consciousness. Below EK-IIf, we see the bosons that transmit and affect the 4 forces of the atom after the Big Bang and the **fermions** affected by the **bosons**.

The *fermions*, the building blocks of the material world, are separate from each other, preferring to keep themselves to themselves, i.e. they are asocial. *Bosons* are social and tend to come together and form groups. Fermions are "*particle*" and bosons are "*wave*". Quantum is both a wave and a particle. The **Bose-Einstein concentration** is also a quantum concentration and creates *consciousness*. This consciousness begins where two quantum meet. The computer-like concrete structure of 10^{10} nerve cells (neurons) in the human body works together with the concentration of quantum in the human brain (such as NUR-U KADIM in Big Bang). We know that thought produced by the brain has similar effects to electric currents. The quantum of vacuum in the human brain of thought is both particle and wave. The thought created by the vacuum develops into a renewed consistency and returns to the vacuum enriched fluctuations (quantum), perhaps through the aura and the chakras. It's like after the vacuum before the Big Bang in the universe, it explodes with the Big Bang, expands, and then assembles with the Big-Crunch and back to the vacuum. Human thought is a pure energy that vibrates at a very high rate by changing itself very quickly, just like the state of energy before the Big Bang, and the formation of that energy of the universe in the Human brain. In the vacuum of the universe before the Big Bang, it is understood that consciousness of the human being is formed in the human brain as well as consciousness of the universe". While less bosons come together more loosely in inanimate life, the number of bosons that can come together in human beings increases, as well as the bond between them. Astral scale, stars, galaxies, atomic scale, like atoms come together and cluster. Because the gravitational field is a boson field. I wonder if the intensification in the middle of the galaxies makes the galaxies' own consciousness? The answer to this question is not yet in the literature! It is the Higgs particles that impart mass to the vacuum; **Higgs bosons** covering the universe, the pre-Big Bang vacuum to the present day of the consciousness of our universe that provides the continuity of the consciousness of people.

The British scientist **Danah Zohar** described **consciousness** as the **quantum self** and used quantum physics in our struggle for peace with ourselves as well as with the world. quantum self according to quantum physics; There is no dichotomy between "**internal**" and "**external**" in people's relations with themselves, with other people, and with the world. Because the inner world of the mind (ideas, values, goodness, truth, beauty, virtue, etc.), the outer world of matter, each other gives birth to each other. The dialogue between **mind** and **matter** is the basis of the creativity of man and the universe. For the **quantum self**, neither individuality (fermionic character), nor relationship (bosonic character) comes first. Because both of them arise from the simultaneous and the same weight quantum substrate, namely **Bose-Einstein quantum density. The quantum self thus provides both the materialistic individualism of the West (materialism), the spiritual collectivism of the East (mysticism), and the birth and development of the human cultural world**. Because the physics of human consciousness and physics of the universe consciousness are the same. Every human being is composed of relationships with their own sub-identities (body and mind) and relations with the world. Like our ever-expanding universe, its population accumulation and cultural production is increasing in people around the world.

III) SUBATOMICPARTICLES AND THEIR VIRTUAL GEOMETRIES:

Big Bang Stars (BBS) in **Annex-IIa, b**: It should also be researched if there are "**weaves**" which are assumed to be spherical with a diameter of approximately 10^{-20} cm, in the high temperature and energy in the first 10^{-11} **second** of Big Bang, as in **Annex-Ie**. **BBS**s can also be a "**Quantum bubble**" with a dimension of 10 in the 10^{-35}**m** scale, for 10^{19} **GeV** energy in Planck time (10^{-45}**s**).

In all of the 3 tables in **Annex-IIe, g, h, i**, as indicated on each of them, single, dual and triple particles were explained.

Condition of each of the two Rhombic prisms in a quantum package, which is composed of **Graviton, Neutrino and Quarks** in the upper center of **Annex- IIa, b**, looks like the **Rhombic Prisms** in the **"Quantums of Universes"** in **Annex-Ia**. In the following atom model of **Louise de Broglie**, Rhombic Prisms can be thought virtually in the quantum packages around a nucleus like photon (light) quantums of electrons in a circular orbit. Of course, quantum waves of the electrons, which rotate in a very small area with a speed of **50000 km/sec,** would not appear as rough waves as in the following Broglie atom model, but as a cloud, in high frequency, nano wavelength and width, as in Heisenberg-Schrodinger-Dirac model. This structure can also help us solve Werner He isenberg's**Uncertaintyprinciple**.

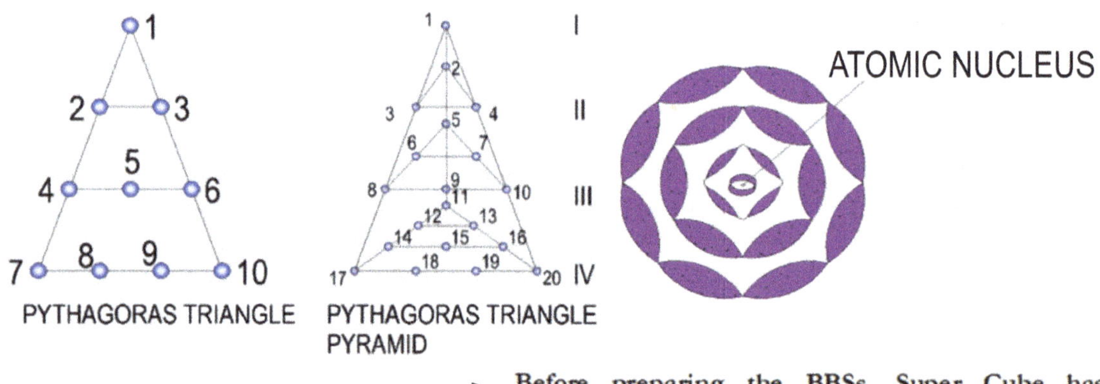

Supercube in ANNEX-II c
Star of David Prism in ANNEX II d

} Before preparing the BBSs, Super Cube has appeared in the 1st phase, Star of David Prism has appeared in the 2nd phase, and BBS has appeared in the last phase.

In Annex-IIg, in Big Bang's **Double Pythagoras Triangular Pyramids Table** on the triangular pyramid, the last particles of our universe before Big Bang can be seen on the upper half, and the first particles of our universe after Big Bang can be seen on the lower part, in Crystallographic Geometrical structure. The specialty of Pythagoras triangle above is that it is composed of 10 parts in two dimensions. 10 is a universal number. Because as can be seen on page1, using from $10^{-\infty}$ to $10^{+\infty}$, we can explain not only our universe, but also the previous and next universes, as well. In the **Pythagoras Triangular Pyramid** which is the three dimensional model of Pythagoras triangle, number of total particles in the Ist, IInd and IIIrd levels is again 10; and this number becomes 20 together with the ones in the IVth level.

In Annex-IIe, from up to down, the number of first particles of universe (10) is equal to the number of particles in the three-leveled triangular pyramid. Total number of particles, which is 20 together with the next 10 Neutrinos, is the total number of particles of four-leveled Pythagoras triangular pyramid.

The existence or absence of **Super partners** given in **Annex-IIa, f** could not be proved, yet. The result obtained from this table is the probability that Super partners, which were created before the first particles of Big Bang, joined the dark matter outside the universe and formed the first particles of dark matter.

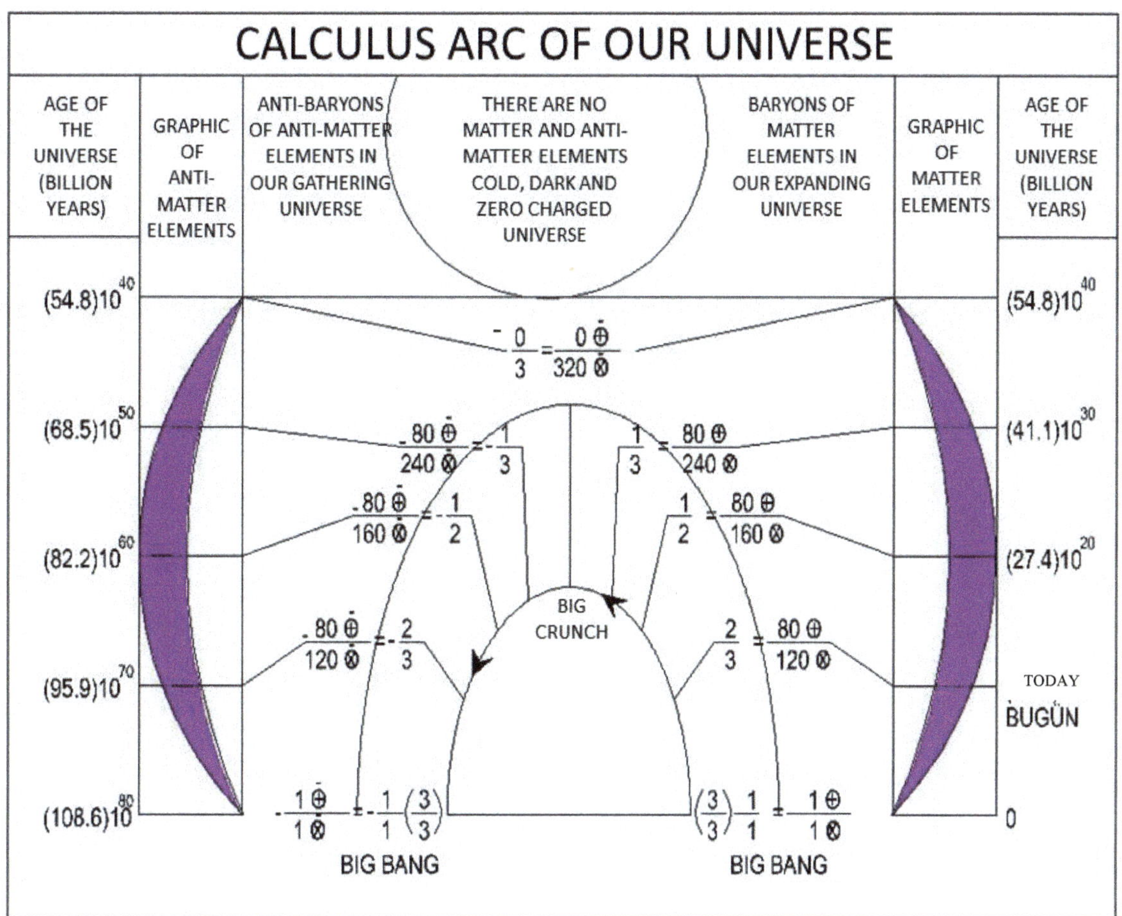

IV) ATOMS:

In **Annex-IIIa**, we can see the number of electrons in the electron orbits and sub-orbits of atoms of the elements.

In **Annex-IIId**, **Double Pythagoras Triangular Pyramids** are given as the code of universe from Sub-atom Particles to the elements. But above, we can see **the calculus arc of the elements in our universe between two Big Bangs**. In this table, the evolution of rates of Proton and Neutron numbers in the matter elements in our expanding universe, are given in billion years.

In **the tables in Annex-Ia, b, c, d**, separation of the development stages of **universes into +1, -1 and 3 denominator, and into 2 denominator simple fractions**, is a right decision, not only because of **3 denominator** electrical charges of Quarks, but also in accordance with the following table, as can be seen in **Annex-Ia, IIa, b, e**.

"**Average rates of number of Protons and Neutrons**" in the atom nuclei of elements present in the current phase of our universe, can be seen in the following table. Today, this average is **80 Protons/120 Neutrons** (or **2 Protons/3 Neutrons**) as can be seen on the left side of the table. As in the table on the previous page, our universe is still 10^{10} **years old, in other words 13,7 billion** years old. Today it is known that when it comes to 10^{30} **billion** years (approximately **41, 1billion**), protons will start to be disintegrated. When the universe comes to 10^{40} **billion** (approximately **54, 8 billion**) years, all of the Protons will be transformed to Neutrons and they will be gathered in dark matter around the universe. Since there will be no elements left in the universe, while the universe collapses by means of **Big-Crunch** because of the Mass Gravity force around the Dark matter, Anti Material Elements will start to be formed, and the universe will go into the gathering stage as an **Anti-Material Universe**, at the same time. It can be seen in **Annex-Ib, c, d** that when the universe comes to 10^{80} **billion** (approximately**109 billion**) years, it will explode again by means of the IIth Big Bang.

ELEMENTLER		PROTN SAYISI ⊕		NÖTRON SAYISI ⊗		TOPLAMI ⊕ + ⊗	⊕,⊗ ORANI
H	HYDROGEN	1	+	1	=	2	1/1
H	DEUTERIUM	1	+	1	=	2	1/1
He	HELIUM 4	2	+	2	=	4	1/1
Li	LITHIUM	3	+	3	=	6	1/1
B	BORON	5	+	5	=	10	1/1
C	CARBON	6	+	6	=	12	1/1
N	NITROGEN	7	+	7	=	14	1/1
EY	OXYGEN	8	+	8	=	16	1/1
Ne	NEON	10	+	10	=	20	1/1
Mg	MAGNESIUM	12	+	12	=	24	1/1
Si	SILICON	14	+	14	=	28	1/1
S	SULFUR	16	+	16	=	32	1/1
Ar	ARGON	18	+	18	=	36	1/1
CA	CALCIUM	20	+	20	=	40	1/1
Te	TELLURE 130	52	+	78	=	130	2/3
Nd	NEODIM 150	60	+	90	=	150	2/3
Gd	GADOLINIUM 160	64	+	96	=	160	2/3
Er	ERBIUM 170	68	+	102	=	170	2/3
HF	HAFNIUM 180	72	+	108	=	180	2/3
Os	OSMIUM 190	76	+	114	=	190	2/3
Pt	PLATINE 195	78	+	117	=	195	2/3
Hg	Mercury 200	80	+	120	=	200	2/3
Po	POLONIUM	84	+	126	=	210	2/3
Uuq	I UNUNQUADI	114	+	171	=	285	2/3

In Annex-IIIb, "Periodical Elements and isotopes according to the number of Protons and Neutrons" are given. Below, on the right side, you will see a small table showing the elements, which came into existence up to today with a Proton/Neutron ratio of 1/1, and **2/3**exactly. In **Annex-Ia, b, c, d** and on the top of previous page, **particles** and **Critical Density** ratios, phases of expanding and gathering universe were determined in accordance with: **-1/1, -2/3, -1/2, -1/3, 0, +1/3, +1/2, +2/3, +1/1**ratios in accordance with the **Dark matter**. In the table above, the accurateness of determining these values can clearly be seen in terms of elements.

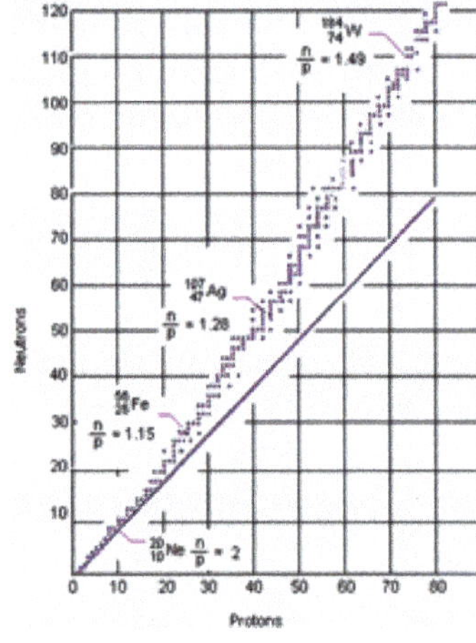

In **Annex-IIIc,** it is presented in the table "**120 Elements base to base Double Pythagoras Triangular Pyramid**" that the creation codes of Anti-Material Elements before Big Bang and Matter-Elements after Big Bang, are included in the **Pythagoras Triangular Pyramid**.

It is interesting that the maximum **120 neutrons** and the maximum **80 proton** ratios in the small diagram above left is seen in the **mercury 200** isotope in EK-IIIb as well. In addition, does it not show that we should actually use the three-dimensional table since the last element of today's two-dimensional element table seen in Annex-IIIa is **118**, while the last element of the three-dimensional triangular pyramid table in Annex-IIIc is **120**? If the elements **119** and **120** are found in the future, the answer to this question will be understood. This can indicate that the tables in Annex IIId and IIg are not wrong.

"**The Numerical Development Table of Pythagoras Triangular Pyramid**" is presented below. In this table, you can see the two dimension **Pythagoras Triangle**, in three-dimension **Pythagoras Triangular Pyramid** format. Mathematical relations are given between the number of points on every level from the top to down, and these numbers.

RABGİA PYRAMİDS TABLE:

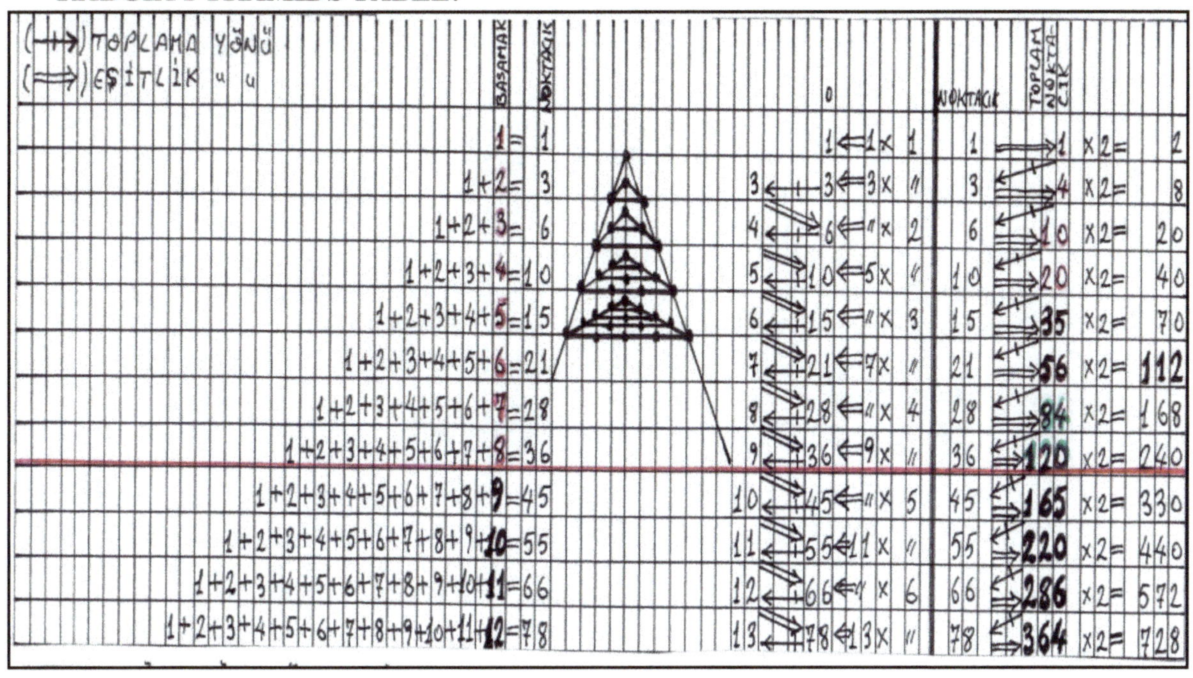

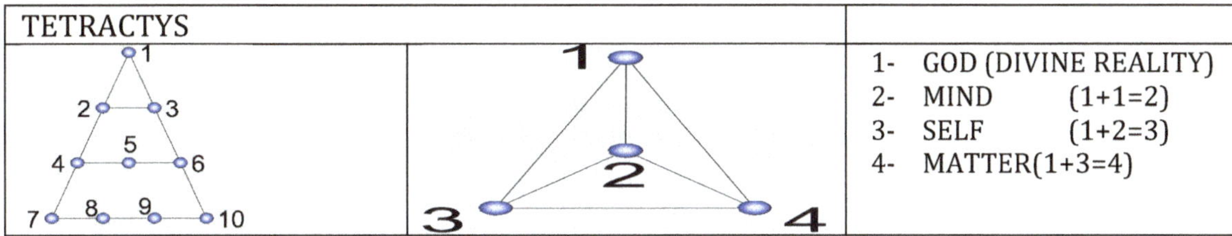

"TETRACTYS" in the above picture is a triangle composed of 10 numbers in four lines (**1+2+3+4=10**). Regarding as the steps of God and an expression of Cosmos, this **Esoteric Doctrine** which has passed from **Hermes Thot** to **Pythagoras**, to **Plato,** to "**İvan-üs safa**" inIslamic Philosophy, to Byzantine philosophy(**Palamism**), to **Kabbalah**, to **Knights Templar**, later impressed the **Modern Science**, **Numerology, Kepler**, **Galileo** and **Einstein**.

Also, four numbers in the tetrahedron represent the 10 numbers in

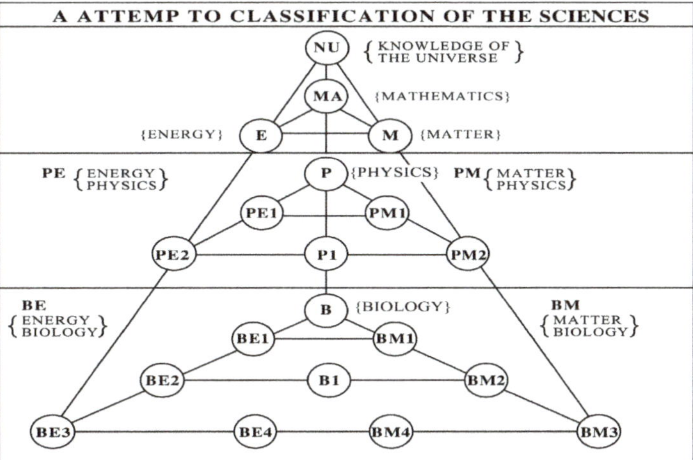

Tetractys. According to 'İhvan-üs Safa', the origin of every number is the numbers from 1 to 4. The other numbers consist of these 4 numbers (RABIA). In the 20th century, **Werner Heisenberg** indicated

in his Modern Quantum Theory that; "**Basic particles will finally take mathematical shapes, but they will possibly turn into a more complicated structure than Pythagoras has once imagined**".

According to the information in this book, the "Table of Classification of Sciences" is given above. "SOCIOLOGY" may be added under the table. In this table "UNIVERSAL KNOWLEDGE" essentially belongs to "ALLAH", who is the owner of both the Matter and Energy, and who covers everything. According to Sufism, that is why the "Excellent Human", having "Beautiful Morals"and the representative of the world, is immortal.

8 levels Pythagoras Triangle Pyramid in Annex-IIId and the foresights which were asserted by **Heisenberg** (P. 94) just for sub-atom particles, occurred in our universe which has expanded first, and which will be gathered later between two Big Bangs, and which has been created in **4 steps and which will be swept away in 4 steps**, (It is also useful to remember **RABIA** here) both for sub-atom particles and elements, in a very simple way despite **Heisenberg** has imagined. (See: **Annex-Ia, b, c, d and Annex-II a, b, e, g, Annex IIIc, d**)

V) MOLECULES, COMPOUNDS, CREATURES and HUMANS:

Molecules are groups of atoms in spherical form, which are constituted by same or different atom beads coming together. **Compounds** are constituted by the groups of molecules in spherical form. **Molecule is the smallest part of a compound, which has the same characteristics of itself**. The number of atoms in the molecules can be from 1 to 1 billion. There are spaces between the molecules of each compounds as big as between the galaxies in the space. Also molecules and atoms are continuously in motion with an increasing speed like the stars of a galaxy. We can **liken billions of known compounds to the galaxies**. As per **Annex-Ie**;

300 thousand years after Big Bang, the first **atoms,** in other words, **Hydrogen atoms** have come into existence.

1 billion years after Big Bang, the first **galaxies**, 9,2 billion years after Big Bang, our **Sun, planets and the World**, 10 - 7 billion years after the Big Bang algaeas (which are the first single celled living creatures) emanated by the connection of Water and Carbon (H_2O+C). But before the first living creature was created; **Organic compounds** like: **Carbohydrates, Fats, Proteins, Vitamins and Nucleic acids** have been formed in the waters in the East of Africa, together with the **Inorganic Compounds like: Minerals, Acids, Bases** and **Salts,** which would form the organelles of that living being. These compounds, then turned to the first single-celled creatures by means of the electrical shock of thunders and lightnings. Today, single celled creatures can be obtained by applying electric shock to these compounds in a container filled with water in the laboratories.

It is also interesting that the African tribe, which were the ancestors of **Homo Sapiens** including the humans in which we are also a member, migrated from this area in the East of Africa to Asia approximately **50-100000 years ago**. Today, the variety of living beings in the Red Sea, may be telling to us much more things which we do not know yet. Our universe which expands by means of Kinetic Motion Energy (Electromagnetic Force) and which takes its energy basically from +, -, 0electrical charges; started the life by means of the same electricity (lightning and thunder). **Single-celled creatures,** later **Multi-celled plants** and **animals** were varied first in the seas and then over land. And today, while the universe is **13, 7 billion** years old, we, the humans who are superior than the animals with the ability of thinking and talking, are living on the world approximately since **50-100000 years**. Today, we know **118 elements** in the universe which was formed by step by step after the first atom Hydrogen, and increasing of Protons and Neutrons in the nucleus of atom (**Annex-IIIa**) and the electrons which rotate around them (**Annex-IIIa**).

As in the evolution of atoms, the other creatures and humans are said to be formed by means of the continuous evolution of DNA in the nucleus of single-celled algae. Today in our body, we have the sub-atom particles from Big Bang, the creation of our universe; and atoms, elements, molecules and compounds which appeared later. For that reason, in the 21stCentury, maybe for the first time, universe makes us who it created itself, comprehend with not only the universe we live inside, but also the previous and future universes. Evolution from the first single-celled living being to the humans; reminds us naturally that there is a probability for more advanced humans to come into existence, after the Homo Sapiens having **100 Trillion cells, which includes us as members**.

Annex no	SCALE	
(NM)	EVOLUTION OF OUR UNIVERSE FROM BİG BANG TO THE HUMANS	
I a, b, II, a, b, f	$\sim 10^{-20}$	SUBATOM PARTICLES
I a, b, II, a, b, f	$\sim 10^{-20}$	SINGLE PARTICLES
I a, b, II, a, b, f	$\sim 10^{-20}$	- BOSONS
I a, b, II, a, b, f	$\sim 10^{-20}$	- GRAVITONS
I a, b, II, a, b, f	$\sim 10^{-20}$	- PHOTONS
I a, b, II, a, b, f	$\sim 10^{-20}$	- W+, W-, Z°
I a, b, II, a, b, f	$\sim 10^{-20}$	- LEPTONS
I a, b, II, a, b, f	$\sim 10^{-20}$	- QUARKS
IIc	$\sim 10^{-15}$	II- DUAL PARTICLES (MESONS)
IId	$\sim 10^{-15}$	III- TRIPLE PARTICLES (BARYONS)
IIIa, b, c, d	$\sim 10^{-10}$	ATOMS
Ie,f	$\sim 10^{-9}$	MOLECULES
Ie,f	$\sim 10^{-6}$	COMPOUNDS
Ie,f	$\sim 10^{-6}$	LIVING BEINGS AND HUMANS
Ie,f	$\sim 10^{-5}$	PROTISTANS
Ie,f	$\sim 10^{-5}$	- SINGLE-CELLED PLANTS
Ie,f	$\sim 10^{-5}$	- SINGLE-CELLED ANIMALS
Ie,f	$\sim 10^{-4-1}$	MULTI-CELLED ORGANISMS
Ie,f	$\sim 1^{-20}$	- MULTI-CELLED PLANTS
Ie,f	$\sim 1^{-3}$	- MULTI-CELLED ANIMALS
Ie,f	$\sim 1^{-2}$	- HUMANS

VI) DNA, ASTROLOGY AND QUANTUM UNIVERSE:

Today; we know that the Universe is an intact, indestructible neither creatable energy passing from state to state, that gained a conscious through Bose-Einstein quantum concentration in the moment of the Big Bang and interacts with each other, moves, changes and vibrates at different frequencies. We also know that there are 4 forces inside the atoms that make up this energy. **(P-27-29)**

We have previously explained that atoms and molecules, compounds in the **Atomic and Molecular scale,** look like the **stars and galaxies in the Astral scale.** Again, in the atoms in the open formulas of living being's DNA, which can be seen in the next page, it is obvious that there is no big difference from astral structures. In the nuclei of our cells, on the **Chromosomes,** which are as big as approximately the thousandth of a needle tip, there is a "**DNA spiral**" as long as a football field when opened as a stairs. On two sides of DNA chain:

While 1 Sugar (**S**) and 1 **Phosphate (P)** lie as a chain, **Nucleic acids,** in other words **Genes** are formed in the middle by 64 different 3-letter codes composed by 4 types of bases and the combination of **S** and **P**. 3 letter codes with each letter to represent the Base, can be liken to the **Words**; and **Genes** which are composed of **P, S** and **2 Bases (C-G or A-T)** can be liken to the **Sentences.** We can say that in **human DNA** there are **30000 Genes,** in other words sentences. Our DNA includes knowledge, which can fit into 1000 books with 300 page each. According to DNA in the nucleus of cells of humans; first of all **20 types amino acid** and **Proteins** which composes the most of our body after water by means of various sequence of these acids. Proteins which are produced as per the DNA of each human, constitute the **body structure, body temperature, thoughts, emotions, motions and also fate** to a certain extent of that human.

The first call of the child (Zygote) is formed by the combination of cell from the father and cell from the mother in the womb of mother. The reasons of possibility of astrological factors' in the formation of steps of DNA spiral in the nucleus of Zygote are as follows:

1st **REASON:** The graphic below is the flat and open view of **DNA, which** is in the spiral form, normally in the nucleus of living cell. In this example, chemicals constituting the DNA (**Nucleotides**) are shown in their open formulas together with accustomed classical symbols. Thus the atoms of DNA look like the stars in the sky. It was found out that in the space, the allocation of atoms of all nucleotides constituting the DNA**, was two-dimensional, in other words superficial. (like the Sun and planets or Graphite crystal)**This structure makes the nucleotides come one on the top of the other as leaves when DNA spiral is closed as an arc, to occupy less space. We know that the plenary structure, which the Sun forms with its planets, around the center of our galaxy and in the Milky way**,** rotates once in 250000 years, indiscreetly, continuously in spiral movements, without ruining this plenary structure. We know that Sun, Moon, Earth, planets and the stars of 12 signs, which compose the Zodiac zone, are in infinite interaction with each other, because of their elementary structures, by means of **Electric, Magnetic and Mass Gravity.**

2nd **REASON:** While our universe was in "Absolute energy" condition (Ancient divine light) before Big Bang, this energy first created the **Radiation Universe** which was another dimension of itself, later it created the **Material Universe,** and then created the **Creatures**, and at last, created the living being which could understand all this evolution and beyond, in other words the **humans.** Thus this "**absolute energy state**" before Big Bang, created living beings, which could comprehend with their own existence after **13,7 billion years**. As a result of modern sub-atom researches; it is proved that matter is in relation with each other in a universal integration; and knottily as a combined single network; and it is defined as the **"Hairy Ball Theorem"**. In this case, **modern science** came to the same conclusion with **Islamic mysticism, Tao, Brahma** and **Dharmakaya**. In the universe, everything is a distinct part in a different dimension of universe. Until everything is transformed in to energy again since the Big Bang and before the next Big Bang, i. e. **NUR-U KADİM**, that is, the Energy of the Universe continuously **create, sustain and end everything on a different scale of time and space as a whole in a different vibration and energy dimension in accordance to a mathematical, musical and divine program.**

3rd **REASON:** There are **Electrical Energy fields and Magnetic fields**, which are associated with health and life in the Far East and Asia and used in Alternative Medicine; which have the 7 colors of Sun around the body of humans; which is composed of 7 layers; which protect the humans and which are defined as **Aura**. Besides, in specific places from top to toe in the human body, with **7 energy centers**, it other words, with **7 Chakras** and **Aura** in the environment, radiate energy in various vibration frequencies and wavelength as a whole, and receive energy from the environment. Showing difference in each of the humans this bioenergy existence, which is transferred in humans by means of birth, **is in continuous rotation towards right for the men and towards left for the women**. It is the same as the electrons rotating in pairs in opposite directions around the atoms. Existing in the brain, in the center of this bioenergy system, which forms the intangible existence of humans, thought, is a pure type of energy which changes itself very quickly and which vibrates in very high frequency. In other words, it is the formation of **Universal Energy in the brain of humans.** It was already known that the thoughts produced by the brain, were showing effects similar to the electric current. **In the 20th century, Russian Semyon-Valentina Kirlian** succeeded in shooting the color photos of electric fields in the humans and other creatures. Thus this knowledge from Pythagoras up to today, was shot by means of modern technology. So, both Universe's, in other words, **Cosmos'**; and Humans' (as **Micro Cosmos)** constituting an inseparable whole as **Matter** and **Energy**; shows that astrology should also be taken seriously.

64 AMINO-ACIDS AND THEIR ABBREVIATED SYMBOLS

	U	C	A	G	
U	UUU (PHENYLALA-NINE)	UCU (SERINE)	UAU (TYROSINE)	UGU (CYSTEINE)	**U**
	UUC (PHENYLALA-NINE)	UCC (SERINE)	UAC (TYROSINE)	UGC (CYSTEINE)	**C**
	UUA (LEUCINE)	UCA (SERINE)	UAA (STOP)	UGA (STOP)	**A**
	UUG (LEUCINE)	UCG (SERINE)	UAG (STOP)	UGG (TRYPTOPHAN)	**G**
C	CUU (LEUCINE)	CCU (PROLINE)	CAU (HISTIDINE)	CGU (ARGININE)	**U**
	CUC (LEUCINE)	CCC (PROLINE)	CAC (HISTIDINE)	CGC (ARGININE)	**C**
	CUA (LEUCINE)	CCA (PROLINE)	CAA (GLUTAMINE)	CGA (ARGININE)	**A**
	CUG (LEUCINE)	CCG (PROLINE)	CAG (GLUTAMINE)	CGG (ARGININE)	**G**
A	AUU (ISOLEUCINE)	ACU (THREONINE)	AAU (ASPARAGINE)	AGU (SERINE)	**U**
	AUC (ISOLEUCINE)	ACC (THREONINE)	AAC (ASPARAGINE)	AGC (SERINE)	**C**
	AUA (ISOLEUCINE)	ACA (THREONINE)	AAA (LYSINE)	AGA (ARGININE)	**A**
	AUG (METHIONINE)	ACG (THREONINE)	AAG (LYSINE)	AGG (ARGININE)	**G**
G	GUU (VALINE)	GCU (ALANINE)	GAU (ASPARTIC ACID)	GGU (GLYCINE)	**U**
	GUC (VALINE)	GCC (ALANINE)	GAC (ASPARTIC ACID)	GGC (GLYCINE)	**C**
	GUA (VALINE)	GCA (ALANINE)	GAA (GLUTAMIC ACID)	GGA (GLYCINE)	**A**
	GUG (VALINE)	GCG (ALANINE)	GAG (GLUTAMIC ACID)	GGG (GLYCINE)	**G**

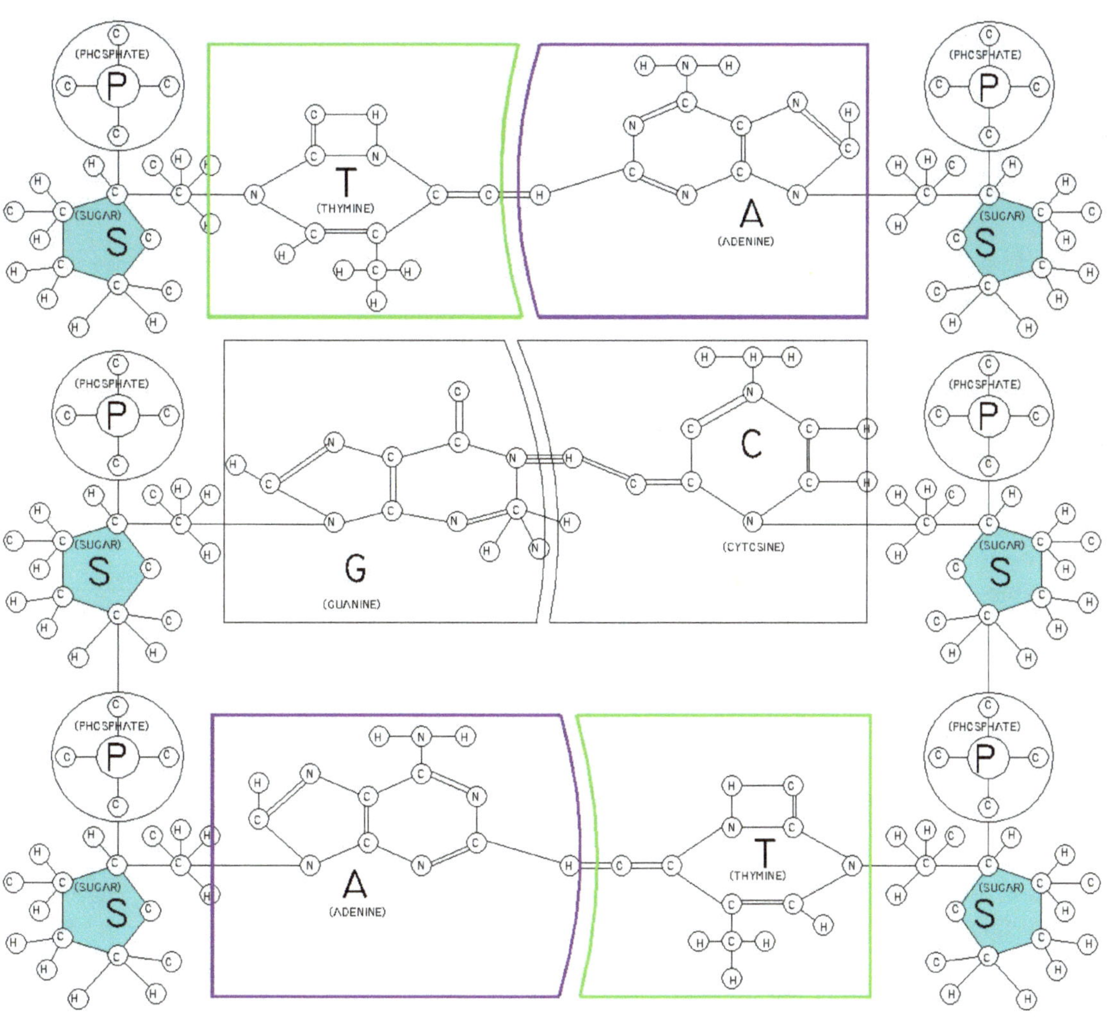

VII) TABLE FOR THE   CLASSIFICATION OF SCIENCES

This table and structuring of community are revealed by means of "**Knowledge**" (intangible) on the right palm of **Mawlawi Dervish**, and "**Matter**" under the left palm, (tangible) duality basis. Also may be for the first time, the classification of sciences took a divine and ideal form in this **universal and communal duality** scheme. There are the Management institutions of **State** and humans on the top center of the table; there are Knowledge institutions and humans **(Knowers)** on the right of state; there are Producer companies and humans **(Makers)** (producers) on the left of state. According to this table, the obligation of state is making the Knowers transfer all they know to the Makers in every respect. State is obliged to spend the tax collected from **Producers** (moneymakers), in other words **Money**, basically for the **Knowers**, in other words, educational institutions and humans, together with many services to be performed for the community; and also for making the intellectual humans, which these institutions have educated, as Producers. Dynamic expression of the system: while the **Money** goes from **Makers** to the state as tax payment, a sufficient amount of this money is being transferred to the **Knowers**. Thus, raised human is joining to the **Makers**, in other words, becoming a Producer is obtained by means of the modern knowledge of each occupation. Thus it can be possible to achieve the ideal of continuously increasing **Quality, Production, Manpower, National Income,** in other words, increasing **Money** (capital).

D) PRELIMINARY MATHEMATICAL INFORMATION PREPARING THE THEORY OF EVERYTHING:

I) SOME DEFINITIONS OF AND INTERPRETATIONS ON MATHEMATICS:

* Mathematics is the systematic of universe.

* Mathematics is a tool, which will be able to lead us to the perfection, beauty and facts of universe as a result of the unchangeable basic theories named 'axiom'. The subject of Matematics is the unchangeable facts of universe.

* Being the language of universe, **Mathematics is an axiom system in which the symbols are numbers and figures (geometry) and letters (nomenclature); and the formulas are words and sentences** (a collection of organs). Mathematics has the secrets of unknowns in itself, in the universe and in all sciences.

* **Mathematics is** the basics of all sciences. It binds the sciences together. It constitutes the systematic of sciences.

* **Platonism:** It is the idea that Mathematics is a study on an independent being apart from humans and cultures; and a study on the absolute facts related with an eternal reality. Today the mind of humans neither could explain the Mathematics perfectly, nor the "**Physical Universe**" and "**Mathematical Universe**", or the relation between them.

* **Plato** accepts that the numbers give specific keys to solve the secrets of nature. Like Plato, many philosophers after him indicated that mathematics have a universal characteristic beyond the cultures, and that the subject of **Mathematics** is the **unchangeable facts of universe**.

UNIVERSAL VALUES - IN METERS	
$10^{-googol}$	-GOOGOLPLEX
10^{-100}	-GOOGOL
10^{-30}	
10^{-29}	
10^{-28}	
10^{-26}	
10^{-24}	
10^{-23}	
10^{-22}	
10^{-21}	
10^{-20}	SUBATOM PARTICLES
10^{-19}	
10^{-18}	
10^{-17}	
10^{-16}	
10^{-15}	
10^{-14}	DIAMETER OF ATOM NUCLEUS
10^{-13}	
10^{-12}	
10^{-11}	
10^{-10}	DIAMETER OF ATOM
10^{-9}	
10^{-8}	ANGSTROM (Å) MOLECULES
10^{-7}	NANOMETER
10^{-6}	COMPOUNDS
10^{-5}	SINGLE-CELLED CREATURES
10^{-4}	MICRON (μ)
10^{-3}	
10^{-2}	
10^{-1}	
1	DEVELOPED LIVING BEINGS
10^{1}	
10^{2}	
10^{3}	THOUSAND
10^{4}	
10^{5}	
10^{6}	SATELLITES OF PLANETS
10^{7}	PLANETS
10^{8}	
10^{9}	BILLION
10^{10}	
10^{11}	
10^{12}	STARS
10^{13}	
10^{14}	
10^{15}	QUADRILLION
10^{16}	
10^{17}	GALAXIES
10^{18}	QUINTILLION
10^{19}	
10^{20}	OUR UNIVERSE
10^{21}	SEXTILLION
10^{22}	
10^{23}	
10^{24}	SEPTILION
10^{25}	
10^{26}	
10^{27}	OCTILLION
10^{28}	
10^{29}	
10^{30}	NONILLION

* **Mathematics:** It is the branch of science, which tries to solve the problems in nature by means of numbers and figures. As "*Mensuration*", it is used as a tool in the other sciences.

* **Mathematics:** It is the secret of the beauty and facts of the universe. **Facts** can be reached as a result of exact series of arguments. The specialty of Mathematics rules is that they are indisputable.

* **MATHEMATICS IS DIVIDED INTO TWO:**

1- Pure Mathematics: Ideas are the world of thoughts and they are intangible.

2- Applied Mathematics: It is about the real world and tangible.

* **Mathematical World** lives in the head; and "**Real World**" lives out of it. The intellectual area, which is determined by the use of mathematics to understand the nature and to study on physical subjects, is called as '**Applied Mathematics**'.

* The development of **Mathematics** starts from intangible basics, and comes to a theoretical level by means of intangible generalizations. Later, to verify the obtained theory, we turn to the tangible problems back again. In short, going from tangible to intangible and coming back from intangible to tangible are the essences of Mathematical thought. As the "**feed-back**" in cybernetics (continuous knowledge exchange).

* **Galileo:** "The Universe is like a book written in mathematical language, and can be comprehended only by those, who know that language. Its letters are triangles, circles and other geometrical figures."

* **Universe** has a divine accord, harmony and balance which are written using mathematics and which could be perceived thinking **mathematics**.

* **Mathematics is** a method for looking for an order from conditions, which seem complicated. So, mathematics will be the only mean to bring an order, a solution to the chaos between the communities and sciences in today's world.

II) BASIC DEVELOPMENTS IN THE HISTORY OF MATEMATICS:

At the beginning, Mathematics has started with the numbers up to 9 in the "**Mohenjo-Daro and Harappa**" cities in the Indus Valley, Yenisei and Orkhon Rivers, which Sumerians have come from. This 9 numbers have been related with the 9 levels of sky in which there was the once-believed Goodness God. The most important development after these **9 numbers, is the use of zero**. It is assumed that zero has been first used by Sumerians in the astronomy calculations. After **700-500 B.C.,** we know that in Mesopotamia a special sign was used to represent zero in the astronomy texts. In **632 A.D.,** in his book named "**Brāhmas phuta siddhānta**", Hindu Mathematician and Astronom **Brahmagupta** explained how 0 would be written together with the other 9 numbers, for the first time. This book was translated in the year **773 A.D.,** in Baghdad, the palace of Caliph Al Mansur with the name of "Big Sinhind in the sight of Astronomers". In the year **830 A.D., Muhammad ibn Mūsā al-Khwārizmī** explained who to use the numbers 9 and 0 in the arithmetical operations in his book named "**Kitāb al-jabr wa-l-muqābala**". By **XIIth century,** after the books of **al-Khwārizmī** have passed to Europe via Al-Andalus, Europeans started to use the zero calculation system instead of Roman numerals. Muhammad ibn Mūsā **al-Khwārizmī** presented **algebra**, which is at least as important as zero, to the humanity for the first time. Why is Algebra important? Because the universe is created based on **symmetry. Al-Khwārizmī** expressed that in numbers as **equation (equality= balance)**. Algebra thus ended up by bringing the possibility to find the unknowns from the knowns on both sides of the equalities. The third important development in mathematics is the knowledge **Pythagoras** has learned from Egyptian Temple priests in the **VIth Century B.C.** Pythagoras called the doctrine as "**matamatas**". It has the meaning that: "**containing all of the information that humans have**". Today, we can call it as "**Theory of Everything**". According to Pythagoras, "All things are numbers". According to him, "**all other numbers can be derived through the mathematical operations such as adding, multiplying and dividing the numbers from 1 to 4. Universe is a harmony of numbers and the creation principle is the opposition. Reality happens by the reconciliation of things, which are opposite to each other. 1 is the first being, it is a point. 2 is a line, 3 is a triangle and 4 is a tetrahedron.** A tetrahedron represents the harmony and virtue; and beauty comes with the harmony. The purpose of life is **cleaning and purifying the soul by means of knowledge.**"

In the 10th century, in the Islamic World, the ideas of Platonists, Pythagorists and Hermeticists (the first Greek who has learned all of the information from the Egyptian temple priests, before Pythagoras) on 'Purity brotherhood' (**İhvan-üs-safa**) in Basra: Starting from number '1', "**Science of**

Numbers" as the origin of all of the sciences and as a supernatural science, led the people to the idea of the unity of God.

In the Middle Age Europe, the belief that there was a numerical harmony between **universe-world and humans** was stated as "Harmonia Mundi" (**Musical Harmony**).

In XVII. Century, in Europe, first **R. Descartes** brought the Algebra in the Geometry. Thanks to Figures, **Pure Geometry, Physical Geometry, "Cartesian Geometry coordinates"** and "**Analytical Geometry**", geometrywas, in Descartes's words, **reduced to algebra. Also Algebra was visualized by means of Geometry. As per Descartes' analytical geometry**, there are one-to-one pairings between the positive (+) points to the right, negative (-) points to the left on a beam, and real, in other words, inordinate numbers (+, -, $\sqrt{2}$). While the geometrical expression of this structure was approved, now **Geometry and Arithmetics were combined as** "mathematical analysis" (**Annex-Ib**). There are one-to-one pairings between the Cartesian coordinates, rowed real measures and points in Beam, Plane and Space. "**Points**" or "**Point clusters**", thus, can be expressed in terms of algebra.

In the XVIIIth century, Newton and Leibniz suggested **the calculus** (Advanced Mathematics). Thus "**algebra**" of al-Khwārizmī, in other words, **numeric expression of equations** united with "**symmetry**". The quantities varying up to many variables, in other words, "**functions**" could then be able to be calculated in **calculus equations**. Also counting with the natural numbers like **1, 2, 3,..** took the mathematicians to the "**granular plurality**" notion and a "**continuity**" notion was reached. Besides, a "**science of studying the properties of shapes of figures which a continuous formation cannot change**" was developed by the up-down, right-left, rear-front connection without using number figures, in other words, "Topology" which Leibniz called as "**Analysis Situs**". **Algebra** is one of the most important universal notions like **symmetry and equality** in **topology**. This definition of topology matches the 'unchanging beings in a changing environment', in other words, "**essence**" definition of "**monat**" and "**essence**" notions in which **Leibniz** defines the God as 'the most perfect being'. Also in Europe, researches started on "**probability theory**" (Calculation of probabilities) based on duality principle like **Yes and No,** and **Futurology** which is the calculation of future as per the data of history. Knowledge input to the living and non-living beings from the environment-**adjustment (control)– performing the necessary actions (knowledge output)** phenomenon was first applied in **Cybernetics and then** in the **computers**. Information exchange in the computers; **positive (+) means yes**, **negative (-) means no**. For the creatures, nervous system can survive thanks to the (+) and (–) current exchange in the cells.

Let's come to the last point, to the theory, which Einstein has started but could not prove in 30 years:

The purpose of "**Combined Field Theory**" and **Quantum mass gravity theory**; was to prove that in 'one in 10^{-43} zeros' of the first seconds of Big Bang, in other words, in **Planck time (Annex-Ia, b, c, d, e)**, when the energy of universe becomes10^{19} **Gev**, 4 forces inside an atom would become a single force as per the geometrical properties (why should this geometry not be the quantum geometry of **Annex-Ia**?) of 4 dimension **space-time** notion. Since **Time** also depends on the distance, it was assumed as the **4th dimension**. In other words, Einstein tried to combine the **Mass gravity fields and Electro/Magnetic fields in a single geometrical structure,** but he could not manage to do it for 30 years. (The two fields have been combined at least visually in Annex Ia). To manage this, when the **Standard Quantum methods** are applied in Einstein's theory, today it is being assumed that it would be because of the exchange of quantums of mass gravity forces, named "**Graviton**". If this theory could succeed, it would be possible to explain the movements of atoms and sub-atom particles in accordance with the Laws of **Quantum Mechanics**. (Our Annex - Ia and other tables may perhaps help to achieve this goal.)

According to the **Quantum Mass Gravity Theory**: At least a 10 dimension space was assumed which could be in only Planck-scale (**10^{-45} m**) representing the sub-atomic particles or all combinations of their interactions. (**Annex-I a, b, c, d, e**) According to Quantum Mechanics, **all of the energy units** are composed of quantized tangible quanta particles. In the tables in the annex, the previous and next universes were given as quanta half (+) and half - loaded. In the **Quanta Electro Dynamics**, the interaction of electric charged (**+, -**) particles is originated from the exchange of quantums of electromagnetic field, which are called as "**Photons**". The most important theory of modern physics in the macro universe is the determinism that occurs according to the theory of relativity, for certain reasons, certain results can be achieved under certain conditions son with the quantum theory in the last stage, albeit on a micro-universe scale led us to dangerous indeterminism. Thus, in the past, something was either right or wrong, and while it could not be both right andwrong, it became a possibility on the quantum scale, right and wrong, with quantum theory. This development is as dangerous as the skeptics who doubt our minds in the history of philosophy. **Page 174** will explain how to return to indeterminism.

III) 6 BASIC NUMBERS IN THE CREATION OF THE UNIVERSE ACCORDING TO THE ENGLISH ASTRONOMER MARTIN REES,:

After **MARTIN REES** pointed to the splendid plan and design of universe and life, let us try to put these 6 numbers which penetrate into the smallest and biggest parts of universe, and which seem to be separate from each other, into place in out table and especially in the "**Calculus Equations of Universes**" tables which it indicates as the "**master plan of the universe**".

1st **Number: D=3:** This number represents the three dimension of universe. It it was 2 or 4 dimensioned, the universe wouldn't exist.

2nd **Number: E=7/1000:** This number is the energy strength of the force which keeps the atom nucleus together. The nucleus of Helium atom constitutes 99,32% of the total weight of **2 Protons, 2 Neutrons** in it. The remaining **0,68%** comes out as heat. Thus when Hydrogen gas in the Sun is converted to **Helium**, 0.007 of its mass is converted to energy. In case this number was **0.006,** Proton would not be connected to Neutron and Universe would only consist Hydrogen and there would be no life. In case this number was**0.008**, fusion would be very fast, there would be no Hydrogen left from Big Bang to-day, Solar System and life would extinct so fast. This number can be associated with **Annex-Ia, b, c, d**.

3rd **Number: N=10^{36}:** This number is the ratio of the force holding **atoms together to the Gravity Force** between the **Atoms**. In other words, it shows that there is a gravity between the atoms, which is stronger than the Gravitation. In case this number was lower, Universe would be more ephemeral. This number can also be associated with **Annex-Ia, b, c, d**. Besides, is it about **Gluons** and **Quark-Antiquarks**, or is it about "**Matter-Anti Material Force**" in other words the 5th force within the **Atom**? This can be researched.

4th **Number: Ω=Real Density/Critical Density**: It is the visible and invisible matter density of universe. In case this matter density was higher, the universe would not enlarge anymore and universe would start to collapse in itself. If this number was lower, galaxies and stars would not be created anymore and universe would be different. The evolution of this number can also be clearly seen in **Annex-Ia**.

5th **Number** (Λ): Being explored in 1998, this number is the **strength of anti-gravity force**, which controls the enlargement of universe. Since this number is very small, it does not affect the structures, which are smaller than 10^{9}**lightyears**. If this number were higher, stars and planets would not exist. Can't this number be the value of 10^{-40} **MeV**, which is the astral scale equivalent of the electromagnetic force, which is the atomic scale minimum (**1 Mev**) in our enlarging universe which can be seen on the upper right side of **Annex-Ib**.

6th **Number: Q=1/1000:** It is the **Complex Irregularity** or in other words, Amplitude of Waving which makes the planets and galaxies come into existence in our enlarging universe. In case this number was higher, the big matter clusters in the space would become giant black holes and absorb the Sun and stars. As in the 5^{th}number, can't it be the electromagnetic force on the upper right of **Annex-Ib**.

It is obvious that these tables need to be developed after putting these 6 numbers of Martin Rees in the tables of **Annex-Ia, b, c, d**. Also it should be investigated for what could be the equivalent of **1/1000**, which is the value of **Q** in the astral scale.

Martin Rees also states that there is a need in the universe for a theory that combines very large and very small. If we call it the theory of everything on **page 168,** "The theory of all things Rabia according to numbers and tetrax of one to four" section tries to explain how to reach a definite solution according to contemporary data in this book and today.

IV) TABLE OF MATHEMATICS (Essay of Dichotomy in Mathematics):

| | INTANGIBLE UNIVERSE | | | | TANGIBLE UNIVERSE | | |
| | INTANGIBLE (PURE) MATHEMATICS (THOUGHTS, IDEAS) | | | | TANGIBLE (APPLIED) MATHEMATICS | | |
	INTANGIBLE ARITHMETICS		INTANGIBLE GEOMETRY		TANGIBLE GEOMETRY		TANGIBLE ARITHMETICS (SCIENCES)
A	NUMBERS	A	a-PURE G. B-ANALYTIC G. c-MENTAL G. (1-NONRATIONAL	A	PHYSICAL GEOMETRY	A	SUB-ATOM PARTICLES, ATOMS, MOLECULES, COMPOUNDS,
a	NATURAL N. b-INTEGERS c-FRACTIONAL N. d-RATIONAL N.						
e	IRRATIONAL N. f-REAL NUMBERS g-IMAGINARY NUMBERS		2-A PRIORI 3-HYPOTHETIC				PLANTS, ANIMALS, HUMANS
h	COMPLEX NUMBERS		4-INDUCTIVE 5-DESIGN G.				
B	UNIVERSAL NUMBERS (π, FIBONACCI, e, etc.)	B	NONDIMENSIONAL FIGURE (POINT)	B	SYNTHETIC GEOMETRY	B	PHYSICS, CHEMISTRY
			1-POINT REFLECTION SYMMETRY				
			2-PROJECTION OF A POINT				
C	ALGEBRAIC EQUATIONS	C	ONE-DIMENSIONAL FIGURES	C	EXPERIMENTAL GEOMETRY	C	ASTRONOMY
a	ONE UNKNOWN FIRST-ORDER EQUATIONS	a	LINES		(A POSTERIORI EMPIRICAL)		
b	ONE UNKNOWN SECOND-ORDER EQUATIONS		1-BILATERAL SYMMETRY				
c	SECOND-ORDER TRINOMIAL EQUATIONS		2-PARALLEL LINES				
d	SECOND-ORDER PARAMETRIC EQUATIONS		3-PERPENDICULAR LINES				
e	LINEAR EQUATIONS		4-HARMONIC PARTITION				
f	POLYNOM	B	CURVES				
g	RATIO, PROPORTION, PERCENTAGE		1-ASYMPTOTES OF CURVES				
		C	ANGLES				
			1-PARALLEL EDGE ANGLES				
D	SETS, SENTENCES	D	TWO DIMENSIONAL FIGURES	D	EXACT GEOMETRY	D	GEOLOGY, GEOGRAPHY
a	ARRAYS (1-ARITHMETIC ARRAYS 2-GEOMETRIC ARRAYS)		(PLANES)				
b	MATRIX AND DETERMINANT		1-NORMAL PLANES				
		a	2-PROJECTION OF A PLANE TRIANGLES				
		b	TETRAGONS				
		c	POLYGONS				
		d	CIRCLE				
E	INEQUALITIES	E	THREE-DIMENSIONAL FIGURES	E	DEDUCTIVE GEOMETRY	E	BIOLOGY, AGRICULTURE, BOTANY
a	FIRST-ORDER INEQUALITIES AND THEIR TABLES		1-TOPOLOGY 2-LENGTH				
b	SECOND-ORDER INEQUALITIES AND THEIR TABLES		3-DEPTH (HEIGHT) 4-WIDTH				
F	RELATION	a	PYRAMID	F	LOGICAL GEOMETRY	F	ZOOLOGY, ANIMAL BREEDING
			1-REGULAR POLYHEDRONS				
G	FUNCTIONS AND CONTINUITY		2-REGULAR TETRAHEDRON	G	AXIOMATIC GEOMETRY	G	MEDICINE, SPORT, PSYCHOLOGY
a	LOGARITHMIC FUNCTIONS		3-TRUNCATED PYRAMID				
b	EXPONENTIAL FUNCTIONS	B	PRISM				
c	TRIGONOMETRIC FUNCTIONS	C	CONE				
	1-PLANE TRIGONOMETRY		1-TRUNCATED CONE 2-ELLIPSE				
	2-SPHERICAL TRIGONOMETRY	d	3-PARABOLA 4-HYPERBOLA CYLINDER				
		e	SPHERE				
H	CALCULUS EQUATION	F	CALCULUS GEOMETRY	H	SOLID GEOMETRY	H	HISTORY, SOCIOLOGY, RELIGION, LANGUAGE, LITERATURE
H	CAL	a	LIMIT GEOMETRY				ART, SCULPTURE, PHOTOGRAPHY, ARCHITECTURE,
B	DIFFERENTIAL EQUATION	b	DIFFERENTIAL GEOMETRY				STATISTICS, ECONOMY, COMMERCIAL LAW
C	INTEGRAL AND DERIVATIVE	c	INTEGRAL, DERIVATIVE GEOMETRY				EDUCATION, TRAINING, PUBLIC ADMINISTRATION
I	THEORY OF PROBABILITY			I	NUCLEAR GEOMETRY	I	INFORMATION SCIENCES
i	ALGORITM, CYBERNETICS, COMPUTER						ENCYCLOPEDIAS, ESSAYS, PERIODICS,
J	MODERN COMPUTER THEORY						PRINTED AND VISUAL MEDIA

The main Mathematics problem of our age is if the Mathematics is **tangible** or **intangible**. The answer to this question is that Mathematics is both tangible and intangible as per the **matter** (tangible) and

energy (intangible) duality of everything in the universe. Hence, I tried to give a general unprofessional sketch of Mathematics in the following table:

MATHEMATICAL CODES:	
1) SYMMETRY IN NUMBERS (EQUATIONS)	
1) $-\pi/-\sqrt{2}../-1,-2/3, -1/2, -1/3 \mid 0 \mid 1/3,1/2,2/3,1/\sqrt{2}, \pi$ (NUMBERS IN Annex IB)	THEORIES OF PYTHAGORAS, KEPLER, NEWTON, EINSTEIN AND THE OTHERS
2) FIBONACCI NUMBERS	
3) LOGARITHM	
4) MUSICAL NOTES (HARMONY) AND MELODY (RATIOS OF FREQUENCIES AND WAVE LENGTHS)	
11) SYMMETRY IN GEOMETRY (BALANCE)	
1) PLATO OBJECTS	
2) GOLDEN RATIOS	
3) TRIGONOMETRY	
4) PI (3, 141…)	
5) PYTHAGORAS TRIANGLE AND PYRAMID	
IN OUR MATTER UNIVERSE:	
I) IN THE ASTRAL SCALE:	
1) SPHERES' (PLANETS, SATELLITES, STARS AND GALAXIES)	- DIAMETER - CIRCUMFERENCE - MASS - SPEED (ROTATING AROUND ITSELF)
2) THEIR ORBITS'	- DIAMETER - CIRCUMFERENCE - SPEED (ON THE ORBIT) - NUMBER OF PLANETS
II) IN THE STRUCTURE OF LIVING BEINGS	
1) IN THE PLANTS	- LEAVES, FLOWERS, ETC.
2) IN THE ANIMALS	- HEAD, ARMS, LEGS, ETC.
3) IN THE HUMANS	- HEAD, ARMS, LEGS, ETC.
III) IN THE ATOMIC SCALE:	
1) SPHERES' (SUBATOM PARTICLES, ATOMS, MOLECULES)	- SPINS - MASS - SPEED (TURNING AROUND ITSELF) - ENERGIES - CHARGE
2) THEIR ORBITS'	- DIAMETER - CIRCUMFERENCE - SPEED (ON THE ORBIT) - NUMBER OF ELECTRONS
IV) SYMMETRY IN THE CRYSTALS (IN 3 DIMENSIONS)	- NUMBER OF AXES - LENGTH OF AXES - ANGLE OF AXES

V) MATHEMATICAL CODES AND OUR MATERIAL UNIVERSE:

As indicated in the table above, although the relations between the above Mathematics and below Astral and Atomic Materials, creatures and crystals, have been tried to be illuminated throughout the history and today by the scientists; a universal explanation for the last words could not yet be made.

We can define the above Mathematics, which is listed below, as the code of **divine symphony of universe. Accord, rhythm, harmony, balance** in all of the notions in the following table of above Mathematical Codes, creates the divine symphony in our universe in the above table of our below Matter Universe. Hoping that it would be exemplary in our World in which we live the rumble of inconsistencies and imbalance, let us give some more information about the **divine symphony of universe**:

1) MATHEMATICS IN THE ORBITS OF PLANET:

In 1778, Johann Elert Bode obtained the numbers 4, 7, **10**, 16, 28, 52, 100, 196, 388 when he added 4 on the series of numbers (0), 3, 6, 12, 24, 48, 96, 192, 384, based on the observation of **Johann Daniel Titius in 1766**. In this series, while the World corresponded to **10**units, the other numbers appeared, in the same sequence, as the radius of planet orbits, excluding Neptune.

2) GEOMETRIES ON THE ORBITS OF MOTIONS BETWEEN THE PLANETS:

- **Two nested pentagons in the formation of planets:** In1948, **Carl Von Weizsäcker** explained the theory that planets were formed by condensation of a particle cloud, as nested two pentagons, in the scheme above.

- **Besides, nested two pentagons appear:**
  b1) In the orbit thickness of **Mercury.**
  b2) In the space between **Mercury and Venus,**
  b3) In the average orbits of **World and Mars,**
  b4) In the space between **Mars and Ceres.**
- **Nested 3 pentagons:**
  c1) In the space between **Venus and Mars,**
  c2) **In the average orbits of Ceres and Jupiter**

In our astral universe of diameter 10-20 billion light years, we know that, there are billions of galaxies and billions of stars in these galaxies, and much more planets and satellites; and that there are 10^{80} particles and 10^{79}Moleculesin the atomic structure of our universe; and that 2 million species are living on the world today; and that 90% of the previous living beings no longer exist now. As we can understand from the above examples, that the existence of our **matter universe is up to the Mathematical Codes which we have indicated in the beginning of this chapter. We can summarize the Mathematics codes** as the Spheres in the geometry and the Plato objects in the Spheres. Now, let's start with the **Big Bang Star (BBS),** which should be included in the Plato objects, in our opinion.

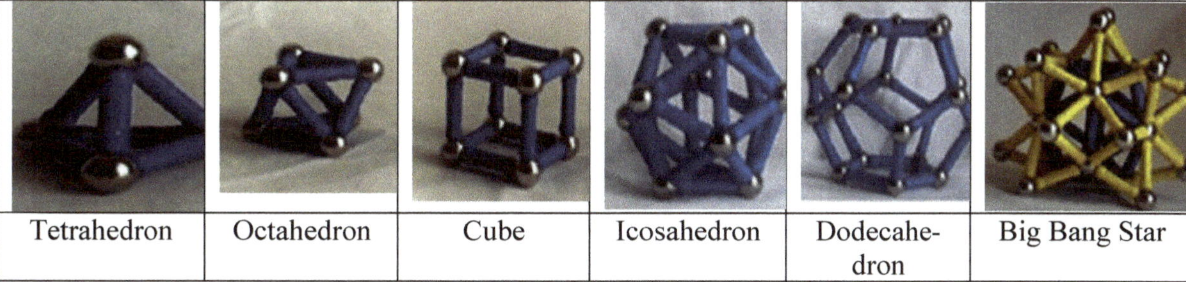

Tetrahedron	Octahedron	Cube	Icosahedron	Dodecahedron	Big Bang Star

3) MATHEMATICIAN EULER'S EQUATION ABOUT PLATO'S REGULAR POLYHEDRONS

Euler presented the equation of relation between the numbers of **edges, vertices and faces** of Plato's regular polygons. This relation can be seen in the following table. We added the Big Bang star in the tables in **Annex-IIa and b,** which we claim to be the sixth regular polyhedron, to Plato's 5 regular polyhedrons. As is seen, Euler's equation is valid for the **Big Bang Star,** as well. Each face of three dimensioned **Plato objects** is composed of the same perfect polygon. Their edges have the same length. Every point of these polyhedrons are in the same distance from the center of regular polyhedron. Besides each of Plato's Regular polyhedrons have 3 spheres which are; **Surrounding sphere** (which is tangent with all vertex points); **Central sphere** (passing through the middle points of each side); and **Inner sphere** (touching to the surface completely). The radius of **inner sphere** in the **Tetrahedron** is equal to 1/3 of the radius of surrounding sphere. **Golden ratio (Phi) √2, √3, √5 irrational numbers** appear continuously in Plato's regular polygons. (**Phi=1, 618., √2=1, 414., √3=1, 732, √5=2, 236**). Golden ratio, which is an irrational number cannot be expressed in terms of proper fractions. Golden ratio is present in the structure of all creatures, and also in the planets of Sun, as well.

REGULAR POLYHEDRON	NUMBER OF FACES (F)	NUMBER OF VERTICES (V)	NUMBER OF EDGES (E)	EULER'S EQUATION: F + V = E + 2
TETRAHEDRON	4	4	6	4 + 4 = 6 + 2
CUBE	6	8	12	6 + 8 = 12 + 2
OCTAHEDRON	8	6	12	8 + 6 = 12 + 2
DODECAHEDRON	12	20	30	12 + 20 = 30 + 2
ICOSAHEDRON	20	12	30	20 + 12 = 30 + 2
BIG-BANG STAR (BBS)	60	32	90	60 + 32 = 90 + 2

MATHEMATICIAN EULER'S EQUATION ABOUT PLATO'S REGULAR POLYHEDRONS

SHAPE OF THE REGULAR POLYHEDRONS	NAME OF THE REGULAR POLYHEDRONS	NO OF FACES (F)	NO OF VERTICES (V)	NO OF EDGES (E)	EULER'S EQUATION: F+V=E+2
	DODECAHEDRON	12	20	30	12+20=30+2
	ICOSAHEDRON	20	12	30	20+12=30+2
	CUBE	6	8	12	6+8=12+2
	OCTAHEDRON	8	6	12	6+8=12+2
	TETRAHEDRON	4	4	6	4+4=6+2

4) MATHEMATICIAN EULER'S EQUATION ABOUT THE STARS OF PLATO'S REGULAR POLYHEDRONS

SHAPE OF THE STARS OF REGULAR POLYHEDRONS	NAME OF THE STARS OF REGULAR POLYHEDRONS	NO OF FACES (F)	NO OF VERTICES (V)	NO OF EDGES (E)	EULER'S EQUATION: F+V=E+2

	DODECAHEDRON STAR	60	32	90	60+32=90+2
	ICOSAHEDRON STAR (BİG BANG STAR)	60	32	90	60+32=90+2
	CUBE STAR	24	14	36	24+14=36+2
	OCTAHEDRON STAR	24	14	36	24+14=36+2
	TETRAHEDRON STAR	12	8	18	12+8=18+2

In the tables above, you can see **Euler's equation** showing the relations between number of face, vertices and edges of Plato's regular polyhedrons. In the table on the previous page, you will see the compliance to Euler's equation on the right side of Plato's regular polygons. And in the table above, you can see **Euler's equation** showing the relation between **number of face, vertices and edges** of Plato's regular polyhedron stars.

5) PLATO'SREGULAR POLYGONS ARE COMPOSED OF TWO TYPES, RIGHT ANGLED TRIANGLES:

a) The right-angled triangle, which is the half part of the equilateral triangle: Tetrahedron, Octahedron, Icosahedron and Dodecahedron on the next page, are composed of these half-equilateral triangles.

b) Right triangle formed by dividing the square diagonally in two: These triangles form the cube. **Tetrahedron, Cube and Octahedron,** they all exist in the crystals of minerals. In the **Crystals (page 36-40),** notions like Symmetry Axis, Symmetry Plane, Mirror View, Rotation Axis are also valid for the Plato objects. On **page 162**, in the center of the given table, we can see the lines of sounds in music, which are determined by two different harmonographs. And on the left and right, there are the tracks, which the planets follow in pairs in the space. The similarity between the lines of music sounds in the middle, and the lines of orbits of planets on the right and left, in other words **common harmonies,** is astonishing. This issue will also be discussed in "**Mathematics in Music**" section (**page 157).**

6) OTHER ISSUES THAT MIGHT BE CONNECTED TO "BIG BANG STAR" (BBS):

a) OTHER PLATO OBJECTS IN BBS:

Geometry is the science, above all sciences, which would tell us the creation code of the Universe. Geometry starts with a point. It continues up to a second point, and becomes a *line*. In other words, duality starts. With a third point at an equal distance to these two points, it becomes an *equilateral triangle*. Thus, 2 dimensions start with three points. As the number of points increase in 2D, it continues as Regular tetragon (Square), Pentagon, Hexagon and Decagon, and ends with a *circle*. In 3D, Equilateral triangle becomes a *regular tetragon* with a fourth point at an equal distance to these three points and here 3Ds start. All 3D objects like Cube, Pentagon Prism, Polygon Prism, Cone and Cylinder, end with *sphere*. There are only 5 regular 3D objects which are known as "Plato objects", in this formation which starts with a Point and ends with the sphere (In a sense, sphere is a macro point).

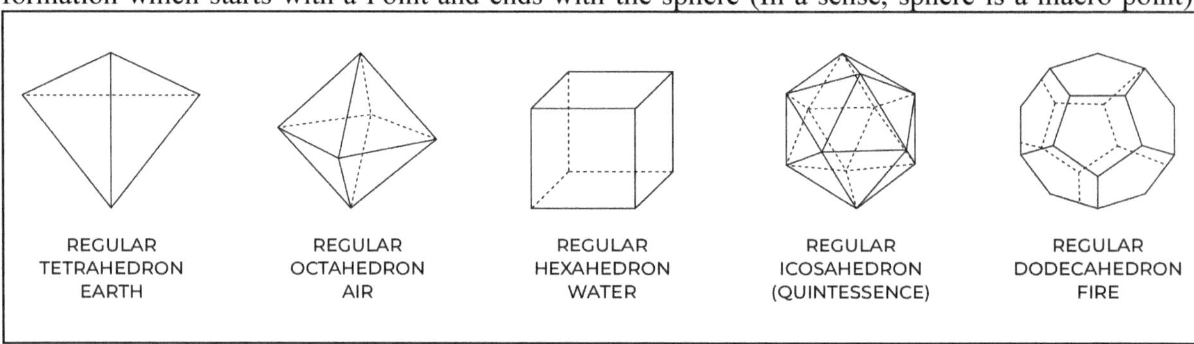

The above-mentioned geometries can be seen in the crystals of elements and some virtual models of subatomic particles. Actually it would be more appropriate to call BBS "Quintessence".

The Plato objects above, are polygons whose surfaces and sides are equal to each other. Below are the Tetra Hedrons (**RABIAS**) at the two ends of one of the 10 pairs of Rhombic Prisms in the 10 pairs of quantum forming the BBS.

Icosahedron is composed of 20 separate triangles;

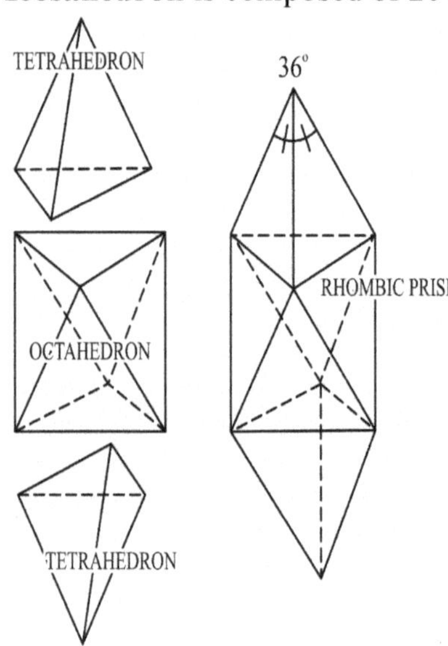

Dodecahedron is composed of12 equal pentagons on the surface of the sphere.

Tetrahedron; upper and lower parts of every rhombic prism composing the **BBS** as in Annex-IIa, b and above, in other words, ends of the stars;

Octahedron: middle parts of rhombic prisms, constituting the **BBS**;

Icosahedron: base on the surface of sphere after removing the tetrahedrons out of **BBS**

Finally, the ends of rhombic prism in **5**s identical groups out of **BBS** constitute 12 pentagons on the surface as in the **Icosahedron**.

b) CALCULUS EQUATION:

(**Annex-Ib, c, d**) Integral (complement) of derivative of a function is equal to the function itself. Instead of Galileo's claim "The book of nature is written in the language of mathematics", there are mathematicians who says "The book of nature is written by Calculus".
Derivative: It is the ratio of one variable to another variable (sometimes it is read as speed)
Function: A variable quantity depending on one or more quantities.
Analytical Geometry: It is the Coordinate geometry.

Analysis: It is the modern development of Differential and Integral calculations.

Integral Calculus: Deals with the "limit" of the sums of small quantities like length of arc named definite integral; and volume of oblique objects.

Differential Calculus: Deals with derivatives. **Integral** and **derivative** are the opposite of each other. Differential Equation (D. E.):

F (x) → Integral x Derivative → Calculus
(Function) (f)   (Limit)

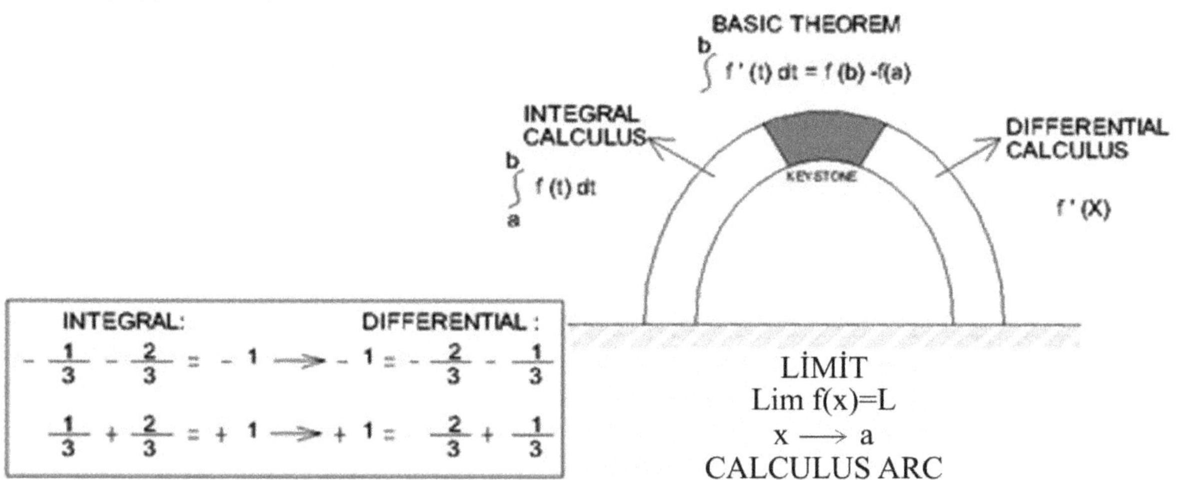

The ratios of electrical charges in each of the Rhombic Prisms in **BBS**, as in the **Fibonacci numbers** are in accordance with the "add the last two to get the next" principle. And they also are suitable to **Calculus**, as can be seen in the tables above and below. Naturally, the momentum of vectors of these electrical charges in all particles of 10 axes of **BBS** should form a **Universal Calculus** while constituting a **Resultant**, which is equal to the sum of momentums as per these axes. Here; **BBS, Calculus** and **Fibonacci numbers** are also integrated.

c) ELEMENT CRYSTALS:

We know that a, b, y angles between **X, Y, Z** symmetry axes in the unit cells of element crystals determine the geometry of unit cell. 10 axes presented in **BBS** would provide an easy to understand classification or the integration of **BBS, Calculus and Crystallography**.

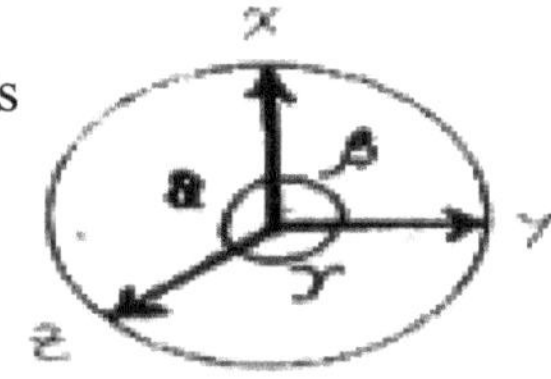

d) GOLDEN RATIOS:

d1) Golden ratio in one dimension: Two quantities are in the 'Golden ratio', if their ratio is the same as the ratio of their sum to the larger of the two quantities.

$$\text{GOLDEN RATE}: \quad \frac{CB}{AC} = \frac{AB}{CD} = 1{,}618$$

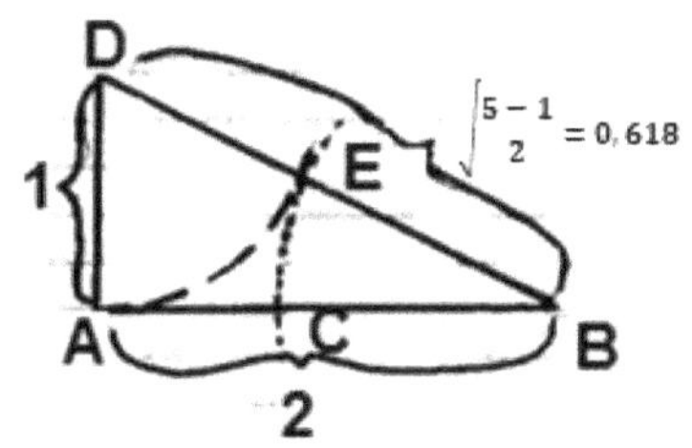

Finding the Golden Ratio using the 1: 2: √5 perpendicular triangle:

An AB line is determined, a vertical line is drawn from point A at the half-length of AB. Point E is obtained by drawing an arc with A.D. diameter and point D in the center. Then Golden Point C is obtained by drawing another arc with diameter EB and point B in the center.

d2) Golden ratios in two dimensions:

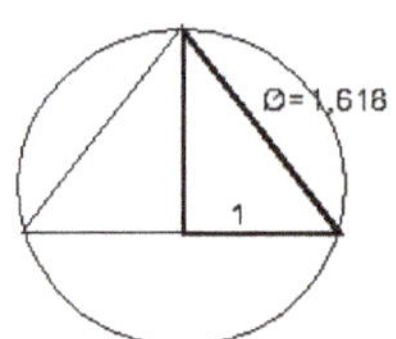
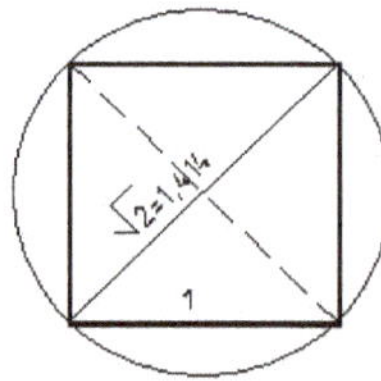
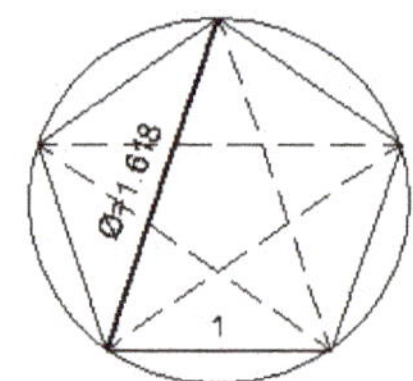
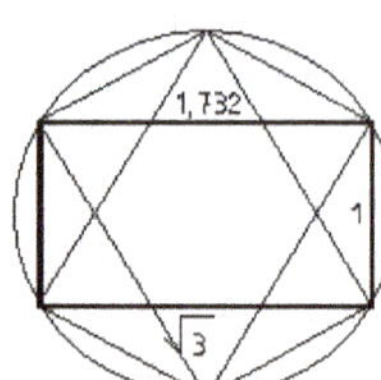
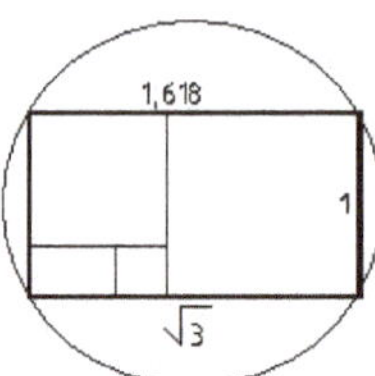

d3) Golden ratios in three dimensions: The secret to creation of all beings in the universe and three dimensional structures of living things on Earth is basically the Platonic Solids and the Big Bang Star, all one dimension, two dimension, three dimension and Golden Ratio geometrical shapes create a helix. This code can be observed in unlimited beings from galaxies in the astral universe to hurricanes, vortexes, shellfish, leaves of plants and wings of butterflies in the world. The golden ratio has a precise mathematical function, which allows transition from one dimension to other and one form to other. While this function creates variety by allowing transition from symmetry to asymmetry and vice versa, it also creates a perfect balance without a slight monotony. Therefore, the Golden Ratio, the dimensional relation between the hole and its parts, is a source of both Rhythm and balance. It would be wrong to say the golden ratio is everywhere in the universe. The golden ration, however, is observed in utterly perfect and beautiful forms and conditions. For example, disorder and casualness in amorphous structures of solids in the universe leave their place to a final orderliness with their periodic and symmetrical characters in solids with crystal structures. In conclusion, it is understood that there is a balanced harmony in togetherness of orderliness and disorder in the universe.

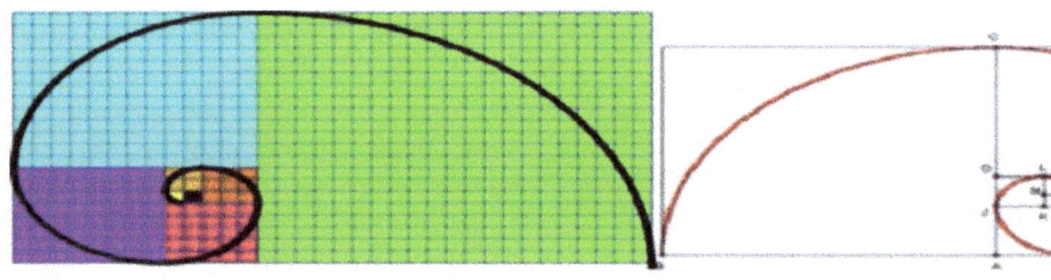

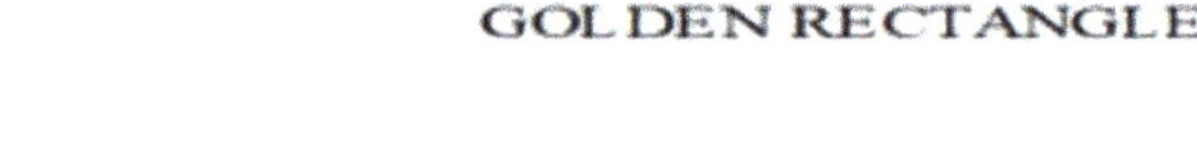

FIBONACCI SPIRAL

GOLDEN RECTANGLE

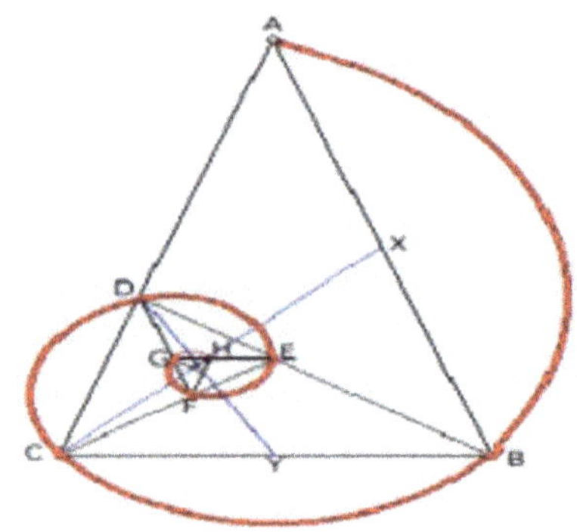

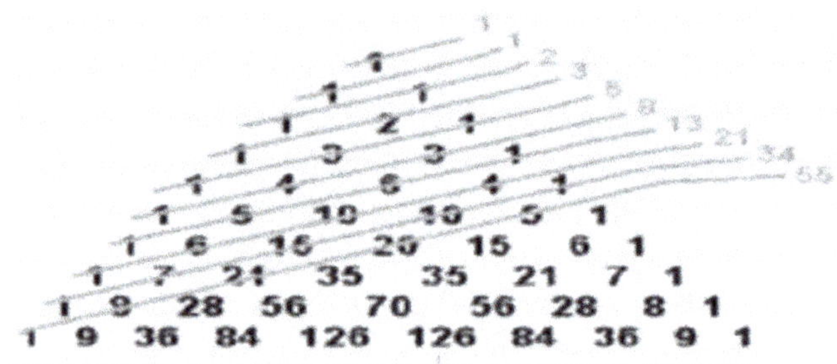

GOLDEN TRIANGLE

PASCAL TRIANGLE AND FIBONACCI NUMBERS

At the zero time of the big-bang, the golden ratio experiment according to the number 3,6,9 of tesla in the spiral structure of the 4 particles of the previous and next universe:

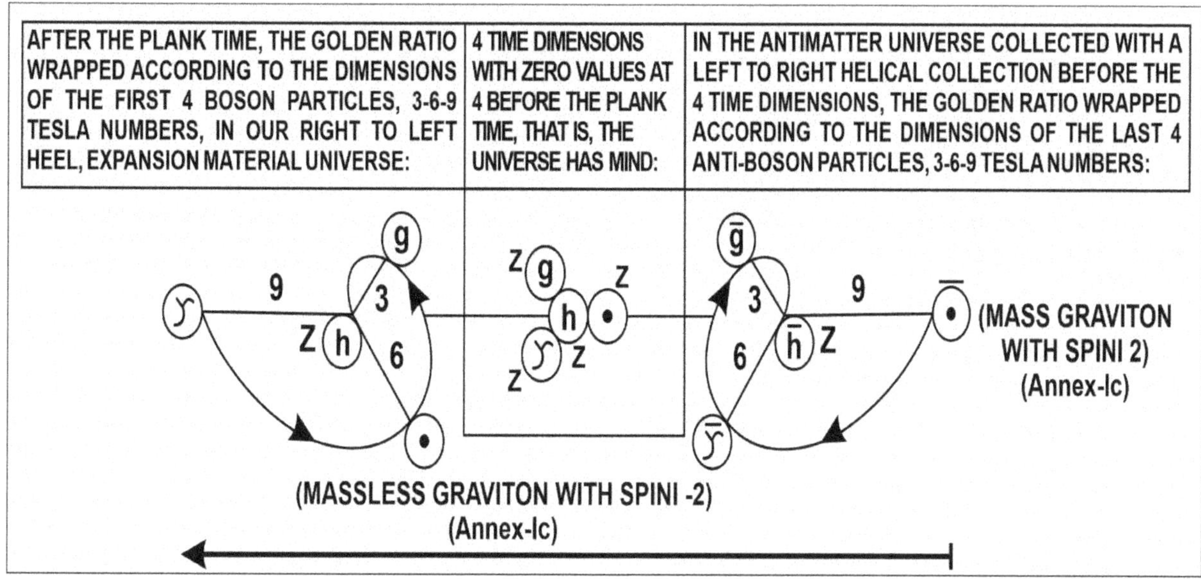

The code of creation of all the beings in the universe and the three-dimensional structures of living things in the world; is the spiraling of one-dimensional, two-dimensional, three-dimensional and Golden-ratio geometric shapes, mainly the smooth faces of Plato and the BBS (Big Bang star). This code is seen in countless beings, from Galaxies in the astral universe to hurricanes, vortices, crustaceans, leaves of plants, wing structures of butterflies in the world. The golden ratio has a precise math function that allows you to move from size to size and form to form. This function allows for the transition from symmetry to asymmetry, asymmetry to symmetry, creating diversity and providing a perfect balance without the slightest single correction. Thus, the golden ratio, which has a dimension relationship between the whole and its parts, is a source of both Rhythm and balance. It would be wrong to say that there is a golden ratio everywhere in the universe. However, it is seen that there is a golden ratio in perfectly balanced and beautiful forms and situations. For example, the irregularity in the amorphous structures of the solid matter in the universe leaves its place in the solid materials with crystal structure with periodic and symmetry features. As a result, it is understood that there is a balance of harmony and order in the coexistence of regularities and irregularities in the universe.

According to the English literary critic **C. Samuel Taylor** (1772-1834); **"beauty is unity in variety"**. Science is the search for exploring the unity hidden in infinite variety in the nature. Poetry, painting and other fine arts are the efforts of people to reach unity in variety. At this point, it would be good to remind the most beautiful definition in Islamic philosophy, **"kesrette vahdet"** which means "unity in multiplicity".

The cuneiform, which began with the **Sumerians** in 4000 BC, was developed with syllables, sounds and letters by the Phoenicians, the Nabataeans, the Himyarites, the Aramites, the Hebrews, the Assyrian and the Arabs in Asia Minor since 1000 B.C. The Arabic **calligraphy** with golden ratios yielded the most beautiful pieces with **Thuluth** (means one-third) script developed after Islam. Especially, Talik script written with a wide end reed pen allows writing all letters in golden ratios and Fibonacci numbers. Calligraphy, i. e. the art of lines, is actually ad Geometry term. The word engineer in Arabic means a person who knows about geometry. Calligraphy is also defined as "moral geometry". There are repetitions in calligraphy as the rhythm in music and the rhyme in poetry.

While the "Golden Ratios" discovered by the temple priests in Egypt thousand years ago before Christ created beautiful art works for humanity in temple and tomb architecture in Egypt, sculpture, painting and temple architecture in Greece, Rome and Europe in Renaissance, they also contributed to architecture and development of calligraphy in Islam after the Greek Era.

Being the expression of harmony and unity in Macro, Normo, Micro and Infra Universes, golden ratio is an irrational number and it cannot be expressed as a proper fraction. Golden ratio is one of the three basic ratios in the regular polygons. It is worth researching that the numbers 1, 3, 21 and 55 in the Pythagoras triangular pyramid given on **page 137**, are also present in this Fibonacci table. It is obvious that the orbit lines of planets on the **page 162** and spiral structures of harmonograph lines of sounds in music cannot be dissociated.

In addition to the ones above, the big circle outside represents the **world**, while the small circle inside represents the **Mercury**, in the following pentagram. Golden ratio can also be seen here. On t
All of the time cycles of **Sun, Moon and World** can be expressed as the basic combinations of Golden ratio. Golden ratio is also present in the orbits of **Mercury**, **Venus**, **World** and **Mars**. Being the mathematical code of the beauties and aesthetics in the universe, **Golden ratio** is present in the 2/5 ratio of the orbital periods of Saturn and Jupiter which are the two biggest planets of **solar system**. Here in the center, you can see the dance of **Saturn** and **Jupiter**; the triangle which can be seen by looking from **Jupiter** to **Saturn**, and from **Saturn** to **Jupiter**; and the sun ($\odot$). On the upper right; there is a 6 vertices star (star of David) which indicates the meeting locations of **Jupiter** and **Saturn** once in 20 years, out of the Zodiac zone. In the figure below, golden ratios can be seen which consist of the locations of **World** (+), **Jupiter and Saturn**, according to the relative speeds of their orbits. There are tens of examples showing that **mathematics, Fibonacci numbers, golden ratios and musical notes** are formed by means of a divine harmony between the orbit radii of Sun and planets; time it takes for them to rotate around the Sun; and the geometries they draw in the space. the right side of the star, **Fibonacci numbers get closer to the golden ratio from up to down.**

0+ 1	= 1	1 : 1	= 1
1+ 1	= 2	1 : 2	= 0.5
1+ 2	= 3	2/ 3	= 0.6667
2+ 3	= 5	3/ 5	= 0.6
3+ 5	= 8	5/ 8	= 0.625
5+ 8	= 13	8/13	= 0.6154
8+13	= 21	13/21	= 0.619
13+21	= 34	21/34	= 0.6176
21+34	= 55	34/55	= 0.6182
34+55	= 89	55/89	= 0.6180
55+89	=144	89/144	= 0.6181
1, 2, 3, 5, 8, 13, ...$\varnothing$ = 10.61803399			

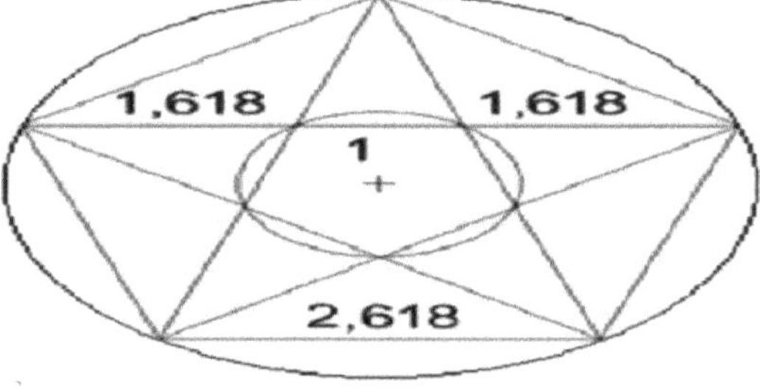

OUTER CIRCLE IS THE SIZE OF WORLD,
INNER CIRCLE IS THE SIZE OF MERCURY

Again, if we go back to the **Dodecahedron** (regular 12 faces) in the geometry of **BBS**; 12 of 20 vertex points form 3 Golden Rectangles perpendicular to each other; and the remaining 8 vertex points form a **Cube** with edge length unit value is equal to **Phi**^{2}. These rectangles can also be seen in the above 6 vertex Stars of David which is the cross section of **Dodecahedron**. 12 vertex points of **Icosahedron** are also defined with **3 Golden Rectangles** perpendicular to each other. Because there is also a 6 vertices star in its cross section. **Golden ratio is 0, (Phi), √2, √3 and √5 in Plato's regular pentahedron. While all of Plato polyhedrons can be formed within each other with Golden ratio; this nested structure can continue forever both towards inside and outside.** Also there is a Golden ratio in the structure of all **creatures**, in the leaves and flowers. Here after all these information, it would not be wrong to claim that **BBS** gathers at least all **Plato objects, Golden ratio, Fibonacci numbers, the codes of the whole universe in Atomic and Astral scale.**

Regular tetrahedrons out of **BBS**, become **Regular Pentagons** in 5s groups; and they become a different star, a **Dodecahedron,** in the whole **BBS**. Perhaps the explanation of a real universe and why the golden ratio is related with the 5s symmetry, and Geometry and Mathematics of **BBS** are hidden here. Here we must indicate that **nested 2 pentagons (page 155)** determine the orbit thickness of **Mercury**, space between **Mercury** and **Venus**, relative average orbits of **World** and **Mars** and the space between **Mars and Ceres**; nested 3 pentagons determine the space between **Venus and Mars**, average orbits of **Ceres** and **Jupiter**. Besides **World** and **Venus** draw a pentagon with their orbits once in 8 years. Kepler has determined the orbit distance between **World** and **Mars** by means of **Dodecahedron**; and the orbit distance between **Venus** and **World** by means of **Icosahedron**. In geometry, as there are only specific number of symbols in harmony like **Plato objects**; in music there are only specific number of notes in harmony.

e) CAN'T THE BIG BANG STAR (BBS) BE THE GEOMETRY OF THE BOSE-EINSTEIN CONDENSATION?

The possibility of the Big Bang star being the Geometry of the **Bose-Einstein Quantum condensation** within the three-dimensional nodes (Annex-Ie) or the human brain, which occurred in the first 10^{-11} seconds of the Big Bang. Because there is no possibility that the first 160 subatomic particles of the universe in BBY should be in place and none of them may be in another part of BBY. It does not make sense to have the condensate in this concentration.

7) MATHEMATICS IN THE MUSIC:

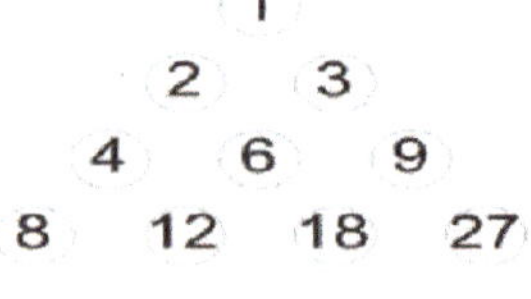

LAMBDOMA II

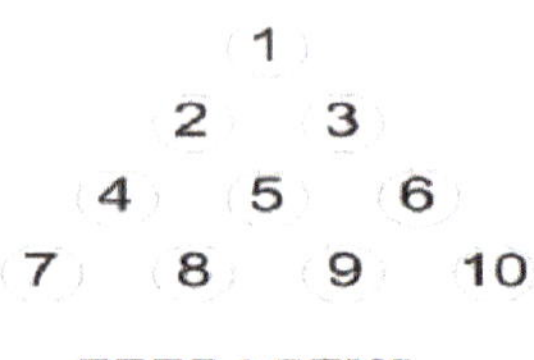

TETRACTYS

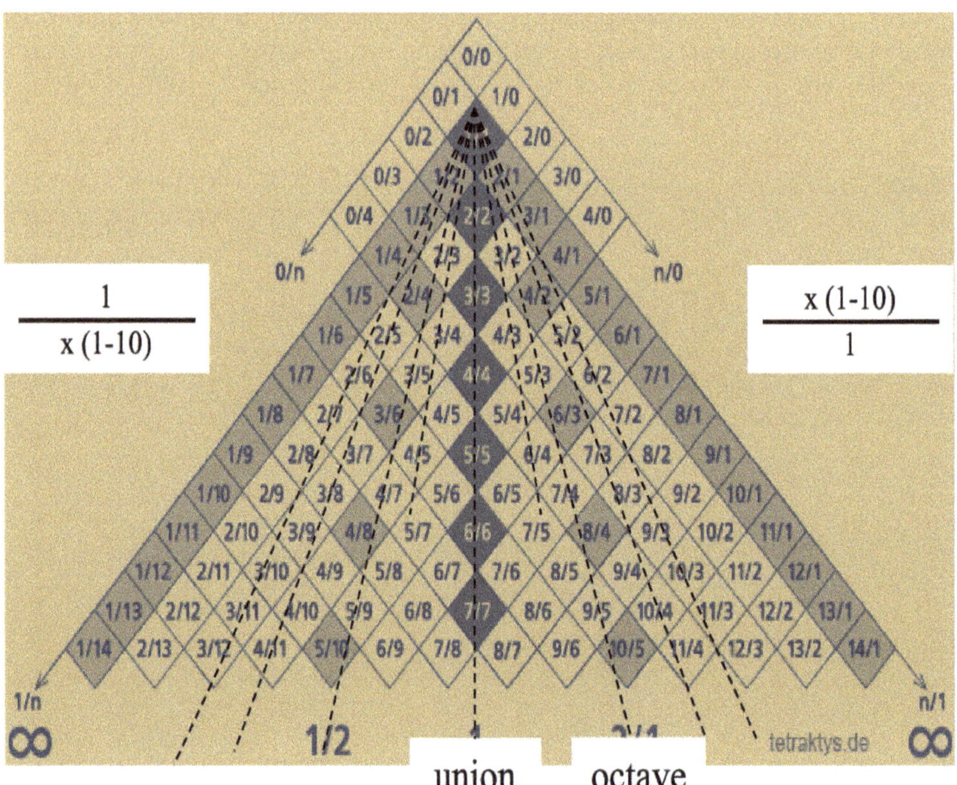

Approximately 2500 years ago Pythagoras found out that when the rates in the frequencies of sounds were composed of simple numbers, an accord (harmony) was obtained and it sounded nice. Since a hammer with a weight half of another hammer makes a sound two times treble than the heavy one, (**One Eighth=Octave=1/2**) and another hammer with a weight ratio of **2/3** were constituting a 5s tone, making a good sound. Strings of various thickness and length in string instruments; pipes of various thickness and length in wind instruments; also the same rule applied for various sizes of cymbals. As a result, they decided that "**All of the nature is composed of the harmony formed by numbers**". After **Pythagoras** found out the relation of sounds in music with numbers; Geometrical views were obtained in the sounds of **Harmonograph** in 1800s, **Kaleidophone** in 1827, **Eidophone** in1880, **Chladni Plate** in 1787, as can be seen on **page162**. **Music** was then called as the number in time (circle), **Geometry** was called as the number in space.

Common harmony: They are the images of double sounds of different frequency and equal volume sonore (wave height) in the same and opposite directions, drawn by a harmonograph.

Lambdoma (λ): Some of the simple ratios are lined up in the above figure in which the harmony in music with upper harmonies from up to down in the right side of **Lambdoma,** in other words, λ; lower harmonies from up to down in the left side. **(Page 156)** While the upper harmonies are lined up from up to down as integers, on the right side of the figure; the lower harmonies are lined up from up to down in halves, on the left side of the figure. On the right side of table, **8/4=6/3=4/2=2/1**intervals from down to up, are the same with each other. These identicalnesses on the right and left, meet on the top in **%** ratio. In total 64 harmonic ratio in this Lambdoma, are as much as the number of **64 Lepton-Quark particles** in our **Super cube** model **(Annex-IIc)**. Besides, number 8, which is the unit number in music, is the same with the number of particles in each **Rhombic Prism** in the virtual models of **Super cube** and other subatomic particles. **(Also 64 different codes of DNA). (Page 141)**

Lambdoma II: Numbers in the Pythagoras triangle on the left of picture in the previous page, are multiplied by two on the left and multiplied by three on the right, and create sounds which separate from the adjacent ones with perfect fifths, horizontally. And this may remind the pentagons on the outer surface of BBS.

NOTES IN THE MUSIC:		
DO λ	RATIOS WITH THE SAME DENOMINATOR	1
RE 8/9	160/180λ	2
MI 8/10=4/5	144/180λ	3
FA 6/8=3/4	135/180λ	4
SOL 6/9=2/3	120/180λ	5
LA 6/10=3/5	108/180λ	6
SI 8/15	96/180λ	7
DO 5/10=1/2	90/180λ	8

Celestial Monochord (Monochord of Universe): Again in the previous page we can see the singularity of creation. In the figure, materials cover a radius more than 10^{40}exponentially, from Quantum waves on the lowermost, to the outer limits of universe on the uppermost; and alignment is as can be seen in the pictures.

Tetractys: The first 3 lines create simple intervals in the triangle composed of 10 numbers (**1+2+3+4=10**) in 4 lines in this Pythagoras triangle on the left of the picture in the previous page. (as in the Pythagoras triangle and regular tetrahedron). It is assumed that the **tuneful numbers in music**are; 16- 12- 9- 8- 6- 4. These numbers can create pairs in various types. 8 notes are the same,below.

On a piano keyboard **7 octachord (octave) = 7 octave** The hearing capacity of humans **11 octachord (octave) = 11 octave**. The highest point of every octave has a frequency 2 times higher than the frequency of the first note. Frequencies start from **16 vibrations in a second** (16 hertz)**, and ends with 20 000** vibration numbers in seconds, which are equal to the lower hearing threshold of humans. Below are the sounds in the frequencies (vibration numbers in a second) which a human ear can hear, and the corresponding number of octaves and scale numbers.

					FREQUENCIES WHICH THE HUMAN EAR CAN HEAR										
Frequency	2	4	8	16	32	64	128	256	512	1024	2048	4096	8192	16384	20000
Frequency	2^1	2^2	2^3	2^4	2^5	2^6	2^7	2^8	2^9	2^{10}	2^{11}	2^{12}	2^{13}	2^{14}	
No of Octaves			1	2	4	8	16	32	64	128	256	512	1024	2048	
Scale Numbers					DO	DO_1	DO_2	DO_3	DO_4	DO_5	DO_6	DO_7	DO_8	DO_9	

In Music: 1/1 (Union) (Duet) in the same fret.

1/2 (Octave) (**Trio**) +2/3(Fifth) =Harmony (as the 1/2, 2/3 phases of universes in **Annex-Ia, b, c, d**)
3 (+) **5** = **8** (3, 5 and 8 are Fibonacci numbers).

As being called as the '**music of spheres**' of the ancient world, planets draw parallel orbits corresponding to approximately 1/2 ratio octave **around the Sun**. As can be seen on the next page, an accord shows up by the lines which are constituted by the orbits of every pair of planets in different times, according to their distance to the Sun and according to their speed.

Above is the **Harmony**, which is constituted by the **Mercury and World**. While **Kepler** was try

DANCE OF MERCURY WITH THE EARTH

ing to find a geometry on the orbits of planets or a solution about the music, he observed that the first 6 planets with the Sun in the center, meant 5 tones in the music. For that purpose, he placed Plato's regular pentahedron objects between the spheres **(page 195, picture 7)** Kepler used the **Dodecahedron** for determining the distance between the orbits of **World** and **Mars**; and used the **Icosahedron** for determining the distance between the orbits of **Venus** and the **World**. Besides, he recognized that all of the ratios between maximum angular speeds of planets are the distances in accordance with the **Harmony in Music**.

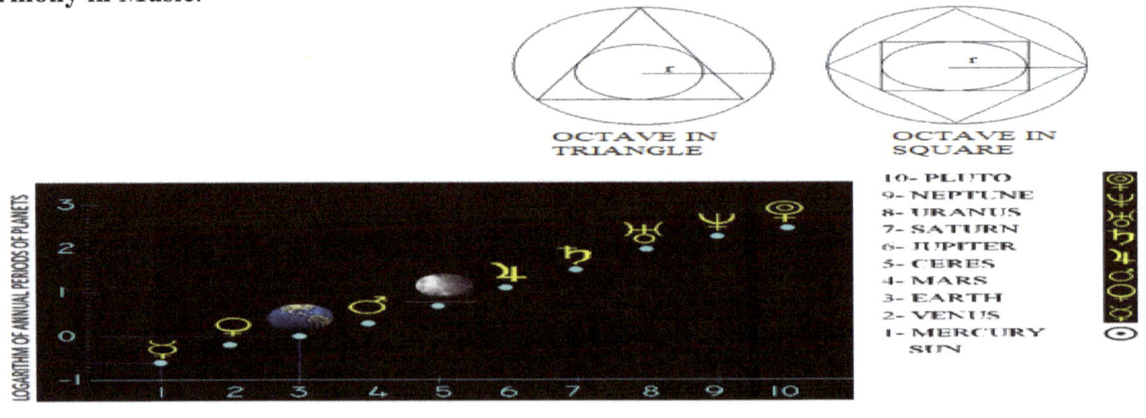

PLANET NUMBERS (QUOTED FROM OUNEDON AND ROY)

Einstein showed that small **Space-Time effects** caused by faster (and as a result, slowing down) motion of Mercury in close locations to the Sun, affect the wobble motion rotations of ellipse orbits which reach up to thousands of years. Thus he supported Kepler's model. In geometry, the method of determining **Octave** which is the halving or doubling of frequency or wavelength, is to draw the inner and outer circles of an equilateral triangle or square, as can be seen above. The diameter or radius of inner circle is half of the diameter or radius of the outer circle.

Uranus (once in 84 years) and **Neptune** (once in 165 years), they constitute parallel orbits around the sun which correspond to an octave (1/2). Cycles of the planets and Basic Logarithmic Graphical Order of their orbits are in the previous page. Sometimes the orbit cycles of planets appear as the simple ratios of each other. 2/5 ratio of **Jupiter and Saturn** is an example. There is Venus on the Sun side of the **World**. Mars is in the opposite direction. **World** meets with Mars 3 times for every 4 meetings of the world with **Venus**. In other words, around the **World** there is a very slow, 3/4 rhythm or music which covers a 4 degrees of notes. While **Uranus** is in rhythm and harmony with **Neptune and Pluton;** when the period of **Uranus** is added to the period of **Neptune**, they present the 1/2/3 ratio, which yields the period of **Pluto**. The orbital period of **Neptune** is two times the period of **Uranus**. (1 octave) The orbital period of **Uranus** is 2/3 times **Pluton's** period. The size of inner circle of **Neptune** is 2/3 times the size of outer circle. And it coincides with a 5 degrees of notes in the music. The ratio of orbits of **Jupiter and Saturn** gives **6/11**. This is two times or as much as the octave of the **3/11** ratio of **Moon and World**. Here it is understood that every planet, like the thickness and length of the pipe in the wind instruments, and the structures of other musical instruments such as the wire thickness and length in the stringed instruments, create different music sounds according to their structural characteristics.

Note: The **mathematics** is at the top of all sciences, but like every being, that is the product of the Atomal and Astral Universe, also the human, the most perfect being of the Universe, is created according to the harmony, melody of the Universe's Mathematical harmony. Because music can only consist of harmonious and compatible sounds in their frequencies, music sounds are perceived by the ears and the brain, creating a very special and inhibiting effect on the human and even in other living beings. For this reason, in the Ottoman period, music was defined as "**İlm-i Şerif**", meaning the honorable knowledge.

There are two classical music prepared according to two different mathematical systems as Classical Western Music and Classical Turkish Music in the World. Both have produced immortal works by separate means. Our wish is that we can achieve harmony between the sounds of music, as well as the Astral and the Atomic Universe. Otherwise, the world community, like every disease structure cannot maintain healthy existence.

ADDITIONAL PICTURES AND TABLES FOR THE MATHEMATICS IN MUSIC:

I) TABLE OF THE APPROXIMATE VALUES OF WAVELENGTHS OF SOUND, COLOR, LIGHT ANDHEA TRADIATIONS; ANDTHE SIGNATURES:

APPROXIMATE COMMON VALUES OF COLORS AND WAVE LENGTHS OF SOUNDS, NOTES, PAINTS, VISIBLE LIGHT, HEAT RADIATION

Left vertical side-labels: "SOUNDS WE CAN HEAR" (along the FREQUENCIES column) and "SOUNDS USED IN MUSIC" (along the OCTAVE column).

FREQUENCIES (Hz)	OCTAVE (NR.)	NOTES		NO	λ and f RATES	λ and f RATES	WAVE LENGTH (nm)	λ and f RATES	λ and f RATES	COLORS	VISIBLE LIGHT (A°)	VISIBLE LIGHT (nm)	PAINTS (μ)	PAINTS (nm)	HEAT RADIATION (μ)	HEAT RADIATION (nm)
2^1 2 2^2 4 2^3 8 2^4 16	1 2															
2^5 32	4	DO	Co 60	0	1/1 1/1	10/10 14/10	700-700 700-700	10/10 a 10/10 f	180/180 180/180	RED MEDIUM RED	0,80-0,75	750	6100-7500	700	0,75 0,65	750 650
		RE	C0 60			8/9 9/8	622/700 700/622	8/9 a 9/8 f	160/180 180/160	ORANGE	0,63-0,63	630	5900-6100		0,60	600
		Mİ	C0 60			8/10 10/8	560/700 700/560	4/5 a 5/4 f	144/180 180/144	YELLOW	0,59-0,59	590	5700-5900		0,55	550
		FA	C0 60			6/8 8/6	525/700 700/525	3/4 a 4/3 f	135/180 180/135	GREEN	0,50-0,50	500	5000-5700		0,51	510
		SOL	C0 60			6/4 4/6	467/700 700/467	3/2 a 2/3 f	120/180 180/120	BLUE DARK BLUE	0,45-0,45	450	4500-5000		0,47 0,44	470 440
		LA	C0 60			6/10 10/6	420/700 700/420	3/5 a 5/3 f	108/180 180/108	PURPLE	0,40-0,42	420	4000-4500		0,42	420
		Sİ	C0 60			8/15 15/8	373/700 700/373	8/15 a 15/8 f	96/180 180/96	PURPLE ULTRA PURPLE	0,004-0,373	373				
2^6 64	8	DO1	C1 61	1	1/2 2/1	5/10 10/5	350/700 700/350	1/2 a 2 f	90/180 180/90							
2^7 128	16	DO2	C2 62	2	1/4 4/1											
2^8 256	32	DO3	C3 63	3	1/6 6/1											
2^9 512	64	DO4	C4 64	4	1/8 8/1											
2^{10} 1024	128	DO5	C5 65	5	1/10 10/1											
2^{11} 2048	256	DO6	C6 66	6	1/12 12/1											
2^{12} 4096	512	DO7	C7 67	7	1/14 14/1											
2^{13} 8192	1024	DO8	C8 68	8	1/16 16/1											
2^{14} 16.384	2048	DO9	C9 69	9	1/18 18/1											
20.000																

NOTE: A HUMAN EYE CAN SENSE WAVES OF FREQUENCY 390-780 MHz

ES ACCORDING TO THE OCTAVE NUMBERS:

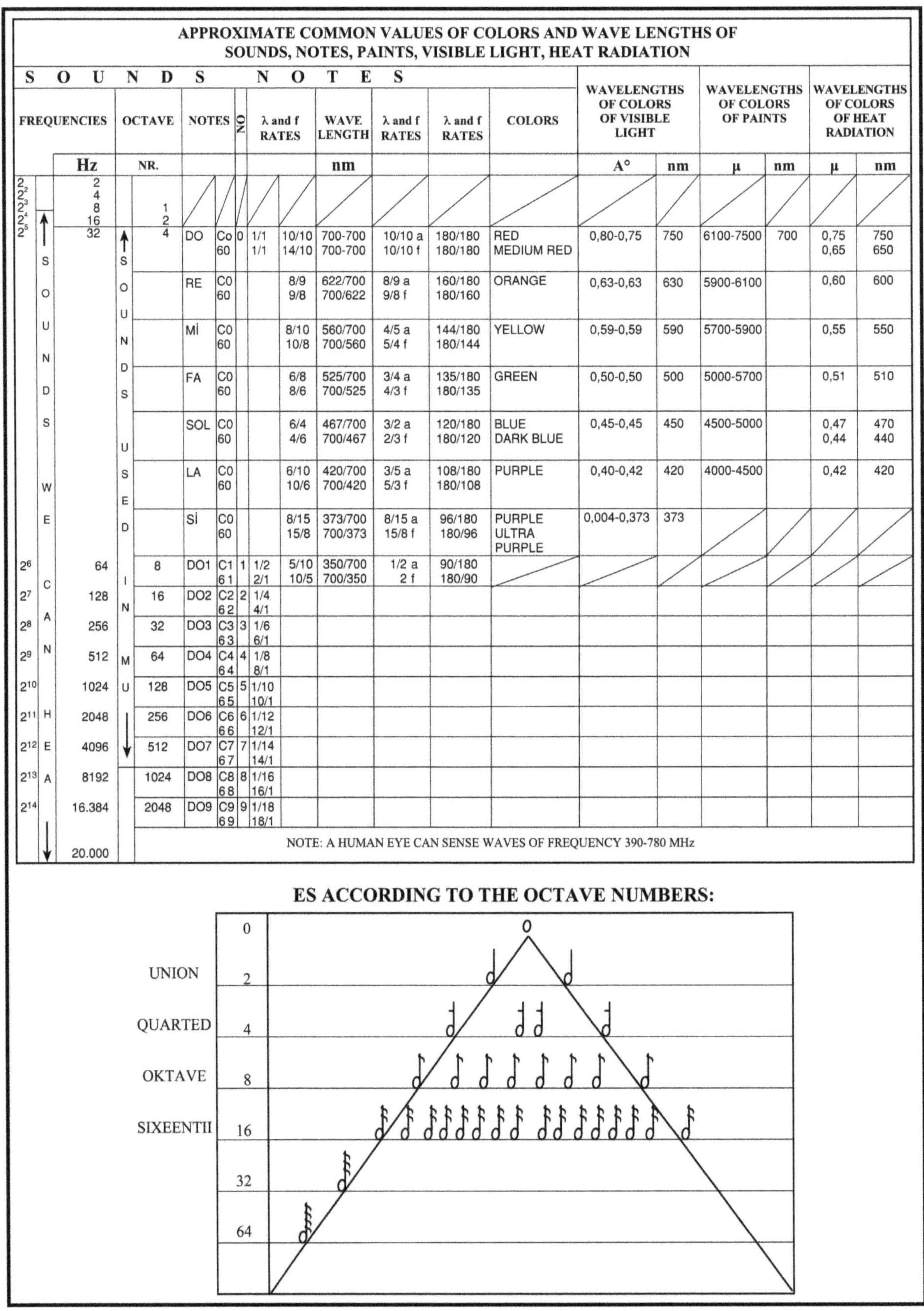

2) TABLE SHOWING THE SIMILARITY OF HARMONOGRAPH LINES BETWEEN ORBIT LINES OF PLANETS AND THE SOUNDS IN MUSIC:

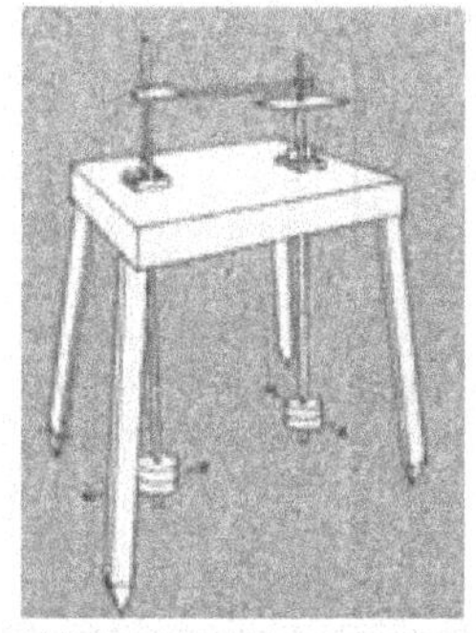

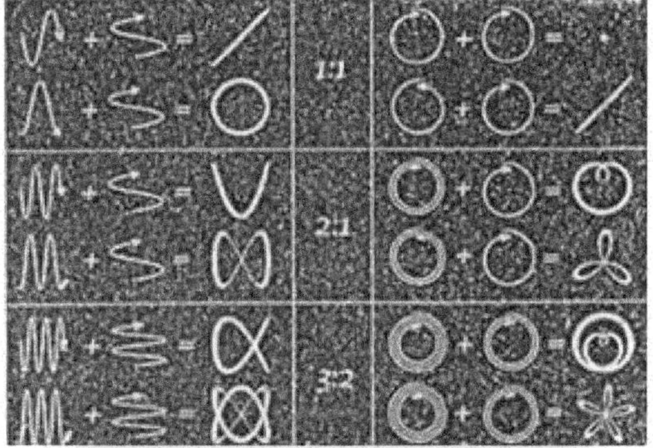

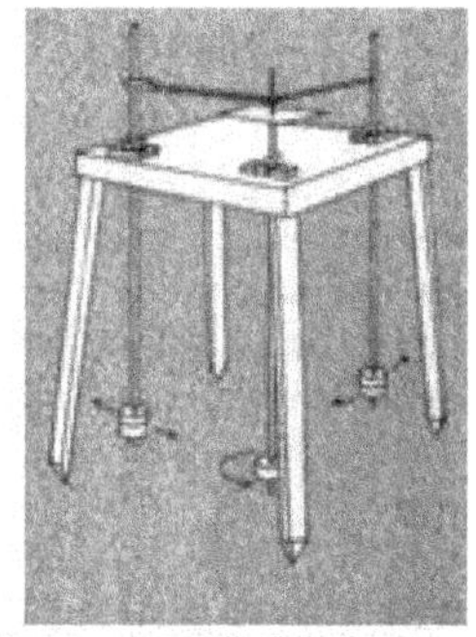

FIGURE 1 – LATERAL HARMONOGRAPH EXAMPLES IN THE LEFT MIDDLE HALF OF THIS PAGE ARE COMPOSED USING THIS HARMONOGRAPH
BELOW IS AN EXAMPLE IN THE ASTRAL SCALE

FIGURE 2 – ROTARY HARMONOGRAPH EXAMPLES IN THE RIGHT HALF OF THIS PAGE ARE COMPOSED USING THIS HARMONOGRAPH
BELOW ARE SOME EXAMPLES IN THE ASTRAL SCALE

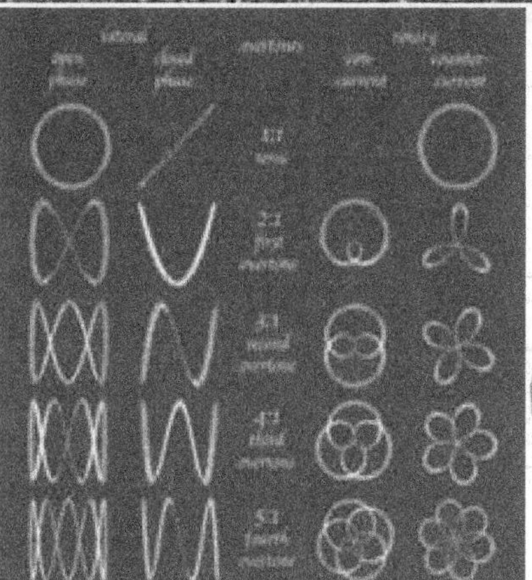

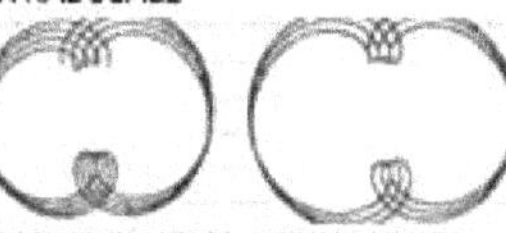

FIGURE 3- THE VISION OF PATH OF VENUS FROM THE WORLD IN 8 WORLD YEARS (OR 13 VENUS YEARS) WHILE THE SUN GOES ROTATES AROUND THE ZODIAC AND WHILE VENUS ROTATES AROUND THE SUN.
THIS PICTURE IS THE ONLY ONE I COULD FIND WHICH IS SIMILAR TO THE LATERAL HARMONOGRAPH LINES (LEFT SIDE).

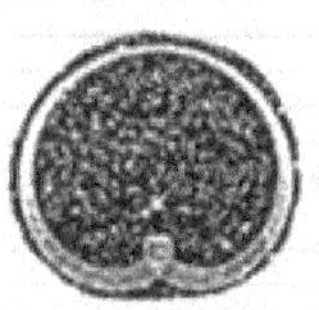

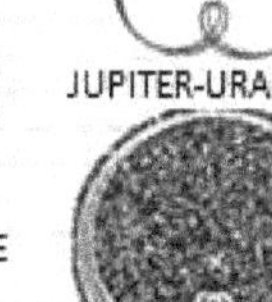

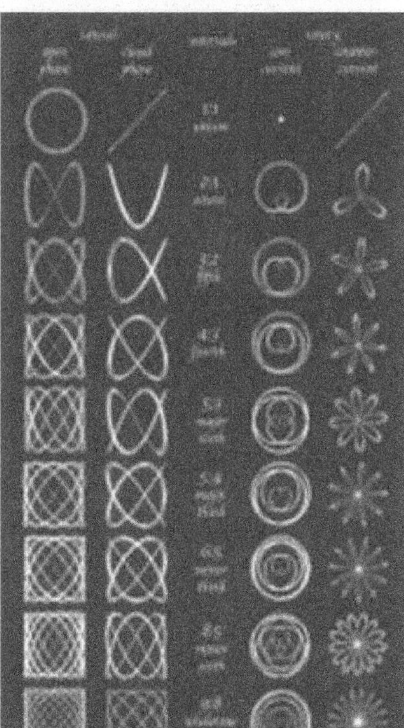

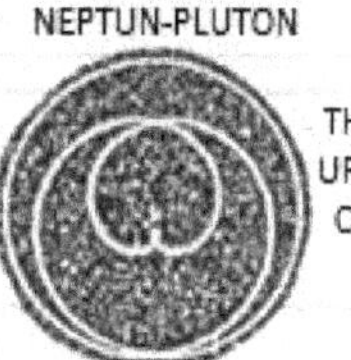

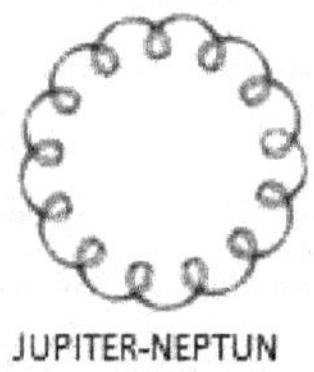

3) TABLE OF RATIOS OF FREQUENCIES IN DOUBLE SOUNDS COMPOSING THE NOTES:

1/16	1/15	1/14	1/13	1/12	1/11	1/10	1/9	1/8	1/7	1/6	1/5	1/4	1/3	1/2 λDO1	1/1 λfDO
2/16	2/15	2/14	2/13	2/12	2/11	2/10	2/9	2/8	2/7	2/6	2/5	2/4 λDO1	2/3 λSOL	2/2	2/1 fDO1
3/16	3/15	3/14	3/13	3/12	3/11	3/10	3/9	3/8	3/7	3/6 λDO1	3/5 λLA	3/4 λFA	3/3	3/2 fSOL	3/1
4/16	4/15	4/14	4/13	4/12	4/11	4/10	4/9	4/8 λDO1	4/7	4/6 λSOL	4/5 λMI	4/4	4/3 fFA	4/2 fDO1	4/1
5/16	5/15	5/14	5/13	5/12	5/11	5/10 λDO1	5/9	5/8	5/7	5/6	5/5	5/4 fMI	5/3 fLA	5/2	5/1
6/16	6/15	6/14	6/13	6/12 λDO1	6/11	6/10 λLA	6/9 λSOL	6/8 λFA	6/7	6/6	6/5	6/4 fSOL	6/3 fDO1	6/2	6/1
7/16	7/15	7/14 λDO1	7/13	7/12	7/11	7/10	7/9	7/8	7/7	7/6	7/5	7/4	7/3	7/2	7/1
8/16 λDO1	8/15 λSI	8/14	8/13	8/12 λSOL	8/11	8/10 λMI	8/9 λRE	8/8	8/7	8/6 fFA	8/5	8/4 fDO1	8/3	8/2	8/1
9/16	9/15 λLA	9/14	9/13	9/12 λFA	9/11	9/10	9/9	9/8 fRE	9/7	9/6 fSOL	9/5	9/4	9/3	9/2	9/1
10/16	10/15 λSOL	10/14	10/13	10/12	10/11	10/10	10/9	10/8 fMI	10/7	10/6 fLA	10/5 fDO1	10/4	10/3	10/2	10/1
11/16	11/15	11/14	11/13	11/12	11/11	11/10	11/9	11/8	11/7	11/6	11/5	11/4	11/3	11/2	11/1
12/16	12/15 λMI	12/14	12/13	12/12	12/11	12/10	12/9 fFA	12/8 fSOL	12/7	12/6 fDO1	12/5	12/4	12/3	12/2	12/1
13/16	13/15	13/14	13/13	13/12	13/11	13/10	13/9	13/8	13/7	13/6	13/5	13/4	13/3	13/2	13/1
14/16	14/15	14/14	14/13	14/12	14/11	14/10	14/9	14/8	14/7 fDO1	14/6	14/5	14/4	14/3	14/2	14/1
15/16	15/15	15/14	15/13	15/12 fMI	15/11	15/10 fSOL	15/9 fLA	15/8 fSI	15/7	15/6	15/5	15/4	15/3	15/2	15/1
16/16	16/15	16/14	16/13	16/12	16/11	16/10	16/9	16/8 fDO1	16/7	16/6	16/5	16/4	16/3	16/2	16/1

E) PRELIMINARY INFORMATION, PREPARING THE THEORY OF EVERYTHING:

In 2005, in Feza Gürsey Physics Institute on the strait side of Kandilli Observatory in Istanbul, after telling "Now I will tell you the curve which Boson particles draw in two dimensions", an Associate Professor Doctor of Physics showed this simple curve and 20 pages of equations approximately to 40-50 audiences by means of a reflecting projector for more than an hour. At last, he indicated that: **"If I have made the same calculations including the numerical values of signs like a, b, x, y in these equations, it would last for weeks. Besides, I am not sure if I could manage to do it"**. This example shows us how to explain the complicated universal facts by means of geometrical figures as everybody would understand. **Universal synthesis would only be by means of Geometry. The geometry, which is the final result of the work in this book, was the "double triangle pyramid" (P.10,175-178).**

As human who can speak and think (Homo Sapiens), we were in Central Asia almost 42 thousand years ago **(P.46-49,193,189,211).** It is said that the continent of Mu was in the Pacific between 20 000-15 000 years ago and the continent of Atlantis in the Atlantic ocean between 15 000-12 000 years ago. We learn information about the continent of Atlantis from Egyptian priests and information about Mu from the "Naacal tablets" that English James Churchward found in Western Tibet. On **page 193** the two-dimensional sacred geometries in the history of people's thought and belief are presented for the first time in this book as a three-dimensional "Double triangle pyramid" **(P.174-178).** It is understood that the concepts of **Rabia**, **Morality** and **Justice** in this geometry, which is the golden age algorithm in the world, have been sacred since Mu time. Information in Mu, to Mayas in Central America (800 BC-1500 AD), to Aztecs, Uighurs in Asia, Scythians, 5-6 000 years old "White pyramids" in China, West Jawa in Indonesia from 28,000 years old "Hilltop Pyramid" to Brahman and Tibetan priests in North India, to Ok-uz Turks in **"ANAV CULTURE"** (VI-VII century) in Turkmenistan, then from Anav to Mesopotamia It reached the Sumerians (3nd century BC) and Babylonian priests of Egypt, and finally Europe with Greek and Islamic philosophers. Now, let's give some Greek and later philosophers below:

HERMES THOT 1300-800 B.C. (P.74) - He was the first Greek, who learned the information that the priests in Egyptian temples, gathered in 7 disciplines **(1-Grammar, 2-Logic, 3-Rhetorics, 4-Geometry, 5-Mathematics, 6-Music, 7-Astronomy)** 3-4 thousand years long. The second Greek was to be **Pythagoras**. According to Hermes Thot, "There is no difference between the little and the big, the inside and the outside. The **light** is the **SOUL** and the **darkness** is the **MATTER**. If the man subordinates to the matter, the divine light in his **SOUL** returns to the place where it came from, and the **SOUL** without the light is left in the darkness, melts down and die out. In the **XII. Century A.D., Mawlana** interpreted this with the **SOUL** over the right palm of the Dervish, whirling around himself and **MATTER** under the left palm.

PYTHAGORAS 570-490 BC (P. 77): After escaping from the temple of Memphis, he established his own temple in southern Italy under the name **Delph**. Later, Al-Khwarizmi would learn the zero **and algebra** from the **Brahman** region **Delhi** in northwestern India, and introduce to the world. According to Pythagoras; the **"Matamatas"** cover the entire information of the universe. According to him, all the matters and non-matter existences are **figures that could be expressed in numbers**. Number is the most dominant thing in the Universe. And the most beautiful thing in the Universe is the **harmony** which is present among the existences and **beauty** is formed through harmony. **Philosophy is the search for unity in the plurality. 1 is a point; 2 is a line; 3 is a triangle; 4 is *an object with four-faces* and it is the divine power; *harmony* represents *virtue*. In Greek *pyro* means fire and *amid* is the center, so *pyramide* means the fire in the center, and most probably the fire at the under the Earth is meant. 10 is *tetraxy* (i. e. the Pythagoras triangle is 1 (+) 2 (+) 3 (+) 4 = 10).**

PLATO 430-347 BC (P-79) later introduced us the **TETRA HEDRON** of Pythagoras as the first regular pentahedron. According to Plato; "The state is an enlarged symbol of an individual's functions, and without justice, the state is degenerated. It is essential to know Geometry to rule people. "

AL-KHWARIZMI - 780-850 (P-81): After returning from India to Baghdad, he was the one teaching humanity how to use 9 numbers and zero in their meanings today for the first time. According to him, the universe founded on the basis of a **SIMMETRY**, which is a geometric definition; he provided this symmetry with two equal number groups and thus brought **ALGEBRA** into humanity. Today, even the space programs and Theories of Everything are prepared with Algebraic Equations. (P-81)

GALILEO 1546-1642 (P-85): The universe is like a book written in mathematical language. Its letters are the triangular, circular and other geometric shapes.

DESCARTES 1596-1650 (P-86): He managed to express the points and point clusters in the space through ANALYTIC GEOMETRY with the CARTESIAN coordinates in an algebraic way.

I. NEWTON 1642-1727 and **G. W. Leibnitz** 1646-1716 **(P-86):** Introduced **calculus**, i. e. the high mathematics. According to this equation, quantities that vary depending on many variables, ie functions began to be calculated. Then counting natural numbers such as 1, 2, 3 led the mathematicians to the concepts of **granular plurality** and **continuity**.

A. EINSTEIN 1879-1955) (P-92) According to the unified field theory or quantum gravity theory, which he defined but could not prove for 30 years, a minimum 10 dimensional space, each of which represents all coordinates of the subatmospheric particles or their interactions and can only occur on the Plank scale (Annex-Ie) (10^{-45} **sec.,** 10^{-35} **m**). Although three **Theories of Everything** were put forward by Einstein to explain everything in the universe, none of them could be proved yet.

F) SOME OF THE FINDINGS THAT ARE PROVIDED IN THIS BOOK FOR THE FIRST TIME:

In addition to the information, which prepared the Theory of Everything, let's list some of the new findings that were revealed for the first time in this book except for these known in the contemporary literature:

The result that emerged in the fourth chapter of the book was the "Double Triangular Pyramid" on page **175-178**, which is the **"Double Triangle Pyramid",** that is, **"Double Rabia",** which is the algorithm for both the Heaven in the World and the Hereafter, for the triangular pyramid at the top for the society and the triangular pyramid for the individual at the bottom. **When the steps of the Pythagorean triangle on page 137; are removed to the levels of 1, 2, 3, 4, 5, 6, 7, 8, 9, 10, 11, 12,** they are given the mathematical table of the number of dots in the triangular pyramids that are formed. At the top of the table with one particle at each of the four corners, **regular tetrahedron**, triangular pyramid, i.e. **RABIA** is seen. The **3-point triangular pyramid has its edges** at the bottom and the total number of dots is **10,** such as the total number of points of the triangle in the two-dimensional structure of the same triangular triangle. Thus, the total number of points of the two-dimensional Pythagoras triangle was **10**, and the edges were obtained with 3-point and three-dimensional triangular pyramid. In Pythagoras, the monks jumped. The total number of points of the triangular pyramid formed in the 8th step of our **12-step table is 120** and it is seen as the total number of the 8-digit (Octave) first triangular pyramid of our Universe after Big Bang in **EK-IIId**. In the same table, **120** sub-atomic particles and 120 elements will end in the middle between Big Bang and Big-Crunch, after which the particles and atoms will be in the middle of the stars. The table on **page 26, 137** shows the same **120** numbers in **120 Neutrons / 80 Proton** rates in the elements of our Universe.

The last element in our two-dimensional known Element table in **Annex IIIa** is seen as element 118. However, the last element in the three-dimensional triangular pyramid of the elements that will occur after the Big Bang of the table in **Annex-IIIc** is the element number **120**. This is more than two elements from our two-dimensional element table. If we find these two elements in the future, this claim will be proven.

In the Table of Subatomic Particles in **Annex IIe**, this number reappears as **120 fermi gluon**. Likewise, in this particle table, the known particles of matter are shown in blue and are given with all the particles to be found. In the **Big Bang Star (BBS) tables** in **Annex IIa and b**, all the particles in the **Annex-IIe Annex-IIe** subtitles with Subatomic **Annex-IIe** have been placed in place of all of the missing and most of them. Cannot be a coincidence. Let ears of the Physicists burn, who underestimate geometry. **BBS** was once given as the sixth of Plato's regular pentahedron.

G) HISTORY OF THE RABIAS FROM THE SUMERIAN, BABYLONIAN AND EGYPTIAN PRIESTS TO THE RESENT:

Actually, the term *ilim* in Arabic is term derived from the word *alem* and it means **sign, form, universe**. So, it can be said that the temple priests in Egypt had defined **ilim**, that is all the sciences as **geometry**. The basic means to understand the **universal truth** through science is **mathematics** and thus **geometry**, which is a summary of mathematics understandable by everyone because it appeals to the eye. The **existence of the universe in the subatomic and astral scales** is the **globules of matter** rotating around themselves and each other. In fact, according to the cosmic energy experts, Energy also consists of Transparent Energy globes. The globules will only be expressed by numbers, i. e. mathematics. It is a known fact that when Element Atoms are forming molecules and compounds, their localization in space in three dimensions, creates concrete **crystals** that can be seen by combining the centers of atomic particles when they become cold and immobile and become solid. It is not yet known whether the **subatomic particles** form similar geometrical structures within the atomic nucleus. Some tables in the Annex of this book have been prepared by further development of the virtual geometries of the particles already revealed by physicists. In this book, the movements and interactions of the particles on the Planck scale cannot be explained with the books full equation and a new everything theory has not been developed yet. Geometry, which physicists underestimate; I have summarized the Virtual Geometry of Subatomic Particles with the concrete Crystals of the Elemental atoms, which are listed above, with the definition of crystallographic geometry and summarized as the RABIA in the light of ancient Geometries and contemporary knowledge. **So that everyone can understand the mathematics simple geometry, which is the essence of all sciences, which encompasses all of our existing and all information was revealed to be above it all.**

On Page 137, RABIA, which is seen at the top of the Pythagorean triangular pyramids table and has its essence in its mathematics and geometry, can be a key to us in the fulfillment of the **"read"** command, the first verse of the Quran. When **"good viruses with artificial RNA"** can be produced based on 4 organic bases (adenine, guanine, cytosine, uracil) in the RNA of plants, the plant diseases and **"good viruses with artificial DNA"**, based on 4 organic bases (adenine, guanine, cytosine, thymine) in the DNA of animals and humans, animal and human diseases can be eliminated without the need for anti biotics and the health problems of plants, animals and people can be solved. With the balance of fire, air, water and soil, the moral health of the world we live in can be saved and "**THE RABIAS OF THE KNOWLEDGE OF PEACE IN THE WORLD AND IN THE UNIVERSE**" can also be achieved. **(P-10, 175-178)**

RABB, means ALLAH in Arabic and Hebrew languages. According to the Egyptian priests and the Sumerian and Babylonian priests in Mesopotamia; the sun god RA (a sole god) meant four directions or 4 dimensions, as in Asian history. Tammuz, one of the sun gods used to be born on July, the 4th. The alternative of KOTH, the sign of the Babylonian kingdom is the sign of RABIA, that is, the sign with the thumb of the hand closed and the other fingers open. According to the ZIONIST Jews, ZION is the same as the RABİA.
Z - Open index finger.
İ - Open middle finger.
O – Open ring finger.

N - Open little finger.
and closed thumb represents 12 Jewish tribes.

The sign of Rabia means opening and closing the doors of both enlightenment and hell. The eye on the top of the pyramid in the US dollar today is RA, which sees everything in the global world, represents forcing God to doom, and the name of the Babylonian Kingdom is KOTH, which is presently ILLUMINATE according to the Jews.

The knowledge of the Sumerian, Babylonian and Egyptian Priests were first developed by the Greeks, then Muslims, then Europeans, westerners, and today we have reached the information age and the space age. We cannot only find the right solution by highlighting the original Rabia or Zion today and ignoring the developments before and after the Priests. In this book, 3D geometric synthesis is the first time that is the result of the analysis of the evolution of knowledge of humanity and the two-dimensional Sacred geometries of the history of religion and belief; (pages 10, 174-178,190,193) became the "Double Rabia crystal" as the geometric algorithm of peace in the Universe and Earth. Thus, for the first time in history, we have reached a comprehensive contemporary solution that will envy the Sumerian, Babylonian, Egyptian and Brahman priests. Since both enlightenment and heaven can open and close in ILLUMINATI, in RABIA or KOTH, there is no need to force God to doom, to prevent the third world war in the world, to close the doors of hell and to open the doors of Heaven, we are no longer in front of us.

The Egyptian and Mesopotamian priests tried to explain that the path to reach Allah through the word RABIA, meaning both the numbers from 1 to 4 and the triangular pyramid, that is through mathematics. We also have to remind and ask that the word Arab can mean an individual, who is on the path of Allah (Rab). Now, without asking any society to account, we all together, can start working to create peace and paradise in the world with the RABIA, THE GEOMETRY OF PEACE ON EARTH AND GOLDEN AGE.

While searching for the creation code of the Universe up to this point, it was the mind that this code could be searched in **Mathematics** and **Geometry**, which is its natural expression. Considering the fact that the **universe**, such as **subatomic and astral spheres**, is still an expanding sphere (perhaps the ellipsoidal egg), it is necessary to start looking for the universal truth in the geometries inside the sphere, namely, 5 smooth faces of Plato. In **Annex-IIa and b**, Big Bang Star (BBY), which I think to be the sixth and last regular polyhedron was given. While the final stage of the geometry inside the sphere in 3 dimensions was BBY, the first stage was Plato's first tetrahedron, the RABIA. Thus, at the top of all sciences, the fact that there is 1,2,3,4 in number and Tetrahedron in geometry in mathematics and mathematics was the main reason that we defined the theory of Everything as the RABIA only. In the figure shown on the front cover of the book and below, the atomic universe seen from the right to the **human, subatomic particles, atoms, compounds and human's GEOMETRIC RABIAS**.

Before we begin to explain the first 4 issues of Pythagoras and its tetraxia as well as the double Rabia crystal of peace in the Universe and the World **(P-10, 174-177,190,193),** let's take the table in the upper part of the front cover of the book in more detail.

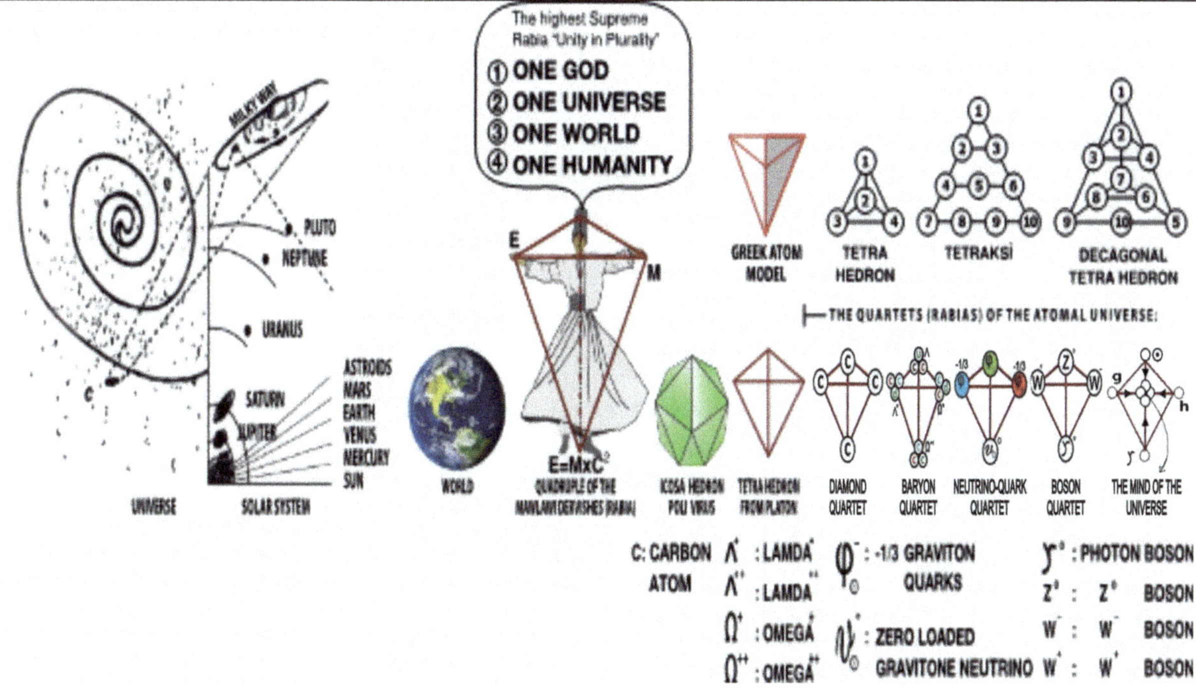

When the **triangle** formed between the hands and head of the **Mawlawi dervish**, representing the **micro-cosmos**, that is, the human, is combined with the one between the feet, the **RABIA** is formed. By repeating the triangular pyramids in three dimensions with the triangle between the hands and the head of the dervish, all geometries, including the sphere and the circle in two dimensions can be formed. 4 also represents 10 numerically according to Pythagoras, that is, the total number of points of the 10 pythagorean triangles. **(1 + 2 + 3 + 4 = 10)** and the numbers following **4** can be obtained from the operation of numbers up to **4** with each other. $10^{-\infty}$ -$10^{+\infty}$ is enough to describe all the universes. The dervish, representing the **wise** and the **perfect** human, has all abstract **spiritual values** on the right palm, and all tangible **material values** under the left palm. **Pythagoras** said that the letter **epsilon (γ)** resembles two ways in front of man, these are ways of **virtue** and **evil**. When **Einstein's E = MC²** formula is set under the dervish, if the **material values** of a human is **1**, then the **spiritual values (virtue)** will be as high as the speed of light (300 million m / sec). So if one knows how many are on the right, how many are on the left, even if you jump down his/her throat, he/she will not look at someone askance. Even though it is late, this truth must be told to everyone. In the table above, let's now briefly define the RABIA, including the geometric pyramids of the subatomic scale, formed in groups of four on the Planck scale, from right to left, after the Big Bang, to the human, in other words, including the geometry of the Mevlevi dervish, namely, the wise. Above all the universe from subatomic particles to the entire universe, Einstein's **E = MxC²** formula below the Rabia of the Mevlevi dervish applies. So **E = MxC²** is valid for both **microcosmos**, i.e. human and **macrocosmos**.

H) THE NUMERICAL AND GEOMETRIC RABIA FROM 1 TO 4 OF THE THEORY OF EVERYTHING AND PEACE IN THE UNIVERSE :
1 IS A POINT ACCORDING TO PYTHAGORAS:

The number 1 is the uniqueness that is the principle of the triangle in all directions in the universe, the uniqueness summarizes the three. The philosophy of Pythagoras and Islamic Sufism are the search for unity in the plurality. In the **Sema Ritual**, the Mawlawi dervishes whirl around themselves and around each other, like the Astral particles known in these times, and the atomic particles known today, to reach **Allah**, the creator of all things through the mystical way.

In each person's brain, the formation of the consciousness of his own personality; **Bose-Einstein condensation** of 4 **bosons**, electrical and quantum in the tangible computer-like system of 10^{20}

nerve cells (**neurons**) in the brain. The Quantum of **vacuum** in the human brain carries both **particle** and **wave** characteristics. The idea of **vacuum** develops into a renewed consistency and returns to Vacuum as enriched fluctuations (Quantum). In the Universe, after the Big Bang before Vacuum in the Universe, it explodes with the **Big Bang** and then expands, and then comes back with the **Big Crunch**. The same Big Bang pre-vacuum is a pure energy like Vacuum. In other words, it is seen that the energy of the universe in the vacuum before the Big Bang is formed in the **human brain**.

The Universe may also have become conscious of itself like the human, through the "**vacuum**", which is the result of the previous universe becoming a black hole and destroying its anti-matter, that is, through **the Quantum Bose-Einstein Condensate**. Is this to a certain extent evidenced by the existence of the God, which **Salvador Dali** predicted in his **nuclear mysticism**? Is the text, where Allah wrote everything before creating and defined as the **Lawh-i Mahfuz** in the Quran and **Nur-u Kadim** in Sufism, scientifically a **Bose-Einstein Quantum Condensate** within the vast energy before the Big Bang? If this is proved in the 21st century, would the **religions** not be unified and the purpose the Mawlawi Dervish approaching Allah through mystical ways by whirling in the Sema Ritual reached by means of **science**? (The mind of universes on P-189) Thus, the gate of Truth, which is the fourth gate opening to Heaven in Sufism, does not open up to all humanity? (Gates of **1- Sharia, 2- Tariqa, 3- Marifa, 4- Haqqiqa**). Would the **human consciousness** and the **consciousness of the Universe**, having the same physical formation, lead the humans to live in accordance with the divine order of Allah in order to solve all our problems?

2 IS A LINE ACCORDING TO PYTHAGORAS:

There is a dilemma between the **illuminated universe** that is expanding and shining with the stars and a **dark universe** consisting of Black Holes and Blackmatter on the Astral scale.

At the atomic scale; as our Universe began to expand with the Big Bang (Annex-Ia), first the **Bosones**, which are the particles that transmit and impact the 4 forces of atom, the Consciousness, the Mind and establish Relation, and then **Fermions**, which are the building blocks of our universe and the particles with which the Relation is established.

MIND was the Relationship and **Matter** was the one to be established Relation with. Bosons have the characteristic of wave, whereas Fermions have the characteristic of particles. Just as the word mind means binding in Arabic, Mind; relational, matter; relationship was established. BOSONS are waves and FERMIONs are particles. FERMIONS, which are the building blocks of the matter of our world, prefer to hide themselves separately, so they are asocial. **Bosones** are social, they tend to come together and create groups. Would it be wrong to say that Western societies have the feature of Fermions and whereas the Eastern societies feature Bosons. While less Bosons are loosely assembled in the non-living things formed after the Big Bang, the number of the Bosons, assembling in the living things increase towards the human and they grow stronger ties. Also the Gravity area enabling the Stars, Galaxies in the astral scale and Atoms and molecules in the atomic scale to gather and build clusters is a Boson Field. So Graviton and Higgs particles are also bosons,

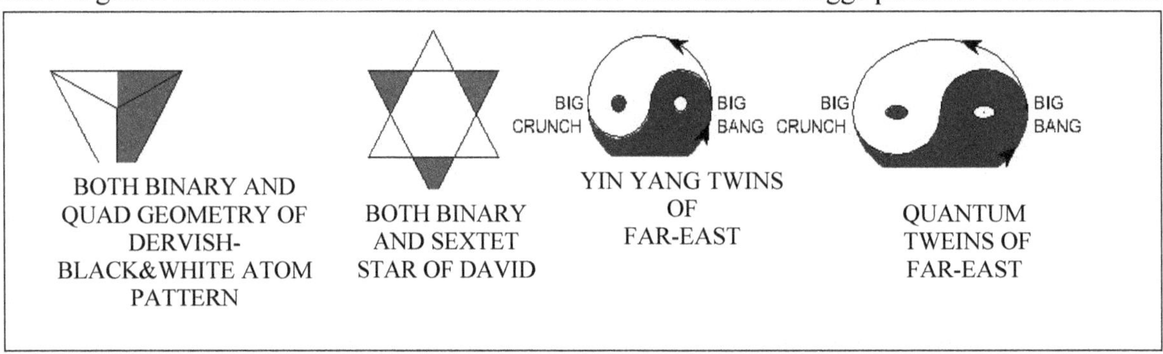

Above given binary examples respectively from the left show; half black and half white triangular pyramid the **greek atom model**; upward white triangle representing **water** (female element), downward black triangle representing FIRE (male element) of the **star of david** show a harmonious duality. According to Chinese duality symbol **yin-yang (P-71, Figure 6),** all the contradictions in the micro and macro level in the universe can survive with **balance** and **harmony**. Since **yin-yang**

means the rise of the Universe, the rotating disc have been given as the quantum of the white luminous Universe that expands with the Big Bang and the dark Universe that will gather with the Big Crunch.

It is another dilemma that the Universe, shown on the front cover of the book and in Annex-Ia, is a shining lumious one inside when expanding, while it is a Black matter outside. In the Universe to be gathered, however, it may be thought that there may be a contradiction as a Dark Anti-Matter inside and a shining luminous Anti-Matter outside.

In the Middle Ages, Leonardo Da Vinci tried to explain in his human model in the center of an interbedded Circle and Square that the Circle represented the Soul and the Square represented the Matter and that Spirit and Matter merged in the Human body. It was very accurate that Da Vinci was able to express this through Geometry after he learned of the Spirit and Matter of man from Andalusia in Islamic Sufism. Geometry was neglected in Islam because drawing images was forbidden.

The Da Vinci symbolized the **soul** with the Circle and the **matter** with the Square. The square and cross-used to represent the Earth as East, West, North and South already in the ages before Christianity. However, putting a naked man into the center resulted both in ignoring the women, and damaged the message, that the "Wise and Perfect human should be balanced between the **matter** (Egoism) and **soul** (Devotion)" which Sufism wanted to explain with the Dervish, representing the **wise** and **perfect** human. Da Vinci's message led to materialism, racism, colonialism, imperialism, and two world wars in Western societies. Today, it is dragging the societies towards the Third World War. The members of the modern Sufism, academicians, and international Islamic organizations should present this correction to the Western scientific world better than me without delaying. Even for the purpose of preventing the 3rd World War, an urgent agenda should be created and discussed at international conferences and symposiums. It is not wrong to say that the Islamic organizations, Pope, Jewish organizations and similar organizations of other religions have the chance to preventhe 3rd World War as a **balance** element.

3 IS A SURFACE (OR TRIANGLE) ACCORDING TO PYTHAGORAS:

So we understand that the contradictions that make up the dualities in the Universe are an inseparable whole, and that they do not have any meaning alone, and the balance (Compliance, Adaption, Harmony, Symmetry, Justice etc.) between controversies is a condition for the former two elements to maintain their existence as a third element.

3 is the compliance, adaption and harmony, morality. The balance, compliance, adaption and harmony of the universe creates; equation in Mathematics, symmetry in geometry and harmony and beauty, aesthetic between the dimensions and frequencies of the colors of the existences.

It provides harmony, peace, justice and balance in the disagreements between people as individuals and society. Likewise, for the people, Adam, Eve and their children, and mother, father and children in the family are the most important trilogy.

In Fine Arts, the artists give beauty to their works as to the colors, dimensions and harmony of the works to their environment, to the extent that they reproduce the balance, compliance, adaptation and harmony in the Universe.

Justice in society, harmony in music, children in the family after parents, golden ratio in fine arts represent the balance that the Mawlawi dervish should provide between the soul and the matter.

We observe that the contemporary Western societies with their materialistic individualistic mentality and the Eastern societies with their morale and spiritualistic mentality become materialistic as they transform into each other. Although a Third World War has not occurred yet, the political, military and economic developments at the global level continue to be agitative. However, the result from the above information: Today, neither individualist and materialist nor spiritualist societies can survive on their own; both of them should integrate by transforming or changing each other in some
extent without applying a force and limiting the other's freedom to provide a third element: **ethics, balance**, **justice, harmony,** whatever we call it. This conclusion is a must also for the Quantum Identities of both the Universe and the Human, which is actually the same.

Message for the individual: **Arif** (Knowledgeable), **Kamil** (The perfect human, who lives according to his knowledge, well-behaved) should provide a balanced life between both matter and spiritual values.

Message for the world society: Both the individualist western societies as well as the spiritualist eastern societies, should unify under a "United World State" on the basis of ethics, justice and knowledge for the survival of the people of the world.

a) TRIADS IN THE DECIMAL NUMBERS: At first, there were numbers in the universe. Because numbers are not only the expression of matter, but they are also the expression of energy. When the energy materialized, the equivalent of numbers, respectively: **elementary particles, Atoms, Molecules, Compounds and then Viruses, Bacteria, Plants, Animals and humans** emanated in the atomic dimension; and **Stars, Planets, Satellites and Galaxies** emanated in the astral dimension, after the Big Bang. Below, the fourth number after three digits constitutes a new unit (Triangular Pyramid):

1 2 3	4 5 6	7	8	9	10	11	12
1 10 100	1000 10000 100000	1 MILLION	10 MILLION	100 MILLION	1 BILLION	10 BILLION	100 BILLION

b) TRIADS IN GEOMETRY: Points take the place of numbers in Geometry. **1 point. 2 points (line), 3 points (triangle). 4 points form the Triangular Pyramid.** If repeated, all 3D geometric symbols can be formed with Triangular Pyramids (including Sphere).

c) TRIADS IN PHYSICS:

c1) EXAMPLES FOR TWO DIMENSIONS IN THE FUNDAMENTAL PARTICLES OF ATOM:

It is the creation of **electron** and **positrons** from the first **photons** in the first seconds of Big Bang as can be seen below:

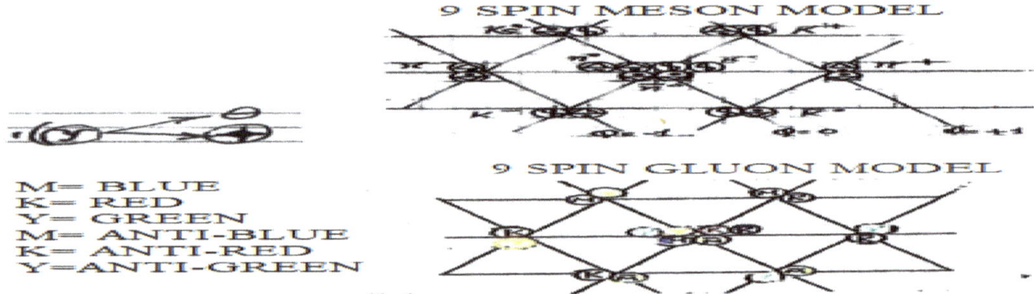

c2) NUMERICAL TRIADS SUBATOMIC PARTICLES

c2a) BOSONS, LEPTONS AND QUARKS IN SINGLE PARTICLES TABLE IN ANNEX-IIE: They are +, -, 0 charged. Quarks in blue-green-red color; **Anti-Quarks** in 3 opposite colors as yellow-orange-purple; also light has 3 basic colors like blue-green-red.

c2b) TWO PARTICLES (MESONS) TABLE IN ANNEX- IIh: All Mesons are +, -, 0 charged. Triangles in the virtual **9 spin Meson model** as can be seen above.

C2c) THREE PARTICLES IN ANNEX-IIi:

-1/2 spin Baryon model which is composed of three quarks baryons in two dimensions.

- In the laboratories, when Zero charged (neutral) Neutron Particles breaks up, they are divided into (+) charged Proton and-charged electrons.

c2d) + charged Protons, zero charged Neutrons and-charged electrons in the atom nucleus are examples of triads.

d) OTHER TRIADs:

d1) Everything is **born, grows and dies**.

d2) Our **perceivable universe** being **3 dimensional**.

d3) In dialectics, there are **Thesis, Anti-Thesis and Synthesis**.

d4) According to Taoism: Tao gives birth to **one**; one creates **two**; two forms **three**; three brings **all beings** of the world into existence.

d5) In Asia, elephant headed stands of 3-pedestal cooker in the center of the tent, which represents the hearth and home, represent **Mother, Father and Children**; and a community which does not depend on hearth and home cannot survive.

d6) According to Buddha, "Happiness is neither in the Physical life, nor in the Spiritual life. It is in the middle of them."

d7) According to **Hermes Thoth** who is the first Greek learning the data in the temples of Egypt before Pythagoras (**approximately800-1400 B.C.**), "There is no difference between the small and big; between inside and outside. Light is the **Soul**. Dark is the **Material**. If humans resign themselves to the matter, the divine light in their souls go to where they have come from. Soul without light is left in the dark, and dies off." "Under the left hand of the Mawlawi dervish, which is the interpretation of Mawlana, there is MATTER (Darkness), there is SOUL (Light) over the right palm."

e) TRIADS IN THE COMPOUNDS:
Blue-Red-Yellow Colors of Paints. Other colors can be achieved from these colors.

4 IS A REGULAR TETRAHEDRON OBJECT ACCORDING TO PYTHAGORAS: (TRIANGULAR PYRAMIDE) AND 4 NUMBERS FROM ONE TO FOUR)

I) TRIANGULAR PYRAMIDE RABIAS EXISTING IN THE UNIVERSE

a) TRIANGULAR PYRAMIDE RABIA IN THE SUBATOMIC PARTICLES

a1) "**Double Pythagoras Triangular Pyramid**" (**P.175**) can be seen in **Annex-IIg**. It is composed of the last 10 particles (10 anti-boson), Mesons with double particles from the first 10 particles (10 boson) and the Baryons with triple particles after Big Bang.

a2) In the table of the **quantums of the universes** in **Annex-Ia**, the first quantum couple formed in the universe from 10 pairs of quantities belonging to BBS was given at the bottom right. Double triangular pyramid, which can be seen below in the center of the first quantum pair which is composed of Double Rhombic prism, the first virtual geometry, is triangular pyramid, in other words, the last of the previous universe and the first **RABIA** of our Universe. which is composed by the last 4 anti-matter particles of the previous universe and the first 4 matter particles of our universe, during the transformation from the previous anti-matter universe to our matter universe. Later, increasing in sequence, rhombic prisms (quantums) constitute 10 pairs of triangular pyramids in the center of BBS (**Annex-IIa, b**) and 10 pairs of triangular pyramids on the outer surface of **BBS**. We can think about the probability of Weavings (in **Annex-Ie**: "History of Our Universe" Table) which have appeared in the first 10^{-11} seconds of **BBS**, being **BBS** itself.

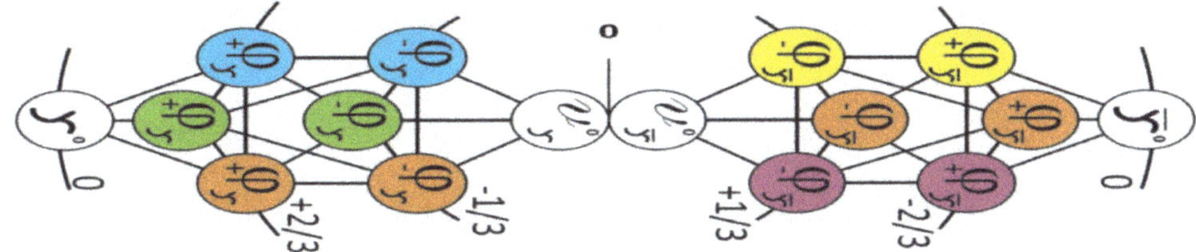

THE FIRST QUANTUMS, OCTAVES (OCATAHEDRONS) OF THE UNIVERSE

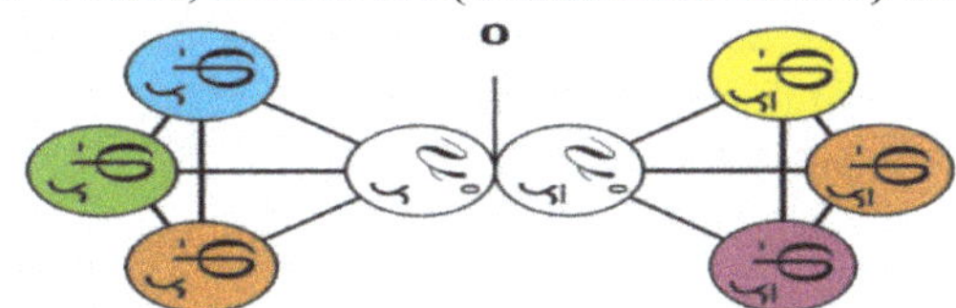

THE FIRST TRIANGULAR PYRAMIDS OF UNIVERSE

a3) The arrangement in the Pythagoras triangle is formed on all four sides of 20 baryon pyramid which is composed of virtual **3/2 spin "souped-up" baryons**, and which can be seen below.

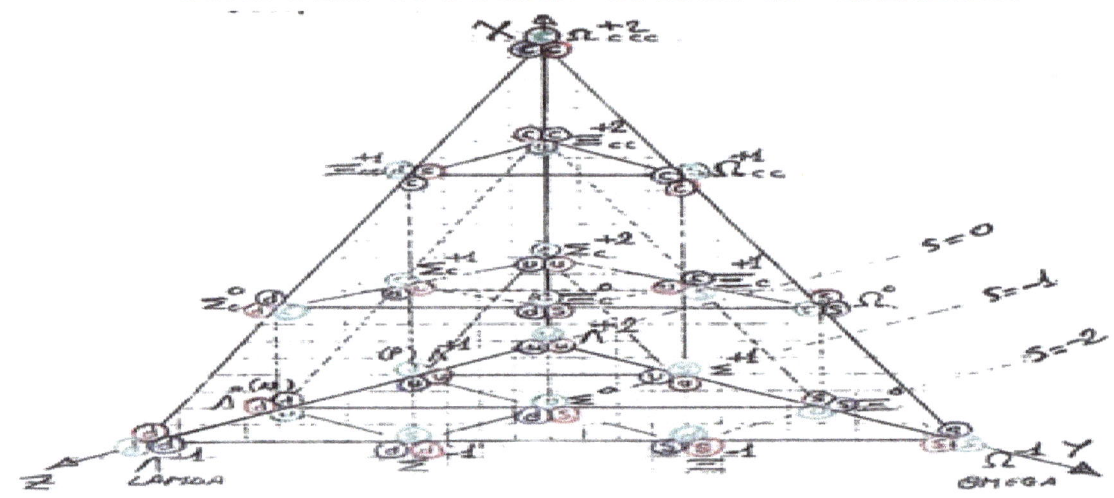

Also the same triangular pyramid structures can be seen in the crystals of the following elements and compounds and in the meanwhile in the ice crystal: These triangular pyramids can be considered in addition to the RABIAS on the front cover of the book.

b) TRIANGULAR PYRAMIDE RABIA IN ATOMES AND MOLECULES:

On the upper right side of Periodic Table of Elements (**Annex-IIIa**), *"Elements with Triangular Pyramid Crystals"* which make more compounds than all other elements (especially complex organic compounds) can be seen below. All of the elements are formed by Helium nuclei. In other words, it is not wrong to call them **helium Rabia**

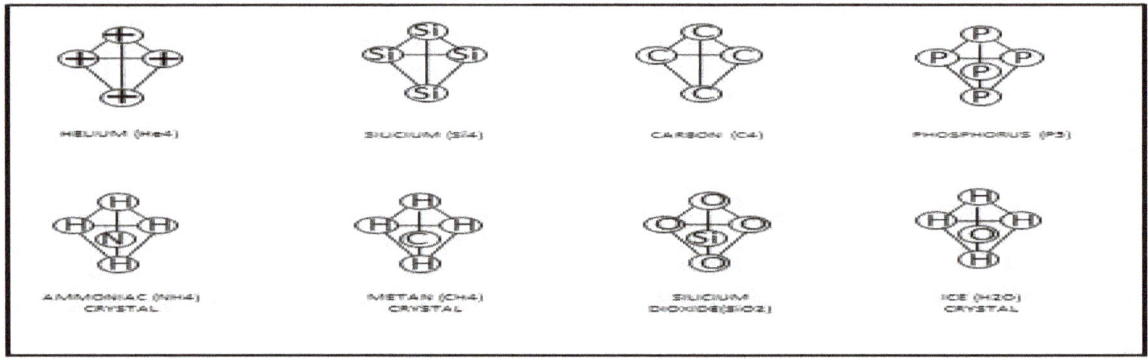

c) TRIANGULAR PYRAMID RABIAS OF CRYSTALS:

On **page 42**, there is a triangular pyramid under the heading c3h of the Trigonal, Bipyramidal 6 (triangular double pyramide crystal). (**Double Rabia Crystal**).

d) THE GEOMETRIC RABIA (P-174-178) IN THE FLAG OF THE PROPOSED "UNITED STATES OF WORLD":

1) **TOP UNIVERSAL INFORMATION (RABIAS)**
2) **GEOMETRIC SYMBOL: Symbol** according to Egyptians and Greeks; To discuss together, to combine together, to gather together, to connect to cover everything. Like the word "akale" in Arabic which means combining the word "akıl -*mind*" in Turkish. The top Universal information can only be combined with geometry that can address all people from the same language.
3) **IDEAL PURPOSE: ETHICS.**
4) **INTELLIGENT INDIVIDUALS**: So according to the Greek philosopher democritos; individuals, who think well talk well do well, behave well.

II) NUMERICAL RABIAS EXISTING IN THE UNIVERSE (4 NUMBERS FROM 1 TO 4)

a) NUMERICAL RABIAS IN THE MUSICAL NOTES

On **page 156,** we see the Pythagorean triangle with the names Tetractys and Lambdoma II (1 + 2 + 3 + 4 = 10). In the table on **page 157,159,** the Octave numbers of frequencies that the human ear can hear are also 4-8-16-32-64-128-256..and so on. it continues as multiples of four.

b) NUMERICAL RABIAS IN THE DNAs OF HUMAN AND ANIMALS AND IN THE RNAs of the PLANTS

On **page 141**, the 64-digit harmonic ratio of LAMBDOMA (λ) on **page 156** is similar with 64 different codes (64 Amino acids) of 4 types of Bases.

c) NUMERICAL RABIAS IN THE OUTMOST SCALE OF OUR UNIVERSE

In the tables in EK-Ib, c, d, our Universe will expand in 4 stages according to the Proton Neutron ratios + 1, 2/3, 1/2, 1/3, 0 from Big Bang to Big-crunch. - after crunch - -0, -1/3, -1/2, -2/3, -1 as charged anti-matter Anti-proton, up to II.Big Bang in 4 steps according to anti-neutron ratios how to collect. These formations on the outermost scale of our universe are realized by the effects of the electromagnetic force (EM), strong nuclear force (SNF) and mass gravitational force (MGF) of the atom **(P.27-29).**

d) NUMERICAL RABIAS IN THE QUANTUM SCALE OF OUR UNIVERSE

In the next chapter, the "Uncertainty Principle which is put forward by physicists in the quantum physics, which is the deepest subject that contemporary science can reach, will be explained. Einstein argued that this uncertainty was caused by a parameter unknown to physicists. In the light of the tables and information in this book, we will try to find the missing parameter that Einstein mentioned on this occasion while describing the numerical RABIAs on the Quantum scale.

In the table in ANNEX-Ie, the quantum particles known to physicists are given in blue and the ones that they do not know are black. In EK-Ic, our Universe is loaded with + 2, 3/2, 1, ½, 0 or 4/2, 3/2, 2/2, 1/2, 0 spins from Big Bang to Big-crunch all quantum particles were expanded in 4 steps, and after the Big-crunch, the charged anti-quantum particles were -0, -1/2, -1, -3/2, -2, or -0, -1/2, -2. / 2, -3/2, -4/2 spin numbers are seen to be collected in 4 stages.

EK-Ib, c, d Expansion of 4 stages in the outermost scale of the universe and its collection in 4 stages were explained according to Proton / Neutron ratios in element atoms. I wonder if quantum particles were formed in the expanding universe in 4 stages according to + charged positron / neutrino ratios, in the collected stage in 4 stages according to - charged electrons / neutrino ratios? In other words, the ratio of Electron or Positron in each quantum particle to Neutrinos differed.

In this regard, the fact that the number of spins of the mass graviton superwoman in the upper right corner of EK-IIe is 3/2 only corresponds to the number of 3/2 spins of the mass graviton superwoman in EK-Ic. It is also seen that the supercharge of the loaded massless Anti-Graviton at the end of the universe to be collected coincides with the number of spins -3/2.

When quantum physicists look for Einstein's missing parameter in the "Uncertainty principle, it may be useful to take into account the above.

d1) THE WAYS OF QUANTUM PHYSICIANS AVOIDING FROM THE INDETERMISM THAT EMERGED DUE TO THE "PRINCIPLE OF UNCERTAINTY":

The determinism that occurs according to the relativity theory of the macro universe, that is, "for certain reasons, under certain conditions, certain results are achieved" led us to indeterminism in the last stage according to Quantum theory. So something used to be either right or wrong, both right and wrong, but now on the quantum scale, right, wrong, has become the possibility. First of the reasons that lead quantum physicists to indeterminism.

In the atomic model of Louise de Broglie, seen on **page 134**, electrons in the circular orbit around the nucleus in the form of quantum packets of 50 000 km per second. As it rotates rapidly, a small area on a quantum scale will only be seen as a cloud in the Heisenberg-Schrodinger-Dirac model. Because quantum physicists could not calculate the position and velocity of electrons around the atom at the same time, they called it the "uncertainty principle". In a second event that led physicists to uncertainty; they sent electrons from an electron gun to a metal plate with two slits on it, the electrons passing through the gaps in the plate passed behind the plate, making wave interference. This time they installed a measuring device to observe the electrons passing through the plate. In this case, the light passed through the gaps in the plate and left a streak-like mark on a second light-sensitive plate on the back, i.e. it became granular. Thus, they concluded that light had both wave and particle properties. However, in my opinion, the measuring device has the feature of matter because it reduces the speed of electrons. Because we know that matter is a slowed state of energy. It is hardly a question of which limit the light slows down to become a matter, ie it becomes a particle.In addition, according to W. Pauli, "In an atom there are never two electrons with the same 4 quanta number." This may be the reason why light has both wave and particle properties.

Einstein, B. Podolsky, and N. Rosen (EPR) have argued that incomplete parameters in quantum physics, which are still unknown, cause this uncertainty. Einstein, too, opposed quantum physics, "God does not throw dice," he said. Einstein was right because the principle of uncertainty put forward by quantum physicists, namely indeterminism, leads the world to fear, panic and chaos. We have to eliminate indeterminism, even on a quantum scale. But if it really is, let's not leave it on the quantum scale and infiltrate it, or look for ways to return to determinism by means of the Rabias, which we will explain below:

I) "UNITED STATES WORLD" ACCORDING TO THE GEOMETRY OF PEACE IN THE UNIVERSE AND THE WORLD:

We are homosapians, who could read, talk and originated from an African tribe that migrated from the south Nile region to Central Asia 42 thousand years ago. We established numerous states until and after the flood of Noah. Today, in 2021, we are a population of 8 billion. We know that if all the bombs in all countries having nuclear weapons, would explode, they could destroy the world hundreds of times. Today, because the big states comprehended that they could no longer fight with Atom and Hydrogen bombs, they have turned to asymmetric warfare with terrorist organizations today, unfortunately there is no possibility that the world will have no future because there is no ethic, justice and peace.

If we want to establish a united world state without a war, the qualification of **being wise** that Central Asian peoples, who have had the most state experience for 42 thousand years, can lead us. If we start from Sumerian king Gilgamesh, who searched for knowledge in 2000 BC, from the deification of the people **(P-51);** In the II. Century B.C., the people loved the former King of Sweden, **Odin, (P-181)** who came from the country of the Turks and the Asses (Alans) who gave their name to Asia, so much that they deified him after his death. Like the Gilgamesh (Bilgamış knowledgeable, holy, noble) epic of the Sumerians, besides his qualification of being WISE, **Odin** also had a qualification of justice and an eight-legged mythological horse. 8 represents infinity in mythology. The message is clear: Odin was saying; "I am forever present with **Justice** and **Wisdom**". However, let us re-evaluate Odin according to the **"SOCIAL GEOMETRY OF PEACE IN THE UNIVERSE AND GOLDEN AGE IN THE WORLD",** :

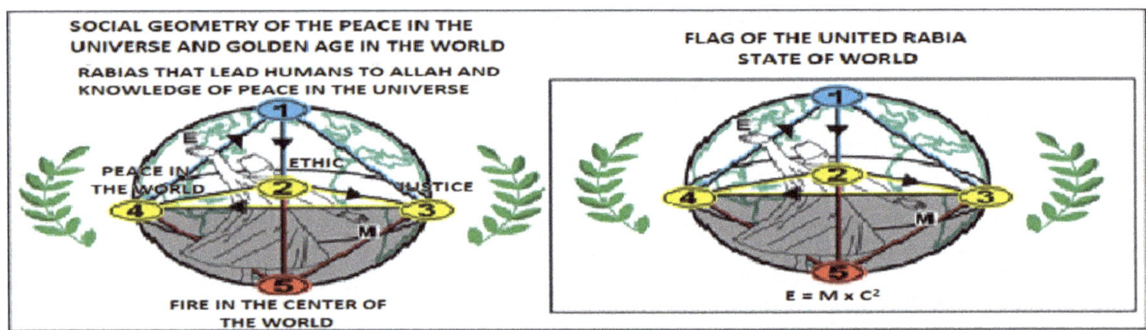

If **Odin** did not have **ethical** first, he could not have **justice**. Thus, according to his knowledge of his own heavenly belief, he believed in the need to be ethic. Because even the Runic script they used was written from top to bottom like the Orkhon inscriptions in Asia. Thus, he must have provided sufficient time in peace in his own community. However, since there is no State that has provided Peace in the World until today, we cannot give an example of history for the **"UNITED RABIA STATES OF WORLD"** to be established. The double triangular pyramid in the diagram shown above on the left overlaps with the double triangular pyramids between Big Bang and Big Crunch in Appendices Ia and IIId of the Book. In Cosmology Institutions in the World, the deep relationship of the double triangular pyramids in these annexes of the book with the subatomic particles seen in ANNEX-IIf can be investigated. 2D sacred geometries in the history of thought and belief of mankind on **page 189,190** appeared here for the first time as an **"Algorithmic double Rabia crystal"** with a 3-dimensional. This two-dimensional Star of David can also be interpreted as a three-dimensional "Nested double triangular pyramid" or Rhombic Pyrisms with a total number of 8 vertices in ANNEX-Ia,IId and **page 151** because 8 represents infinity Its implementation is only possible with the approval of Cosmology chairs at World Univerties today.

I had been wondering, what would come out in the end when I was trying to synthesize the information I had bunched together for 56 years aiming a better World and humanity and in order to guide people for Universality, to find the secrets of the Universe, and to reach a contemporary, easy-to-understand synthesis.

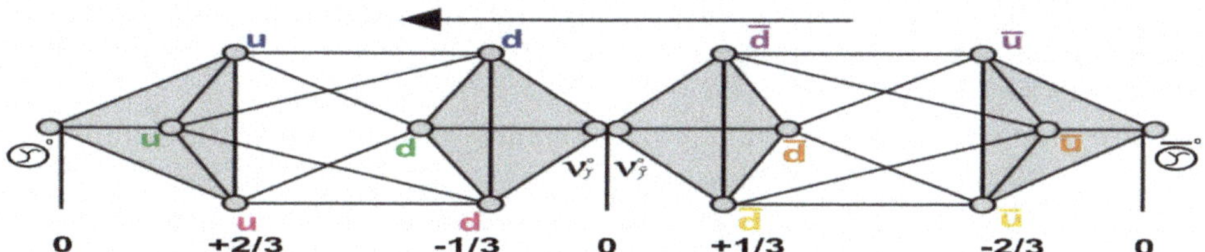

In the first part of the book, contemporary knowledge about **UNIVERSE**, and in the second part, about **HUMANS** is given. In the third part, it is concluded that the synthesis of universe and humans is the **"Triangular pyramid"** (**RABIA**) which represents the human (i. e. micro cosmos) and which is also the geometry of Mevlevi Dervish as per the contemporary knowledge in the previous two parts. The result of a more comprehensive study in the fourth part of the book, based on the additional tables at the end of the book, which is not in the contemporary literature and is the algorithm of the salvation of the Earth; It became a **"double triangular pyramid"**, that is, **"DOUBLE**

All the elementary particles given in Annex-IIe were determined on the 10 pairs of virtual rho mbic prisms in Annex-IIa and b, and BBY (Big-bang star) was obtained. In the first virtual double rhombic prism of BBY in Annex-IIb above, the last anti-boson particle of the previous universe, our own the universe's first particle of light turns into a boson.Then, at the atomic scale, all the other single particles given in Annex-IIe, double particles in Annex-IIh, triple particles in Annex-IIi, atoms in Annex-IIIa,b will be given in order. Thus, the molecules, compounds to be created in the universe and also on the astral scale, quasars, galaxies, stars, planets and their satisfied life are finally life and people we know on planet earth.

This two-dimensional star of David can be interpreted as a three-dimensional **"İntertwined double triangular pyramid"** or a **"Rhombic prism"** formed after the **Big - bang** in the Supplementary tables of the book._According to the Egyptians and Greeks**, SYMBOL** means to discuss together, to unite together, to gather together and connect in a way that covers everything. The highest Universal knowledge can only be combined with geometry that speaks to all people in the same language.According to Einstein, the best solution is the simplest and according to Newton the truth is hidden in simplicity.

If we can manage to establish a "United States of the World" to save the world from the third World War, with this "Double Triangle Pyramid" on the World Flag, we can start the **Golden Age,** which the Sumerians miscalculated and predicted to begin in December 2012, perhaps in 2022. The golden age, on the other hand, becomes the **"Red Apple"** that we have been looking for throughout human history and will bring us paradise both in the World and in the Hereafter. In this geometry, 3 basic concepts at the corners of the triangle at the base of the upward triangular pyramid; **JUSTICE**(3) was given for **PEACE**(4) in the world**, ETHICS**(2) for justice, and **ALLAH** cc.(1) was given at the top of the triangular pyramid above for morality._Atheists may call it **RABIA** because Rabia is the essence of mathematics, both numerically and geometrically or like Pythagoras, they can simply call it Mathematics or Energy._Only if there is Morality, Justice and Peace, the Earth can be transformed into a paradise with virtuous women, men and children._At the lower corner of the downward triangular pyramid and under the foot of the Mevlevi dervish, the **FIRE IN THE CENTER OF THE WORLD** was given. Thus, from the continent of **Mu** to the present day, the 2D sacred geometries given in this book, for the first time in history, have brought people to Allah. It has been revealed as a 3D **"Double Triangle Pyramid"** geometry.

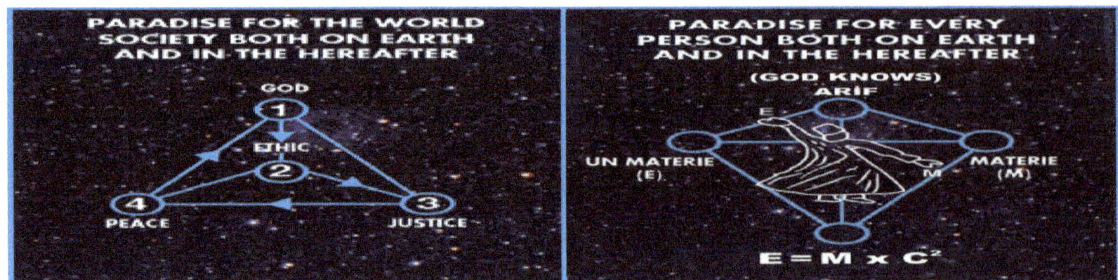

Rabia has Egyptian priests at its source. Thousands of years before Egypt, the Central Asian and Sumerian culture had an inverted L-shaped ÖG stamp. When the 4 ÖG stamps representing the 4 ages, each of which would last 6500 years and would end with **Fire, Air, Earth and Water**, were combined, it became the **Swastika**, the OQ stamp, which represents **"turning to God by turning"** in the 4 ages. This stamp turned into returning dervishes with Ahmet Yesevi , Mevlâna and Hacı Bektaş-ı Veli in Islamic Sufism and has survived to this day. When the swastika went to Germany with the treasure that H.Schliemannn took from Turuva, it unfortunately became the symbol of the Nazi party of Hitler, who was very opposite to its real meaning.

However, placing the flag of the World State to be established, in the upper right corner of the flags of the world, can be submitted to the public vote, even without the establishment of the World State. Thus, humanity is directed to **ETHIC** [In the same sense, **EDEB** in Arabic means of base, rule, foundation, the essence of the 666 verses in the Quran and is to believe in the existence of Allah, that is, faith **(İMAN)**] without losing time, the main problem of our age. However, providing **ETHICS**, **JUSTICE and PEACE** should be the subject of the world's constitution,

Since Allah granted the intelligence only to humankind in the world, every human being is responsible for the existence, health and happiness of other people and all living things as well as their own existence, health and happiness. If people do not bear this responsibility, they cannot survive on Earth Because the existence of every human being depends on the existence of other people and living things on Earth. If only bees disappeared in the world, humankind could no longer survive. In Islam, there is nothing except the concept Allah. Thus, since Allah wanted us to live as a part of the Universe in the magnificent integrity and beauty of everything in the Universe, He gave us a small piece of his own intelligence. Please, let's start using that intelligence before it's too late. The **"double triangle pyramid"** that we put forward for the purpose of controlling people can also ensure that **Artificial Intelligence robots**, which are worked on very intensively today, can be controlled by ethical programming.

A human converging to **ALLAH** becomes virtuous **(ETHIC)** the virtuous human becomes equitable **(JUSTICE),** and equitable people ensure **PEACE** in the World. People who make peace converge to **ALLAH** again. If this chain breaks, the world will come to an end. . Because Allah cc in the Universe there are Mathematical rules that provide the balance and peace between the contrasts it creates. Here for Earth and Humans these rules are summarized in Geometry. Peace reflects beauty to the eyes, music to the ears, health to the body and happiness to all people in the world.

Different information gathered throughout history in the Cortex, the upper membrane of the human brain, which is not found in animals, has been a source of stress and conflict throughout history. Now, let's reconcile these differences with the **"double triangular pyramid"** geometry given here and turn our Earth into Heaven. Then to us moral people, Allah's No one should doubt that he will grant Paradise in the Hereafter.

Our responsibility; The peace that already exists in the Universe should be gathered under the "WORLD OF UNITED LABOR STATES FLAG" seen above, in order to ensure peace in the world, as an example of order. In the sphere in the flag, the double triangular pyramid can be sculpted and placed in squares in large cities. The Universal geometry in this flag emerged from the continent of **Mu**, (P-188) by combining the Rabia knowledge that he found and gifted to the priests in Mesopotamia and Egypt temples in the millennium BC with the example of Gilgamesh and Odin, according to this geometry.

The 4 definitions given in the flag are at the top of the words and definitions that we always speak in the world, in universities, universities, libraries. The downward triangular pyramid will bring heaven in the hereafter for every individual, and the upward triangular pyramid will bring heaven on

Earth for the World society. İn the beginning of the 21st century we live, let us explain the Rabia in this Universal geometry that will save us from the Third World War in order:

1) UNITY: Allah (There are different names given to him according to Mevlana)

The **knowledge of the World** that brings people to the unity of Allah is Rabia

2) DUALITY: It represents **morality**. According to Einstein's formula $E = MxC2$ and fism, one cannot choose only material values (ego, needs) or only human, spiritual values (Energy). He is obliged to balance both. Thinking without a Morality is like thinking of einstein's formula without E. It is contrary to nature, makes societies sick and leads to extinction. Thus, as the basic condition for maintaining our existence on Earth, Morality emerges. Nobodyy has the right to complain about chaos on Earth without teaching people as of the education at the primary schools about the position of ethic in the flag of the united states of world **(P-175)** and in social and universal geometry. Not even to mention the chaos, the world started to putrefy, Viruses come up and that will not stop. We do not know what will happen to us yet. This World must be the World of moral mothers, moral fathers and moral children. **E** (Energy, human) on the right palm of the Mevlevi dervish in the double triangular pyramid, and **M** (Matter) on the left palm.

3) TRINITY: The common ethic (2), justice (3), peace (4) triangle in the middle of the double triangle pyramid. Light enlightens the outside of man, is moral rather than enlightens the soul of man. The term **adalet** (justice) means putting someting into place in Arabic and its opposite is **zulm** (persecution). Thus, its meaning can be expanded to composing with appropriate notes in music, or **writing thuluth** with appropriate arabic letters, creating works in accordance with golden proportions, and managing the state with competent people.

4) QUADRANT: Peace (Kut); represents harmony in the universe, harmony, melody, beauty, happiness in the world and hereafter.

Everything in this book, which has been prepared in 56 years, are given to lead people to Allah, and the geometry that will easily lead people to Allah without reading long books has been put here as the flag of the "United States of World". As Mevlâna says, even if there are other names to God, all the books, all the ways will come to him. **(P-65)**

According to **this semiotic (S-96,104) geometry**, if the establishment of a World State is submitted to the referendum and approved in the existing countries and the existing countries become states and according to the Table of Classification of Sciences in Annex-IVa, if people are categorized and organized in the groups of professions and if every individual connects **to the artificial intelligence Blockchain program in the central quantum computer** through his/her cellular phone according to the occupational group, all problems in the world can now be solved with the right algorithms, human society would gain a monolithic unity and produce happiness like a bee colony produces honey even if the world population would be 28 billion, instead of 8 billion.The fact that the flag seen above is placed in the upper right corner of the flags all over the world today, without the establishment of a World State, is directed to ethic without losing time.

Likewise, human geometry for peace in the world; The philosophy of the Mawlawi dervish, seen Likewise, the geometry of man for peace in the world: The philosophy of the Mevlevi dervish, which Is also seen above on the left and on the front cover of the book, and above on the right the geometry Of the state for peace in the world (Annex-IVa).

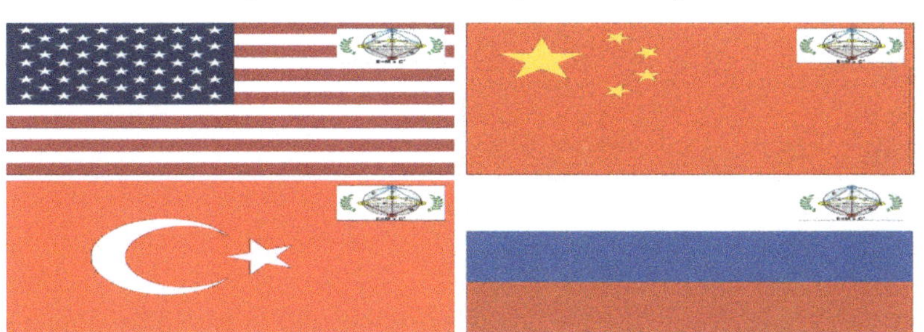

a) DEFINITION Nr. 1 TO ENSURE THE PEACE IN THE WORLD "RABIAS THAT LEAD HUMANS TO ALLAH AND KNOWLEDGE OF PEACE İN THE UNİVERSE"

In the 21st century, in which we live, physicists put forward all three theories of theory, but they could not prove it. We can. now start with the RABIA, without waiting for a theory of everything they can prove and to escape a third world war.

I) THE NUMERICAL RABIA IN THE WORD "SILM" IN ISLAM:
There are 4 main types of peace in Islam: (P-60)
1. **PEACE AMONG ALL CREATIONS IN THE UNIVERSE**;
2. **PEACE AMONG THE WORLD NATIONS AND STATES.**
3. **PEACE AMONG THE FAMILY MEMBERS**
4. **PEACE BETWEEN THE SELFISHNESS (NEEDS) AND THE MIND OF THE INDIVIDUAL.**

II) RABIA FOR SURVIVAL OF HUMANITY, UNITY IN PLURALITY AND PERPETUITY WITH ALLAH:
1) **ONE GOD**
2) **ONE UNIVERSE**
3) **ONE WORLD**
4) **ONE HUMANITY**

Absolute truth that will unite the religions and science in the world are the Rabias (quad riples), which are both the knowledge of peace in the Universe and the essence of mathematics. Maybe Allah wrote everything in a poetry form consisting of 4-line verses when writing in the pure energy state before the Big Bang before creating the Universe, that is in the Ancient Divine Light. (The mind of universes on P-189) Rabias are the call of Mawlana "Come whatever you are!", the expression "kesrette vahdet" i.e. "unity in plurality", and "KUT", that is, the ETHIC, which brings the happiness in the World and the Hereafter to mankind as stated in "Kutadgu Bilig" (P-82). Likewise, the Rabias are the way to achieve the ideal of "healthy, happy, perfect, true Wise people who integrate with everything in the universe and their own existence" (P-54-57), which passes from the Asian Vedas to the Brahmans and then to the Yoga understanding of Hinduism. People who have reached this ideal, will live are no longer with the perception that they are different, with the awareness that they are a part of the Universe, the Earth and Humanity, which gathered in unity of Allah. Unfortunately, we do not have such a common understanding that everyone accepted about "Ethic", which is the condition of maintaining the existence of humanity in the 21st century. Let's combine all the religions and sciences in the world with the "Rabia ethic", which is the gift of the Egyptian temple priests and Islamic Sufism, and let's integrate humanity with the Universe and in this year of 2019 let us start golden age, which was calculated by the Mayans and Sumerians to start in 2012 though we are a little late.

III) THE FOUR REQUIREMENTS OF BEING HEALTHY FOR PEOPLE:
1) **CLEAN BODY AND NATURAL ENVIRONMENT.**
2) **BALANCED NUTRITION.**
3) **BALANCED SLEEP.**
4) **ACTIVE LIFE.**

IV) THE FOUR REQUISITES OF BE HAPPY FOR ALL LIVING THINGS:
1. **HAVING FOOD.**
2. **HAVING CLOTHING OR FURS AGAINST THE HEAT OR COLD.**
3. **HAVING AN ACCOMMODATION.**
4. **HAVING A FAMILY,** i.e. having a partner from the opposite gender

The word "*ocak*" in Turkish is derived from "**üç ok**" meaning three arrows. There is a three pillar Sacred family cooker on fire in the middle of the tent in Asia. (Brahmans call it Mandala and it is called *mangal* in modern Turkish). **The three pillars of the cooker are elephant heads**. The pillars are probably their trunks.. Elephant represents wisdom in mythology. With the Fire in the middle of the tent, both will be warmed up until summer and the food will be cooked. Sun is the source of life.

Hakan (ruler) of the society has a symbol of Sun (a dot in a circle). Japans might be Yakut (Sakha) Turks. *Yakut* means ruby, ruby is red, there is the red Sun in the middle of the white flag. *Hirohito* is deemed to be the son of the Sun. Considering the cooker again, three elephant trunks on three corners of the cooker represent the Mother, the Father and the Children. Wise mother, wise father, wise children.

V) THE FOUR REQUISITES OF BEING A LADY AND A GENTLEMAN FOR THE SURVIVAL OF HUMANITY:
1. **RESPECT FOR THE ELDERLY** is the responsibility of the younger ones.
2. **LOVE TO THE YOUNGER** is the responsibility of the elder.
3. **DEVOTION,** as much as needed, where necessary.
4. **TOLERANCE,** anything that does not end with tolerance cannot be humanistic.

The term **ahlak** (ethic) is the plural of the word "**hulk**" in arabic, it refers to the personality that rivals good qualities, that is, first of all, attachment to God (**Divine piety**, that is to God's commands and prohibitions), then to the **natural environment, society, the state, the family**. In schools, young people should be taught to respect themselves first, as those who do not respect themselves can<not respect others.

VI) THE RABIA FOR BECOMING A PERFECT LADY AND PERFECT GENTLEMAN FOR THE HUMANS:
1) **GOOD KNOWLEDGE.**
2) **GOOD WORD.**
3) **GOOD ACTIONS.**
4) **HİGH ETHİCS.**

VII) THE NUMERICAL RABIA IN SUFISM ABOUT THE FOUR GATES TO PARADISE:
1. **SHARIAT**, Quranic Verses (Ayat).
2. **TARIQAT**, various comments of Qoran.
3. **MARIFAT,.**
4. **HAQIQAT:**

VIII) THE NUMERICAL RABIA OF A IDEAL STATE:
1) **REPUBLIC**
2) **DEMOCRACY**
3) **EQUALITY**
4) **LİBERTY**

IX) THE NUMERICAL RABIA OF PEACE IN ALL COUNTRIES:
1) **ONE FLAG**
2) **ONE NATION**
3) **ONE HOMELAND**
4) **ONE STATE**

X) THE NUMERICAL RABIA OF PEACE IN THE WORLD:
1) **ONE WORLD FLAG**
2) **ONE WORLD NATION**
3) **ONE WORLD CITIZEN**
4) **ONE UNİTED STATE OF WORLD**
   (OR UNİTED RABİA STATE OF THE WORLD)

The initiation of the Golden Age is possible by establishing a "**United States of World**" like the United States or a "**United Rabia States of World**" with geographical names, provided that the sovereignty of each state belongs to its own country unconditionally. Justice and Wisdom, the characteristics of Odin, who is seen on the right above, can be represented by the double bladed sword of the Islamic hero Ali. In a world governed by justice and wisdom, if the human, the most perfect being that is created by Allah, can be ethic also, he may establish a social order that fits both the perfectness of itself or the universe.

If a Eurasian United States is established first in the Middle East, then in the European Union, Russia, India, Pakistan, Afghanistan, Japan, then China and the United States will hopefully reconcile and form a Democratic World United States. Let everyone, including the big capitalists, come and govern the world from Anatolia.

b) DEFINITION NUMBER 2 "ETHICS" TO ACHIEVE PEACE IN THE WORLD:

The reason for all the problems in the world today; is that the people of science and religion have not been able to present a common definition of ethic to society. The most striking result of the analysis and synthesis that took my 56 years from the beginning of this book to the present was the "**SOCİAL GEOMETRY OF PEACE IN THE UNIVERSE AND THE GOLDEN AGE IN THE WORLD**": **(P.174,177)** Peace on Earth can now be turned into an Paradise on Earth. **Justice** is required for **peace**, for justice, people have to be **ethic**. The conclusion, we have achieved through **reasoning**, is that there are "**Rabias, the knowledge of the peace in the universe**" that will direct people to be ethic. Because the word mind, that is "**akale**" in Arabic means "**combining**". Only the mind that combines the knowledges of **Rabia** to eachother, can lead to ethic. As given already, the term philosopher is defined as the person who combined the knowledge of the era on a reasonable basis and created a system. Here, all the information given in the previous chapters of the book have been combined with the mind and reached the "**universal truth**", that is, the **Rabias**. Likewise, the "**Bose-Einstein quantum concentration**" that allowed the universe to reach its own consciousness at the moment of Big Bang and giving every individual to find his/her own conscious through reasoning, overlap with the expression "**I have blown a drop of my own mind into you**" in the Qur'an. I believe that we have to stand up upon these new and striking results and celebrate with fire works in the name of humanity.

What needs to be done is; Among the members of Buddhists, Brahmans, Hindus, Taoists, Christians, Jews, Muslims and other religions, it is accepted that the COMMON RABIAN MORAL, which is the mathematic expression of ethics in the World, the happiness in both in the World and Hereafter, is striving for the sacred ethic, i.e. KUT in terms of Asian Uighurs, is accepted **(P-82** Yusus Khass Hajib). Likewise, according to Hermesthoth **(P-74),** light is **SPIRIT**, darkness is **MATTER**. If a person submits to matter, the divine Light in his soul will return to where he came from, his soul will remain in the dark and will melt away. The exalted soul of man becomes all light, all beauty, all power, all reason, and this is immortality.The **ethic** that will be acquired through **"the RABIAS, the knowledge of the peace in the universe"** will be able to cover not only all the religions but also the sciences. For this purpose, an **"Institute for Researching, Development and Promotion of the Ethics of Rabia"** within the institutes of the cosmology in the world.

According to Pythagoras, the letter **γ** (Epsilon) looks like the parting of two ways in front of a human. The ways are **badness** and **virtue. Mind** is the principle of unity. Prophet Muhammed (PBUH) says, "I was sent only to complete the **high ethics**, and nothing else. Those who encourage each other for patience and compassion are the people on the right side, as for those who deny our evidences, they are the one on the left side, fire will fall upon them." In Islam, Prophet Muhammed (PBUH) represents the **perfect human**, who is complete with **good knowledge, good orders, good action, high ethic.** In the Mawlawi sect, the perfect human is represented by the Mawlawi dervish. According to another definition, **ethics** is the whole of **human knowledge, thought, behavior and emotions**. People who are right and measured in all four elements have high ethics. **EDEB** (*Civility)* (Arabic pedestal, rule, basic, meaning, summary of 666 verses in Kur-an and believe in the existence of god), which is the peak point of morality, is the fourth rule that brings us to perfect people according to Islam and Sufism. **(P-61,180) It is the beautiful image in the behavior of the ethical person** and is over everything else. Let's try to separate the civility and morality from being a general concept in order to make it better understood today and to break it down a little bit as seen below.Some inverse factors of **civility,** which is the fourth condition of being a **perfect man or woman** and need to be balanced according to the Mawlawi dervish:

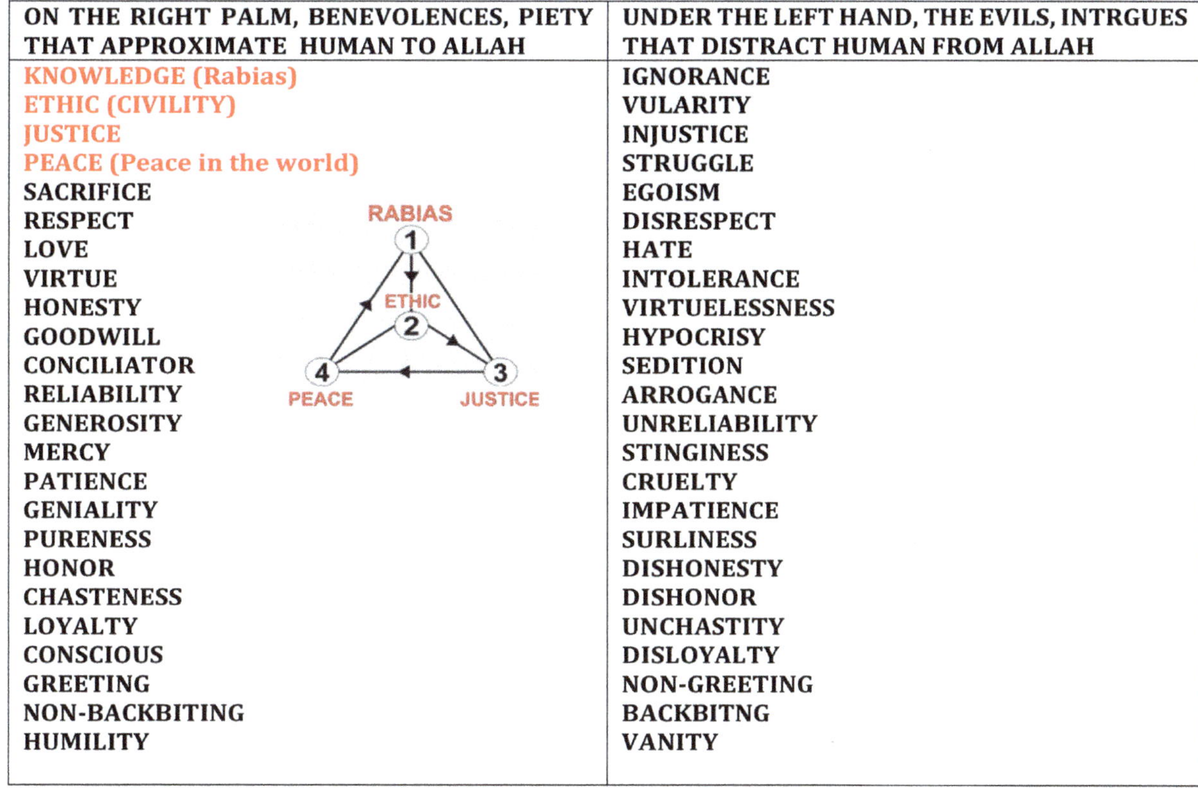

ON THE RIGHT PALM, BENEVOLENCES, PIETY THAT APPROXIMATE HUMAN TO ALLAH	UNDER THE LEFT HAND, THE EVILS, INTRGUES THAT DISTRACT HUMAN FROM ALLAH
KNOWLEDGE (Rabias)	IGNORANCE
ETHIC (CIVILITY)	VULARITY
JUSTICE	INJUSTICE
PEACE (Peace in the world)	STRUGGLE
SACRIFICE	EGOISM
RESPECT	DISRESPECT
LOVE	HATE
VIRTUE	INTOLERANCE
HONESTY	VIRTUELESSNESS
GOODWILL	HYPOCRISY
CONCILIATOR	SEDITION
RELIABILITY	ARROGANCE
GENEROSITY	UNRELIABILITY
MERCY	STINGINESS
PATIENCE	CRUELTY
GENIALITY	IMPATIENCE
PURENESS	SURLINESS
HONOR	DISHONESTY
CHASTENESS	DISHONOR
LOYALTY	UNCHASTITY
CONSCIOUS	DISLOYALTY
GREETING	NON-GREETING
NON-BACKBITING	BACKBITNG
HUMILITY	VANITY

We said **justice**; there could be no justice without **ethics** and without eliminating the **fight**, there can be no **ethics;** and the defeated may turn to corruption. Salvation lies in the "**Rabia of survival of humanity**", likewise **"Rabia of Peace "Silm" in Islam" on Page 178,166,167 and** the world peace, **which can be achieved through other Rabias** on the head of the Mawlawi dervish on the front cover of the book. Thus, in this book, we see that PEACE, which we can understand that it is at the top of the information we have collected over 200 pages, has already been revealed from the beginning with the 4 meanings of the name of Islam, namely Rabia. **(S-60,178)**

However, if people are not clearly told what the **ethics** is, there is no fault for those who are immoral. Thus, it was revealed in the above table that **knowledge** is above **peace**. It is understood that the mind precedes the knowledge, since it is necessary to have **mind** for a person to be knowledgeable. And According to Islam, Allah gave it to human as a mule from his own mind. The first verse of the Qur'an means **"Read!"**, that is, not only read the Qur'an, but try to understand all of what I have created in the Universe with your reason, that is, have all the Knowledge of the Universe. Today, we know that mathematics is at the top of all sciences, and since the priests of Egypt and Mesopotamia reveal that the essence of mathematics is the **Rabia**, it can be concluded that the word **read** in the Qur'an also means that the **Rabia** covers everything.

When the word **read** and the word **mathematics** are removed from a very broad concept, all the remaining basic concepts become easily understandable in a universal integrity that everyone can understand, both divine, scientific, clearly and truly.

In Islam Allah c.c. it is not a god that was found to be worshipped. In Islam, there is nothing outside the concept of Allah. It covers everything from the outermost boundary of the universe or universes to the quantum. As Mawlana states, as **civility** means faith in Allah, the Rabias starting and ending with the civility, and all the other elements of the above-mentioned civility, we tried to collect and all aspects of civility and ethics that we tried to compile above and which we listed or not, are the things that the believers will largely follow any way, as long as a general acceptance is established in the world. Our survival problem can be largely eliminated, because Bekabillah on the head of the Mawlawi dervish seen above on the left. It means that people who possess high spiritual and human qualities in Islamic Sufism should turn to Bekabillah (P-**179,167**) after they stand up for Allah (Qiyam-billah). Thus, it is aimed to explain that the survival of humanity, that is, the existence of mankind, can only be achieved by having moral and human values, that is, with good morality.

If we woud complete the words of İbrahim Hakkı of Erzurum **(P-86)** "Let's see what the Companion (Guardian, Sustainer) makes What ever It makes, It renders it beautiful", with the phrase "It renders its 'Encab' (=divine nobility) also beautiful", one can see that the immortal 'Encab' (=divine nobility) is in an upper rank to to the beauty as well. A higher nobility is no longer conceivable. May Allah grant people to have the good 'encab'. In the words of today, may Allah grant the mankind the rank of divine nobility (encab), defined in terms of Sufism as being one of the Prophets of God (Ricâlullah, Nücebâ) beyond what is beautiful in this way, while living beneficial and good to the World.

The personalities of the people are generally similar to the names given to them, I believe that the foundations of the Ottoman state, which could survived for 6 centuries, was laid down by Edebali (meaning master of civility or supreme civility). He was the teacher of both his daughter and Osman, his son-in-law, and with his advices, he impacted the sultans who were the successors to Osman. This advice is inspiring even today. He named his daughter, who would later become the wife of Osman, "Rabia Bala". So he was aware of Rabia. In addition, "Bala" meant both supreme and child. Edebali was born in the city of Merv in Turkmenistan. In other words, from the region where the Sumerians went to Mesopotamia and Odin to Sweden. Gilgamesh was already the Sumerian king. From the meaning of Bala, meaning both supreme and child, it can be concluded that a child is regarded as a supreme being that God gave to parents in this regional culture, since the child is born. Today we are ashamed to live in an era when newborn children are left in trash cans. Sumerologists claim that Sumerians are also the source of monotheistic religions. Further research can be done by Sumerologists in the future.

Some examples of Edebali to Osman: **O my son! Now you are the Emir! You are strong, powerful, smart, wise, but if you don't know where and how to use them, you will be blown away in the morning wind. The strength of a person will exhaust one day, but knowledge survives forever. The light of knowledge penetrates even through closed eyes and enlighten them. An Emir should know to be patient. A flower does not bloom before its time. If you lose your faith, you turn from greenery to deserts or arid lands. Respect your mom and dad! Know that blessings are with the elders. Let the state live so that people live. When an animal dies, it leaves the saddle, when a human dies, he leaves his work. Those who do not acknowledge their past cannot know their future. Do not forget from where you came so you may not forget where you go.**

This advice should actually be perceived by America, Europe, Russia and even China, who are the masters of atomic bombs. Because like the king of Spain, who chased Muslims from Spain,

Mao burned down the ancient books and founded communist China. If humanity will have a future, everyone should know his/her origins; and this is not possible by burning books. People living in Europe and America today should know that came from Central Asia after the ice age ended 12 thousand years ago; they should know that the knowledge that allowed them to travel in the space is gathered from the Sumerian, Babylonian and Egyptian temples and transmitted to the Greeks, then in the cities such as Baghdad, Aleppo, Damascus, which they are destroying today, in Central Asia and in the Middle East, and then in Europe created the Renaissance and brought us to the space age. Now, the only place to escape and the last chance of humanity to avoid a possible 3rd World War; is a World peace and union based on Rabia knowledge, Ethics and Justice.

On the left palm of the dervish depicted on front cover of the book is the **"matter"** (**M**), and on the right palm is **"God, meaning, ethic, civility, knowledge"** or **energy (E), because all four are intangible**. Such moral values give the kindness of being human to the human with the survival algorithm of humanity. **Ethic** in human is just the same as **energy** in the universe. Thinking the **human** without ethic is like thinking Einstein's formula $E=MxC^2$ **without E.**

The Statue of Liberty, which was bestowed by France to America for political purposes in 1886; and located in New York today, was an ancient but unwanted statue paid by the Ottoman sultan Abdülaziz in 1865 to be placed at the entrance of the port of Port Said in Egypt and to represent an Egyptian woman (maybe Rabia) who lived during the Pharaohs era. The torch in the right hand of the statue is the symbol of freedom that illuminated the world according to the French people, and the 7 pointed crown on its head were supposedly interpreted as 7 continents or 7 seas in the world. The French changed the face and arms of the statue, but left the crown, however the interpretation was wrong. The correct interpretation might be that they represent the 7 planets thought to be revolving around the Earth in the era of pharaohs, and the 7 sciences that priests worked on in temples (grammar, logic, eloquence, ditch, mathematics, music, astronomy) **(P-76).** Hermestoth explained very clearly, what the torch was on the right hand of the statue. The light particle of the first particle of matter in the Universe we live in has been described earlier in this book as **photon**, but if we call this torch liberty, there may be liberty against the French kingdom, but there is no freedom in opposition to Allah, it is like a riot and the world we live in today emerges.

Freedom, being one of the three concepts of Freedom, Equality and Justice and which was not implemented after the French Revolution anyway, was misunderstood. The Freedom against the supreme power that created the universe and human beings, is both on the contrary to the mind and science. The arabic term **"kul"** means captive and **"edep** (civility)" is being a captive of the rules. Thus, being the captive of **"nefs** (desire)" is not freedom but captive. Only one who is the captive of the desires cannot be Free, if not, at least we cannot get out to the streets today, but tomorrow we will not be able to open the window

In P-**180**, liberty was given in the right place in the **"RABIA OF IDEAL STATE"**. The correct interpretation of the statue; Regardless of whether the the statue represents an Egyptian or Roman lady, according to the knowledge of peace in the world or the contemporary sciences represented in the crown of the statue, the torch must repre sent the light that illuminates both the outside of the human and the light that illuminates the inside of the ethic person, that is, the nur. We live in a universe of light that is expanding and is healthy thanks to energy (spirit), the balance of the matter, please, let's not make our Earth sick and leave people in the dark with materialism.

In the table given as **"GEOMETRY OF HUMAN FOR PEACE IN THE WORLD"** on **page 177**, the formula of Einstein is given under the Mawlawi dervish on the front cover of the book and the World flag on **page 174-177** and Rabias, which are the knowledge of the peace in the universe that will bring us the ethic. According to this formula, if the material values under the left palm are 1, the speed of light (C) is approximately 300 thousand kilometers per second (~ 300 million meters); 1 on the left is not multiplied by 300 million, 300 million squared and then multiplied by 1. On the right palm in Sufism, spiritual, ethical values in the singularity of Allah are gathered in the heart and the heart represents love. Thus, when the divine light of the spirit that Allah entrusted to man is in the heart, and when the ethical man dies, divine light of the spirit reaches Allah before the light of his soul disappears. Now, let's invite all the people of the world to put their right hands on their hearts and say, **"We all have the same ore, atomic bombs, hydrogen bombs are scarce, here's the Big Bang."**

Today, we started looking for a planet with water in order to escape from Earth. But why not start the golden age with the geometry of peace on Earth now.

J) THINGS TO DO PERSONALLY AND SOCIALLY

Now we have to bring the Earth and Eastern and Western communities to a **Balanced (Fair)** integration and perfection by means of; scientific knowledge of 21st century, like a winged angel's touch with her scepter on the world to change it to **Heaven;** as a result of all our knowledge; in the light of **Mathematics, Physics, Cosmology and Mysticism;** what the quantum individualism of **universe has created**; taking the perfection and integration of universe as an example; together with the Quantum Individualism of 21st century **humans.** What are told up to now, show that the Quantum Self of the universe gives us this responsibility. Otherwise, it is obvious that the **balance**, which is the necessity for the opposite poles to survive, may be lost on the world where we live. In the light of knowledge we presented up to now, we have to integrate this **Balance (Justice, peace)** with the **Fermionic Materialist Individualism of the West,** in other words **Materialism**, and the **Bosonic Mental Communism of the East as an in separable whole through the balance**. In fact, this universal message, which also exists in Mawlawi Dervish, can be actualized within a **democratic** and **justful "United States of World"**, which will configure both the individuals and the communities of the world.

We are at the beginning of the 21st century. The European Economic Community is now transformed into the European Union; they prepared a quite comprehensive **European Union legislation**. This union and its legislation should be firstly combined with the United Nations and then with all of North Atlantic Pact, Nato, the Balkan pact, all of Russia's and China'a similar pacts based on the "SOCIAL GEOMETRY OF THE PEACE IN THE UNIVERSE AND GOLDEN AGE OF WORLD" in line with the motto "One World, One Dream" of the 2008 Olympic Game – China.(P.**175**)

The aim of all the information given here; was to look for a Universal message to guide people. The extent to which the goal is reached will determine the readers' discretion. As far as I know only in one or two of 10 universities in America, have a **cosmology department**. The common characteristic of more than a hundred Islamic scholars who created the European Renaissance thanks to the knowledge that they had derived from **Sumerian, Babylonian** and **Egyptian** priests, from **Vedas, Brahmans, Greek philosophers** and developed, was that they were acquaintant with all the sciences of their time. Unfortunately, none of the universities in Turkey currently have a Cosmology Department. The Islamic mysticists must be turning over in their graves. In cosmology faculties in the United States, they may know things about the relation of the humans with the matter, which they call **positive sciences**, better than we do. Are the **spiritual sciences** negative ones? It is difficult for the Western world to understand the **spiritual, moral, conscience, humanistic** issues because they do not have as much social and historical accumulation as we do. Such accumulation charges us a historical responsibility, we have to transfer this accumulation to the West. The fact that Indogermans emigrated from North-West India, that is the region, from where the Sumerians and Turks came to Anatolia may be the source of the life style in accordance with the rules, which we unfortunately forgot in a great extent in Turkey nowadays. The fact that the concept of civility in Islam means "**rules**" may make those who regard disobedience to rules as a superiorioty today, more civil. The thing to do is; for the cosmology faculties to be established in both Islamic universities and western universities to find Universal truth both in material and spiritual (humanistic) sciences like the Islamic Sufis did. If they approve the Rabia in the meantime, maybe we can start the **Golden Age** together. Thus, the work done so amateurly should be continued in the **faculties of cosmology** with scientists who have philosophical qualifications in the world. A cosmology magazine can also be published. A research on a world language, which is better and easier from Esperanto can be carried out. The topics studied can be updated by time by a top committee composed of the most competent philosophers selected from the institutes of cosmology in the world. **Textbooks on Rabia Peace, Ethics, Civility and Cosmology can be prepared for primary schools, secondary schools, high schools and universities.** Gifted children can be noticed at an early age and educated by philosophers in the faculties of cosmology.

As a matter of fact, the definition of the word "**University** ", explains in a sense that the universities should deal with all the sciences related to the **Universe**. Today's Universities have been dealing with all sciences separately in various faculties within their organization. What they do not bother is trying to gather all the sciences under the roof of cosmology and put forth the big picture. At least some universities are starting to enter this path with cosmology departments.

Let it not be misunderstood; according to an Anatolian saying, "Where there is no sheep, they call the goat Abdurrahman Çelebi". This book was prepared in the hope that a person without an academic qualification could gather all the important information of the present age and reach a universal synthesis, although it did not fall short of it. **In order to reveal the big picture in real terms, it is necessary to integrate the following subjects in the faculties of cosmology:**

"numbers, geometries, music, quantum interpretation of electron orbits (atomic model of Louise de Broglie), virtual rhombic prisms, triangular pyramid of Pythagorean triangle, gluons, quantums, resonanas. The force of the atom, the force of 4, the 4 dimensions and other dimensions over time, while there is a restriction of the velocity of light in the transmission of kinetic (motion) energy in the expanding universe, in both the expanding and collected universe of gravitational force, between the graviton outside the universe and all the gravitons in the universe; continuous instantaneous interaction and control according to the master plan (Nur-u Kadim) in the state of energy Super-symmetry, quantum Bose-Einstein concentration, "M-theory", space-time, instanton, everything theories, quantum mass gravity theory, trigonometry, Plato's regular polyhedrons, golden proportions, geometry in music, artificial intelligence, cubit and quantum computers, entanglement and other related topics, especially the tables given in the Annex of this book: "the quantum of universes" and "universals calculus equations" where necessary in a syncretic conception of computers, an "automation program" in the form of calculus equations and visual loading and thus trying to illuminate yet unknown universe from these known facts. Transformation of the state structuring in ANNEX-IVa into an artificial intelligence e-state model, perhaps with quantum computers " Maybe quantum computers will solve this in the future.

It will be hopeful that all cosmologists will come together for the sake of world peace and for the moral philosophy of the societies in our world where we are talking about the signs of doomsday. Otherwise, a nuclear war between America, Europe and Russia, China, and Iran, as a result of the movement that began gradually with Israel, America and Syria, could make the world uninhabitable. We are in april 2020 now; all the world's navy gathered in the eastern Mediterranean. Is that why Allah has created the most perfect human being? The information tried to be explained so far can prevent this cockfighting. When the Covid-virus epidemia is over, it is likely that the USA will pick a war, which is not guaranteed to be won, in order to save the Dollar and oil. So the only way to save both the USA and the world should be the establishment of the "United States of World". Because, as in history, people do wrong because the truth is not explained. While it is possible to turn the world into heaven with universal truths, irrational madness continues. **Edeb** (lit: civiliy) comprises the concepts of rules, acceptance, approval, acknowledment.

Accordingly, the word edebiyat (lit: literature) can be defined as ¨the art of observance, acceptance, approval, acknowledgement; that is, the art of expressing the necessity to resign oneself to the rules¨ Even if the world population would be reduced to 500 million, if people do not live according to the geometry in the flag of the United World (Page **175**), they will quickly destroy each other, including the same family members, in that society. However, with the 5G technology, which are posessed both by China and the USA, a chip to be inserted into the body of every person (maybe in the form of a tattoo) and this would record any biological changes and thoughts in the face of events, there is no alternative for people other than living with morality. Every one of the 8 billion people are instantly under record but the "wise" chosen by the peoples of the world must be at the center.

On the basis of " THE SOCIAL GEOMETRY OF PEACE IN THE UNIVERSE AND GOLDEN AGE IN THE WORLD" in Page 175, we hope that after the establishment of the " UNITED RABIA STATE OF WORLD", we will melt the all of warships in the eastern Mediterranean and the world and transform them into rails for high-speed trains and oil and natural gas pipes.

Where there is life there is hope, finally, according to both the **Sumerians** and the **Mayans**, let us explain how the starting date of the **golden age in 2012** is calculated on the following table: The axis of our world is making a full cycle called **precession**, approximately **every 26 thousand years** (25 thousand 760 years).The **precession** draws a cone at the north and south sphere poles. When **26 thousand years** are divided into four, **four ages** of **6500** years are formed as seen below. Below are the dates of the **7 ages** of **Sumerians** and **5 gaes of Mayans**. Due to the difficulty of converting the Mayan and Sumerian calendars to today's calendar, 2012 may be miscalculated. I hope the truth will still be the years we live.

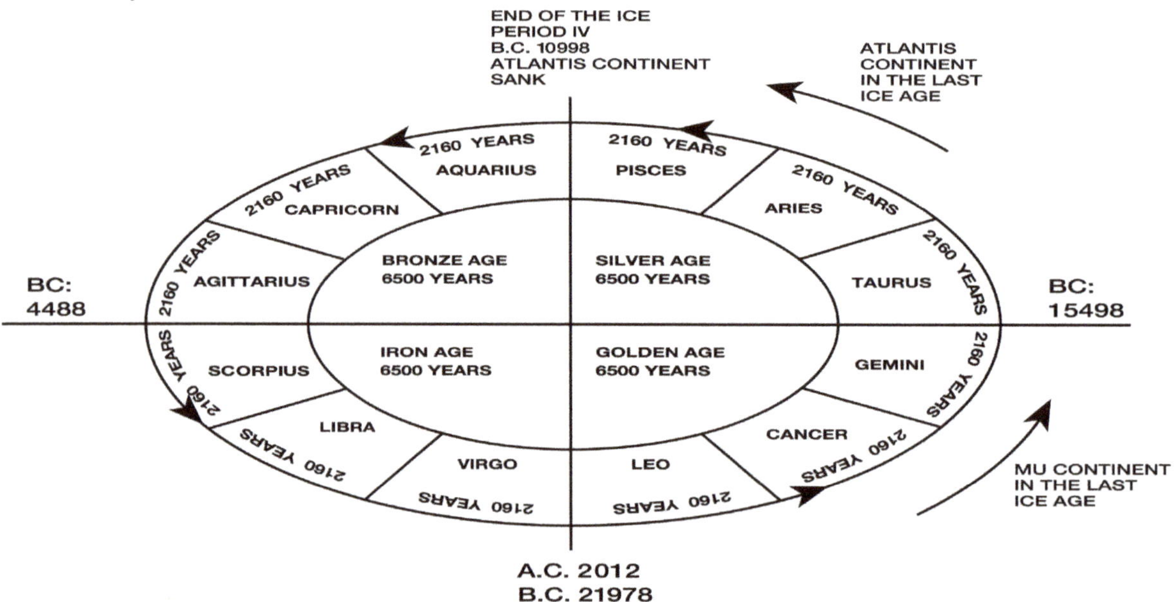

AN AGE ONCE IN 3661 YEARS			AN AGE ONCE IN 5125 YEARS			FOUR
ACCORDING TO SUMERIANS			ACCORDING TO MAYAS			ELEMENTS
3661x7=25627 (7 AGES)			5125x5=25625 (5 AGES)			OF LIFE
12632	B. C.	END OF 1ST AGE	18475	B. C.	BEGINNING OF 1ST AGE	
+3661	B. C.		+5125	B. C.	(ENDED WITH FIRE)	FIRE
8971	B. C.	END OF 2ND AGE	13350	B. C.	BEGINNING OF 2ND AGE	
+3661	B. C.		+5125	B. C.	(ENDED WITH WIND)	AIR
5310	B. C.	END OF 3RD AGE	8225	B. C.	BEGINNING OF 3RD AGE	
+3661	B. C.		+5125	B. C.	(ENDED WITH WATER)	WATER
1649	B. C.	END OF 4TH AGE	3113	B. C.	BEGINNING OF 4TH AGE	
+3661	B. C.		+5125	B. C.	(ENDED WITH EARTHQUAKE)	EARTH
2012	A. D.	END OF 5TH AGE	2012	A. D.	BEGINNING OF 5TH AGE	(DECEMBER 23RD)
+3661	A. D.	(GOLDEN AGE)			(GOLDEN AGE)	
5673	A. D.	END OF 6TH AGE				
+3661	A. D.					
9334	A. D.	END OF 7TH AGE				

We must be able to tell the Sumerian and Mayans' golden age before it is too late for those who prepare the world for doomsday, the third world war with the Dark Avatar legends. Otherwise, when-Mehdi comes, he may not find anyone in the world. Our wish in the present year 2020, the fate of our world; BC: 250'de Sodom and Gomorrah sinking in the Dead Sea in Israel, not as the fate of the Sumerians and the Mayans as the result of the beginning of the Golden Age.

Finally, Let's invite the officials in Turkey to establish a "**cosmology institutes, faculties**" under TUBITAK and in the empty space behind the library under construction in Istanbul Rami barracks, likewise in Erzurum-East Anatolia in the name of Erzurumlu İbrahim Hakkı **(P-86)** and in Van in the name of Said Nursi **(P-92),** for the activities of the academics to have a say about cosmology in Turkey and where they can organize symposia. Just as the former Islamic Sufis know all the sciences of their time in order to find the truth in the madrasas, in today's cosmology faculties and institutes of world universities; studies can be carried out to establish a "**world Rabia society and order** that is able to balance the material and spiritual values that are worthy of the divine perfection in the matter and energy balance of the universe. This is the only way that all the thinkers, scientists and even the prophets' efforts of the Sumerian and Egyptian temples can be achieved. An example of this institute can be created in Madrid, Spain with the Yunus Emre Institute and the Spanish government, in the region of the former Andalusian madrasas in Granada and Cordova.

In addition, it may be suitable to construct a multi-puropose building, where stars and galaxies are projected to the ceiling of the dome, symbols and pictures of the most important cultures in the history of the world can be seen on the three sides of the stage in the middle of the amphitheater., the audience can sit on the rising steps and sema rituals may be performed during Mawlana Remembrance Week within the Opera House to be built in Taksim Square or in the fairground to be constructed at Atatürk airport in Istanbul. After the ceremony, these groups can continue their performances in the festive atmosphere in the semi-touristic World Cultures Fair to be held at the fairground. After the sema ritual is over, sample family groups consisting of mother, father and children, selected from all countries of the world, in their local costumes, father and children around the mother, children and mother around the father, also a different rituals by turning around themselves can be performed. During the week of Mevlana, they can continue their shows in the semi-touristic "**World cultures fair**" and in the city, with the motto "**peace in the universe, peace in the world**". Thus, the importance of the family in world societies is emphasized as well as being a family among the world's people. During the Mawlana week, local and foreign scientists who have a say about cosmology in the world can organize symposiums in the Mawlana Hall. World folk music festivals and competitions can also be held at the fair area to be established at Istanbul Atatürk airport.

Though we might not be allowed to say, we also wish from the big capital owners who try to rule the world; if they want to have a say in the future United States of World as capitalists and save the world from a possible nuclear war or to be destroyed by viruses, biological weapons, today they will earn their hearts by investing in the people who have been **destroyed**, burned, and forced to emigrate from their homes in the Middle East and the world and should strive to establish the balance between the matter and essence in their lives from now on. Now the world should not be parceled by races, but the world of Ethics, Justice and Peace, belonging to all the states that are named by the geographical regions united as "United States of World". On the **page 166,** in the history of Rabia's, it was explained in a humble way why there is no need to force God to doom and how peace and heaven can be created on Earth. Would it be bad if the temple of Suleiman could be reconstructed without demolishing Masjid al-Aqsa and these to temples would be connected to facilitate visits of the tourists, and if Muslims, Jews, Christians, and members of all religions could perform Sema ritual in these temples alternately every Mevlana week?

Now, starting the peace and golden age in the world based on **ETHİC, JUSTICE**, and the world, thanks to Rabias, which is the essential-knowledge of the universe described in the first part of this book, and the human and mathematics described in the second part, is the ultimate chance of surviving on Earth and responsibility of the most perfect being created by Allah against the creator. Because divine peace in our Universe on the Astral and Atomic scales consists of balance, order, harmony, compliance, equation in mathematics, symmetry in geometry, tuneful melody and harmony in music, eye-pleasing beauty in art, golden proportions, morality in sociology, justice in law.

We can only clean the **SOİLs, WATERs, AİR** of the world that we have polluted through clean, moral, dignified people. Otherwise, the fourth element of life, **FİRE** will somehow ask us for an account. There is no other world anymore. We have to clean our Earth before it's too late.

K) AFTERWORD:

So far, it has been explained again and again that the algorithm of salvation of the EARTH and HUMANITY is the "Theory of Everything", that is, the "Double Triangular Pyramid" (Double Rabia). Let's start by trying to understand what happened before the time of Plank according to contemporary physics, as seen in the table below, in order to understand that the theory of everything RABIA also applies to the entire UNIVERSE.

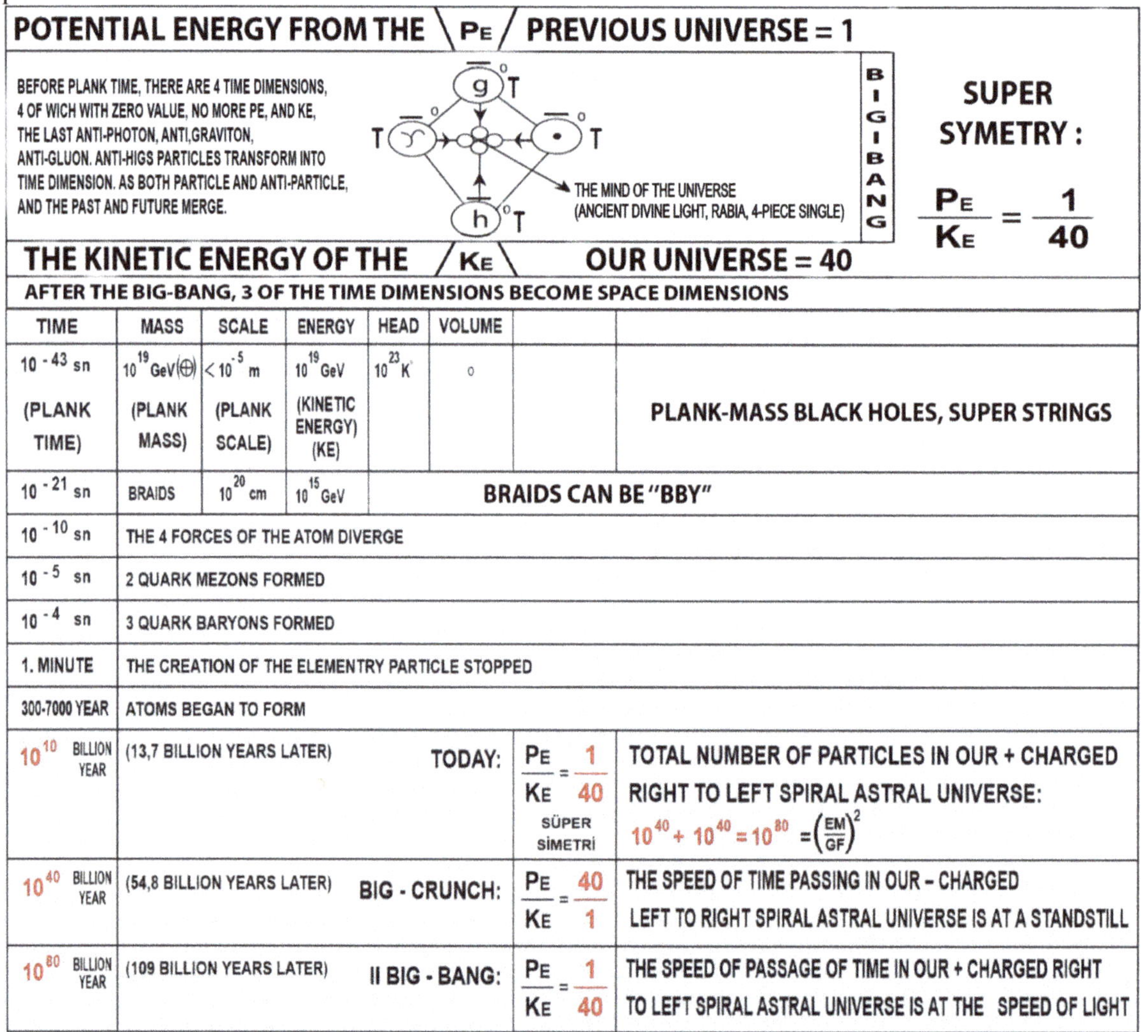

$$\text{SUPER SYMETRY:}\quad \frac{P_E}{K_E} = \frac{1}{40}$$

TIME	MASS	SCALE	ENERGY	HEAD	VOLUME	
10^{-43} sn (PLANK TIME)	10^{19} GeV⊕ (PLANK MASS)	$<10^{-5}$ m (PLANK SCALE)	10^{19} GeV (KINETIC ENERGY) (KE)	10^{23} K	o	PLANK-MASS BLACK HOLES, SUPER STRINGS
10^{-21} sn	BRAIDS	10^{20} cm	10^{15} GeV			BRAIDS CAN BE "BBY"
10^{-10} sn	THE 4 FORCES OF THE ATOM DIVERGE					
10^{-5} sn	2 QUARK MEZONS FORMED					
10^{-4} sn	3 QUARK BARYONS FORMED					
1. MINUTE	THE CREATION OF THE ELEMENTRY PARTICLE STOPPED					
300-7000 YEAR	ATOMS BEGAN TO FORM					
10^{10} BILLION YEAR	(13,7 BILLION YEARS LATER) TODAY:	$\frac{P_E}{K_E} = \frac{1}{40}$ SÜPER SİMETRİ	TOTAL NUMBER OF PARTICLES IN OUR + CHARGED RIGHT TO LEFT SPIRAL ASTRAL UNIVERSE: $10^{40} + 10^{40} = 10^{80} = \left(\frac{EM}{GF}\right)^2$			
10^{40} BILLION YEAR	(54,8 BILLION YEARS LATER) BIG - CRUNCH:	$\frac{P_E}{K_E} = \frac{40}{1}$	THE SPEED OF TIME PASSING IN OUR – CHARGED LEFT TO RIGHT SPIRAL ASTRAL UNIVERSE IS AT A STANDSTILL			
10^{80} BILLION YEAR	(109 BILLION YEARS LATER) II BIG - BANG:	$\frac{P_E}{K_E} = \frac{1}{40}$	THE SPEED OF PASSAGE OF TIME IN OUR + CHARGED RIGHT TO LEFT SPIRAL ASTRAL UNIVERSE IS AT THE SPEED OF LIGHT			

a) PRE- PLANK TIME ACCORDING TO MODERN PHYSICS AND POST PLANK TIME ACCORDING TO THE TABLES IN THIS BOOK:

Egyptian temple priests found that the essence of mathematics was **Rabia**.Justice and Wisdom, which are indispensable in state administration, were given in this book in the examples of Odin **(P-181)**, Gilgamesh **(P-51)** Wise kaan (Bilge kaan) According to the above table and the other tables that I have prepared in this book, let's begin to explain that the essence of the Universes and the true and ultimate Wisdom are Rabia:

It will be prudent to start from the fact that the 4 anti-boson particles of the previous universe as seen at the top, in the middle, and at the right of the table in **Appendix-IIe** transform into 4 time dimensions with the too high attraction pressure of the mass as seen above and form the "**Mind of the universes.**" As can be seen at the top of this page, this mind is the mind of both the previous and the future universes as it unites with the past and the future at that moment. Here, the question that comes to mind is whether or not Bose and Einstein understood that the quantum concentration they hazardous substance found was a 4-Quanta concentration, that is, **Rabia?**

The fact that all four particles are zero at the zero point of the big-bang can be explained by the fact that the particles at the corners of the triangular pyramid are combined in the center.

The interpretation of the same formation with Rabia is given below.

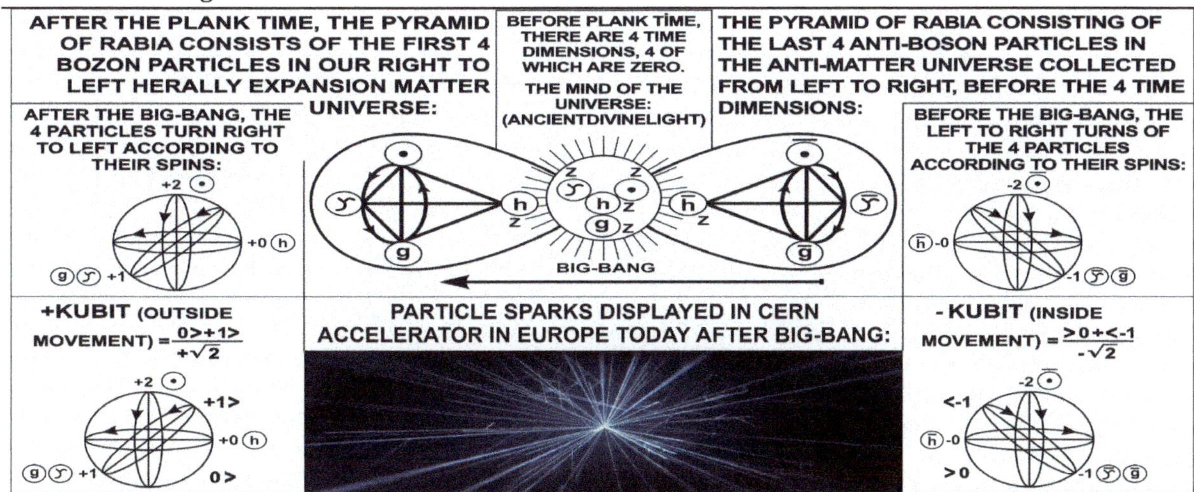

 Above is the Boson triangle pyramid, which is formed by the transformation of the last 4 particles of the previous universe into 4 time dimensions at the zero point of the big-bang, and the transformation of 3 particles into space (space) dimensions in our universe on the left. Also, the boson and anti-bosons at the corners of the Rabia pyramids seen above are each of the atoms. It represents one of the 4 forces in it:

Bosons of the strong nuclear force : (h) (Higs) ,   (h̄) (Anti-Higs)
Bosons of the weak nuclear force : (g) (Gluon),   (ḡ) (Anti-Gluon)
Bosons of gravitational force : (•) (Graviton),   (⊙) (Anti-Graviton)
Bosons of the electromagnetic force : (ϒ) (Foton),   (ϒ̄) (Anti-Foton)

Thus, the 4 forces in the atom were combined here, and the "Theory of Everything" was finally proven for the first time. In addition, large-scale entities in the astral universe can be calculated according to Newton's law of gravity according to the binary (0,1) BIT system in computers. However, these calculations are in our expanding matter universe: For the particles (Quantums) in the nucleus of the atom, in computers according to Quantum Mechanics, above left. It can be done according to the +KUBIT (+Quantum Bit) system seen. In the anti-matter universe, which is collected inwardly before our universe, calculations can be made according to the -KUBIT (-Quantum Bit) system, which is seen on the right above, in computers according to Quantum mechanics.

I also believe that the Big-bang double pyramids of Rabia, seen above, will help solve 125 unsolved problems of natural sciences, published by Science magazine in 2005. We see that the concepts of Unity (Unity), Justice (balance), and Peace, which kept the Ottoman state alive for 600 years, are also present in the Universe. The "Rabia geometry" seen on the back cover of this book, which will bring us heaven both in the world and in the hereafter, can enable us to establish a single world state that will integrate the people of the world with the Universe. This can also be called the 2023 interpretation of the teaching of the Roman Stoic Marcus Aurelius, which recognizes "Humanity as a Nation and the Universe as a Homeland."

Here, after the Big-bang, the fact that the Higs particle has passed back to our expanding Universe as a time dimension suggests that there may be these high-mass Higs particles at the center of Black Holes in our universe, I think. In fact, there may be anti-highs particles in the center of the anti-Black holes that will form there after the black holes in our universe emptied the material they collected in the next universe as anti-matter after the Big-crunch. The last remaining anti-Photon, anti-graviton and anti-Gluon particles before the second Big-bang may have transformed into 4 time dimensions together with anti-Higs, that is, the **"mind of the Universes"**, with the high gravitational power of anti-Higs and completed their evolution.

The total number of points in the top 3 pyramids at the top of the table "Rabia pyramids" given on **page 137** is 40, 20 on the right and 20 on the left.

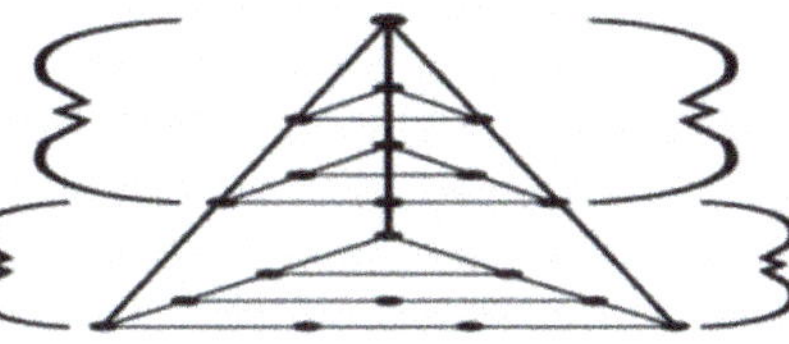

At the top of **EK-IIe**, each of the last 10 remaining ANTI-BOSON grains from the universe collected before the Big-Bang on the right gave birth to 10 BOSON grains on the left side, and when each of these 20 Boson and Anti-Boson grains formed 1 Neutrino grain, a total of 40 new particles were created. Each of these 40 particles will form 40 QUARKS, 1 Quark. The 40 Quarks seen in Annex-IIe will form 120 BARYONS **(EK-IIId)** and then 120 ELEMENT atoms inside the stars under the influence of the 3 GULUON colors given at the bottom of the table (40x3=120) **(EK-IIIa,c)**. Then MOLECULES will be formed from the union of atoms, COMPONENTS will be formed from the union of molecules. Finally, on Earth, the planet of the Sun, COMPOUNDS, IN THE WATERS WERE FORMED FIRST BY SINGLE-CELLED ORGANISMS and then by MULTICELLULAR ORGANISMS, PLANTS, ANIMALS and HUMANS by the electric effect of lightning. In **Annex-IIc**, the 64-particle SUPER CUBE created with some of the QUARKS and LEPTONS may have the code for the DNA and RNA **(P-141)** of living things. Thus the 20. In the century, Werner Heisenberg's **(P-94)** claim in "Modern Quantum Theory" that the elementary particles of the atom would eventually take mathematical forms, but that they would take on a much more complex structure than assumed by Pythagoras, proved to be untrue. It turns out that only the universal code is the three-dimensional **pyramids of Rabia** instead of the two-dimensional pisagonal triangle .

Likewise, in the control of the universes that explode and expand in Annex-Ia and Ig, then gather and explode again, under the control of the **"Mind of Universes"** before the time of Plank, all elements of subatomic particles, from quanta to galaxy clusters, up to the outer limit of 500 billion light years, are constantly interacting with each other and in a state of change, expanding and gathering 10^{80} It seems that it is like a single living organism with a total number of particles and is magnificent in the consciousness of its own self, like every human being.

b) BEFORE THE PLANK TIME ACCORDING TO EINSTEIN:

In order to prove the theory that Einstein called "**Unified Field Theory**" with equations, Plank wanted to go before his time, but he could not prove his theory for 30 years. Due to the fact that the mass of the Sun is so much from the Earth, 1 second passes in the Sun and 11.5 days pass on the Earth. Therefore, Einstein wanted to combine the dimensions of the atom, including the time dimension, according to the geometric properties of the 4 forces of the atom. I think Einstein could have proved his theory if, before he began to prove this theory with equations alone, he had seen the tables that I have put forward in this book, in a way that appeals to the eye of the existing contemporary proven knowledge and of Rabia's geometry. Although I do not fall into my place, I explain this topic below in the hope that it may inspire contemporary physicists.

In the 1930s, British physicists Sir A. Eddington and Paul Dirac **(P-94)** believed that the number 10^{40} was the code for a comprehensive relationship that connected the micro-world of the atom to cosmic structure. In the table above and in Annexure-Ig, the numbers $10^{to\ 40}$ are indicated in red. In the **"Creation Monocourt"** given on **page 157**, the lowest limit of the Universe was given as superimposed attributes from the quantum wave to the highest limit of the Universe from to 10^{40}. Likewise, it is the Weak nuclear force (ZA) in the nucleus of the atom and its particle Gluon (g) that reveals the right to left + charged, left to right - charged spiral structures mentioned above and above EK-Ig.

If Einstein could prove his "**Unifying Theory**," the room would lead us to the owner of the Universes **(the Mind of the Universes).** According to the table under Annex-Ig, the **KÇ** and **ZA** forces of **PE** and the **EM** and **GÜ** forces of **KE** basically lead us to **the** duality of **PE** and **KE**. Although the intensities of the 4 forces of the atom changed at different scales in the time of Plank, the fact that it approached the same around the Plank scale, as well as the union of the past and the future, and the uniqueness, showed that it could lead us to a Bose-Einstein Quantum condensate where **PE and KE** before Plank time also coalesced, that is, to the sole creator, the owner of the Universe. Because it would have been proven that the "**Bose-Einstein Quantum Condensation**", which is also present in the Big-bang and in the human brain and allows each person to become conscious of his own self, has enabled the owner of the Universes to become conscious of himself. Thus, the statement in the Qur'an, "I have given you a katre (drop) from my own mind," would also coincide with contemporary science. The "Owner of Universes" mentioned above (**The Mind of Universes**); As the past and future converge, it turns out that both our expanding Universe owns both the previous Universes that have gathered and steadily destroyed themselves. If its existence remains unscientifically proven, that is, the **PE** (Potential energy) of the collected universe and the **KE** (Kinetic energy) duality of our expanding universe can be examples of conflict for people on Earth. In other words, if there is conflict in the universes that have already gathered and expanded, conflict can be perceived as inevitable within people on Earth. So, what you or I do, whatever you do in this life anyway will stay with you, there is no one who asks us the thought of no one who is now asking us 3. It brings us closer to the world war. We know that if all the atomic bombs in the warehouses of the states that have atomic bombs exploded, they would have the power to destroy our Earth several hundred times. It is also clear that far fewer hydrogen bombs will be enough. Are those who rule the world not aware or not?

The proof of Unified Field Theory by equations is no longer just a physics problem, but 3. It is the question of the salvation of the Earth, which has come to the brink of the abyss of world war, if we do not go any further. Let's invite physicists again to prove Einstein's **"Unified Field Theory"** in the light of the tables I have given in this book and to illuminate the 10^{-43} seconds before the Big-bang, that is, before the Plank time.

According to Islamic Sufism: The Qur'an contains 4 gates in the Qur'an and the 4th door is the 4th gate of **the TRUTH;** It is accepted that the most distinguished people (**Anjab**) who have the Divine **nobility** in the rank of **Ricalullah** (those who reach the unknown secrets of Allah, that is, the Divine truth) can be reached. If they succeed, not only physicists but also all the people of the world can now become people who have reached the level of **Ricalullah and Anjab** (who know the secrets of the Universe). In this way, the only owner of the Universe, in which Jews, Christians and Muslims alike believe, is proved by science.

c) DOUBLE RABIA IN THE BIG-BANG TABLE GIVEN ABOVE OVERLAY OF THE PYRAMIDS WITH THE MERKABA STAR:

According to the Big-bang information revealed by the world physics community today, we see two Rabia pyramids above, which rotate against and opposite each other. It was surprising to see the same pyramids in the star of MERKABA or MERHABA (today, the old meaning of the word salute in Turkish, let the soul and body be the light), which dates back to 13 000 years ago. It means MER (light), KA (spirit), BA (body). The mind unified KNOWLEDGE in the human brain is represented by the downward pyramid, and the emotional LOVE in the human heart is represented by the upward pyramid. When KNOWLEDGE in the brain and LOVE in the heart unite, the human body gains balance, health and happiness, and the holy light MER is formed in the human brain with electromagnetic waves. In the MER, it creates a colorful AURA, up to 16 meters around the body, like the Mandala, the flower of life. Aura, which also occurs in plants, can also be photographed with computerized cameras. The result is ignorant LOVE. KNOWLEDGE without love makes people and society sick. Only if the two come together, a balanced, healthy and happy human and society integrated with the universe can be formed. The history of the Kaaba or KABA, which is the place of cross for Muslims today, dates back to Adam and Eve as a holy place of visit. Likewise, the city of Mecca, in which it is located, is the only city in the world with golden ratios due to the ratio of its distances from the north and south poles.

The main reason for the Chaos we live in the world today is the word AKIL, which means combining Akale in Arabic. In other words, Mind is a verb, not an adjective, and gives us the task of combining the information in our brain.From Mu continent (S-193) to Taoism (S-59), Brahmans (S-55), Buddhists (S-53), Hindus (S-57), Jews (S-58), Christians (S-60)), Islam (S-60), it is our duty to make a contemporary unification, or synthesis, that embraces all faiths. According to the information above, it was our responsibility to make the first synthesis on the male-female relationship.

According to a false interpretation of Christianity, this relationship was seen as a cursed great sin even between parents in the family. According to the yoga belief of Hinduism, it is considered sacred.The privacy of the relationship between mother and father under the roof of the family institution is indisputable according to Islam.This relationship is essential and sacred to the health and happiness of both parents and children. In fact, in terms of the family being an example to the environment, it also positively affects the health, happiness and intellect of the society.

ADDITIONAL PICTURES AND DRAWINGS FOR THE PART FOUR:
A) HISTORY OF THE THOUGHT AND BELIEF OF HUMANITY AND HOLY GEO-METRIES D_FROM THE BEGINNING TILL TOD

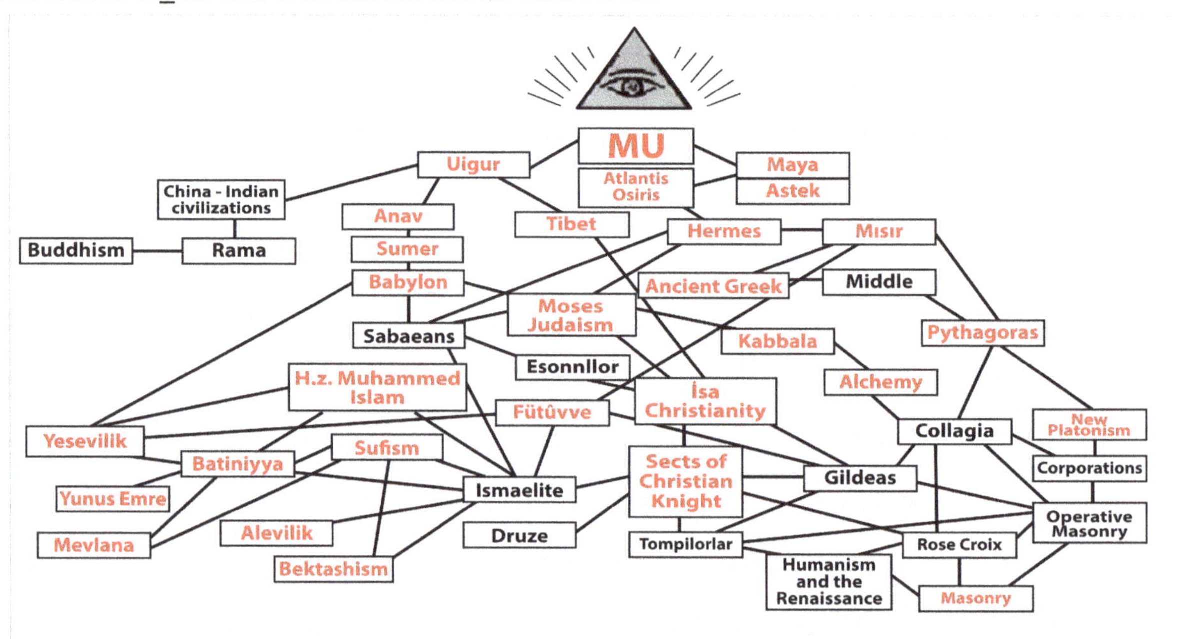

B) MU COSMIC DIAGRAM :
(According to the 15.000 Years old Naacal Tablets)

C) MU ROYAL SYMBOL:
(MU – Meaning Sun Empire. This symbol can be seen by the Indians, , Maori, Nevada and Mexican natives, and Guatemalans.)

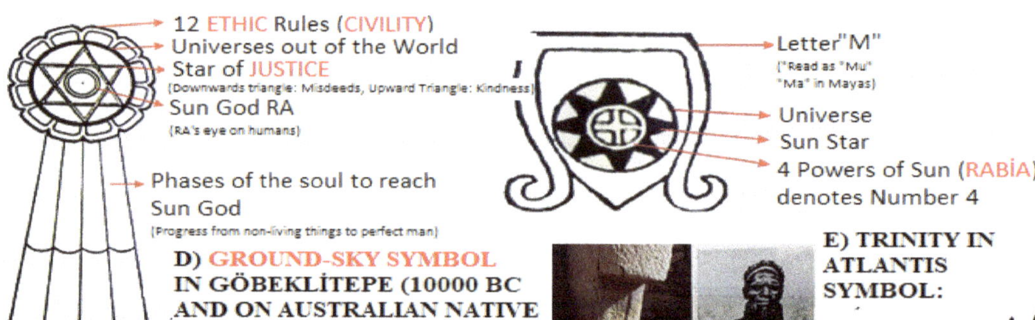

D) GROUND-SKY SYMBOL IN GÖBEKLİTEPE (10000 BC AND ON AUSTRALIAN NATIVE ABORIGINS

The down and upward triangles in "MU" are in arc shape here

E) TRINITY IN ATLANTIS SYMBOL:

15 000 BC?

F) NAACAL TABLET

G) ETRUSCAN GRAVE FOUND IN GLOZER/FRANCE (IV. Century BC)

DOUBLE PYTHAGORIAN TRIANGLES CONSISTING OF QUADRAT POTTERIES: USED NOWADAYS AS THE SYMBOL OF EYES IN ANATOLIAN RUGS. THAT IS, IT PROTECTS HUMAN FROM EVIL SOULD LIKE CINTEMANI AND AMULENT. THE SAME GEOMETRY IS SEEN UNDER THE BOOT OF OF THE SCYTIAN PRINCESS

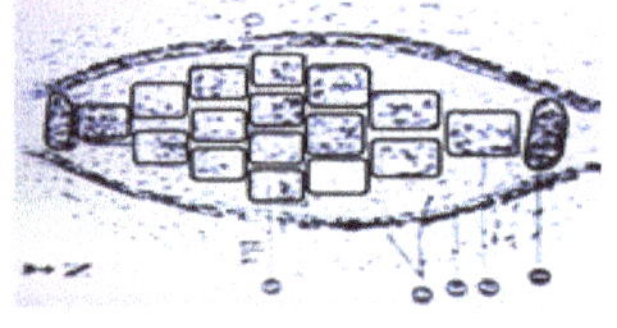

H) MAYAN CALENDAR: I) AZTEC CALENDAR
İ.Ö. 15 000   İ.S. 14-16 yy

J) DOUBLE TRIANGLES ON THE DRESS OF THE PRINCESS IN THE SCYTIAN GRAVE FOUND IN KAZAKHSTAN AND DOUBLE PYTHAGORIAN TRIANGLES UNDER HER BOOT
V. Century BC

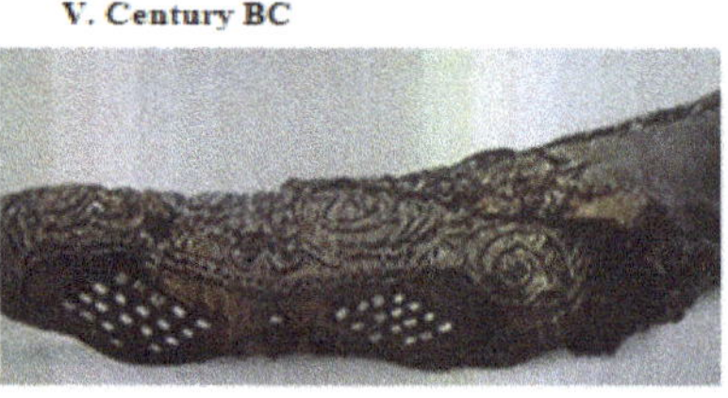

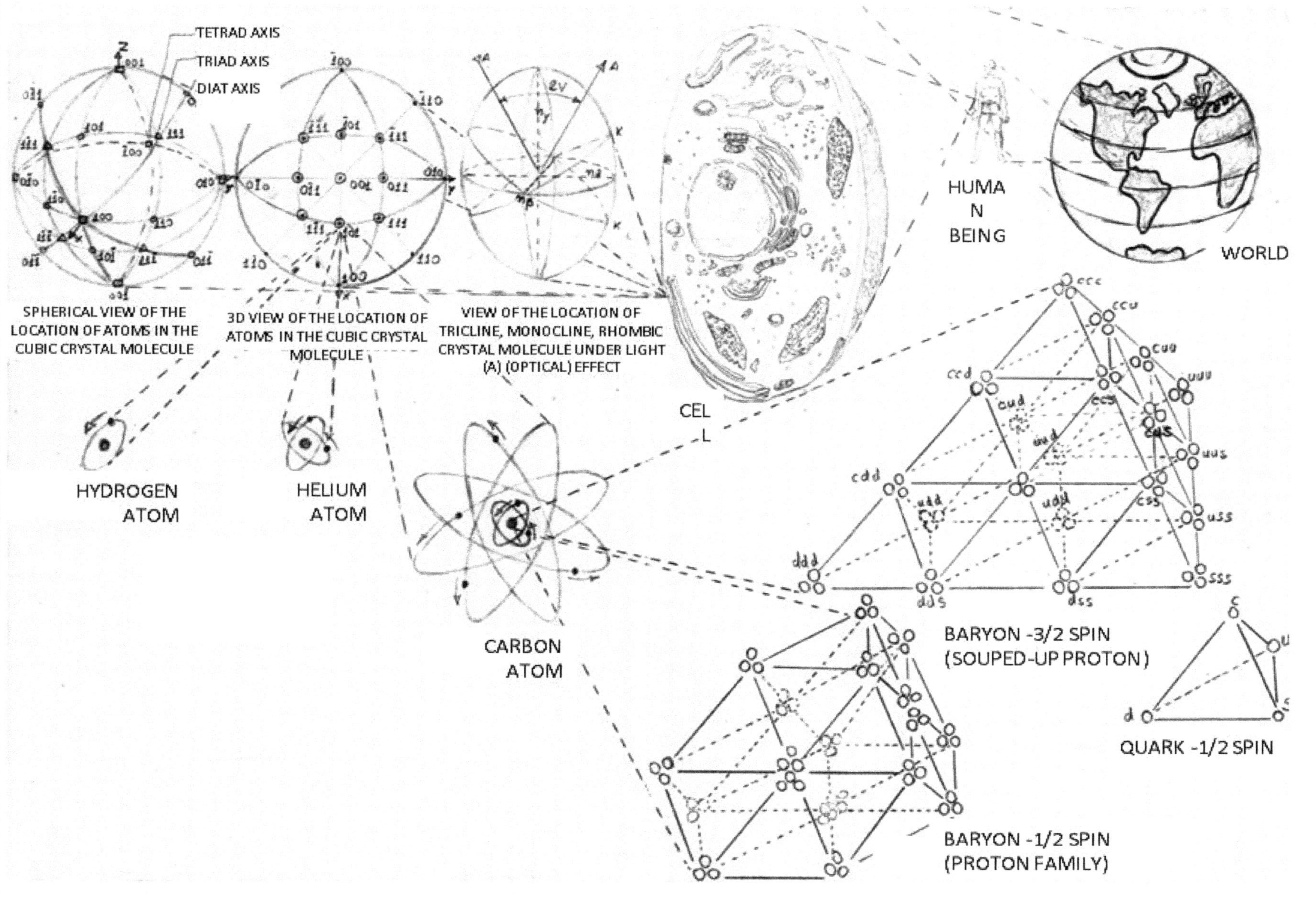
TETRAD AXIS
TRIAD AXIS
DIAT AXIS
SPHERICAL VIEW OF THE LOCATION OF ATOMS IN THE CUBIC CRYSTAL MOLECULE
3D VIEW OF THE LOCATION OF ATOMS IN THE CUBIC CRYSTAL MOLECULE
VIEW OF THE LOCATION OF TRICLINE, MONOCLINE, RHOMBIC CRYSTAL MOLECULE UNDER LIGHT (A) (OPTICAL) EFFECT
CELL
HUMAN BEING
WORLD
HYDROGEN ATOM
HELIUM ATOM
CARBON ATOM
BARYON -3/2 SPIN (SOUPED-UP PROTON)
BARYON -1/2 SPIN (PROTON FAMILY)
QUARK -1/2 SPIN

D IN PART FOUR
ADDITIONAL FIGURES AND DRAWINGS ON THE "NEW THEORY OF EVERYTHING"
BASED ON THE TABLES AND GEOMETRIES IN THE ANNEX:

TETRAHEDRON:

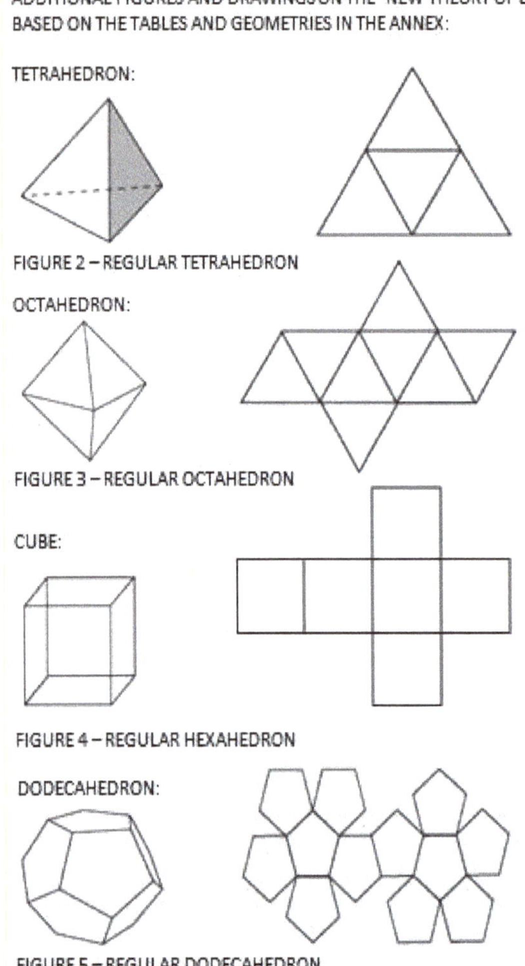

FIGURE 2 – REGULAR TETRAHEDRON

OCTAHEDRON:

FIGURE 3 – REGULAR OCTAHEDRON

CUBE:

FIGURE 4 – REGULAR HEXAHEDRON

DODECAHEDRON:

FIGURE 5 – REGULAR DODECAHEDRON

ICOSAHEDRON:

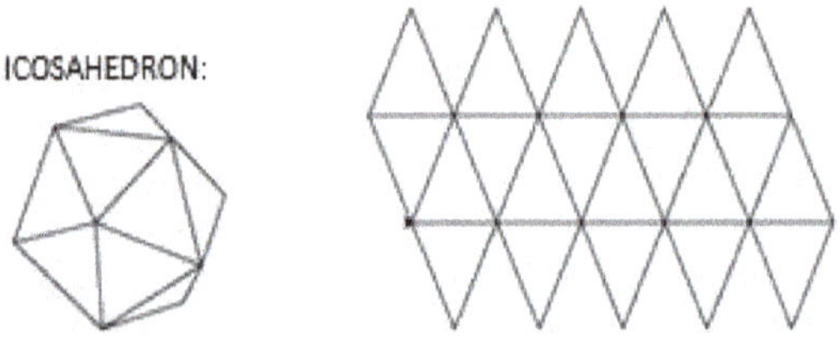

FIGURE 6 – REGULAR ICOSAHEDRON

ASTRONOMER KEPLER (1571-1630): HE FOUND OUT THAT THE PLANETS OF
SUN HAVE A MUSIC AND ALSO DETERMINED THE NOTES OF EACH PLANET.
"WELT-HARMONIE" (HARMONY OF THE WORLD)

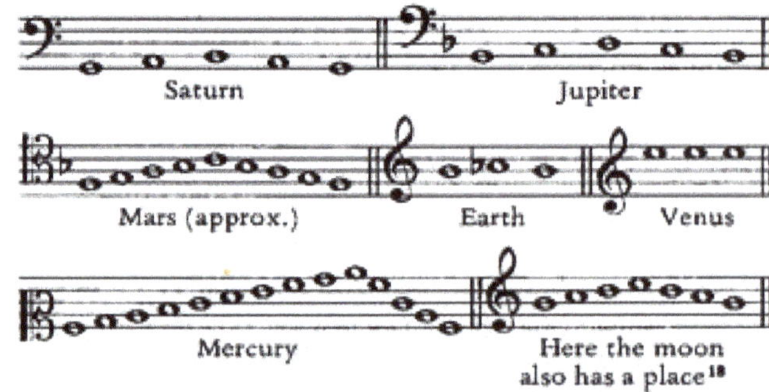

FIGURE 1 – ASTRONOMER KEPLER'S "HARMONY OF
THE WORLD" – MELODY OF EACH PLANET

FIGURE 7 – ASTRONOMER KEPLER HAS PLACED THE
POLYHEDRONS OF PLATON IN THE SPHERES OF 6 PLANETS IN A
NESTED FORM AS CAN BE SEEN ABOVE

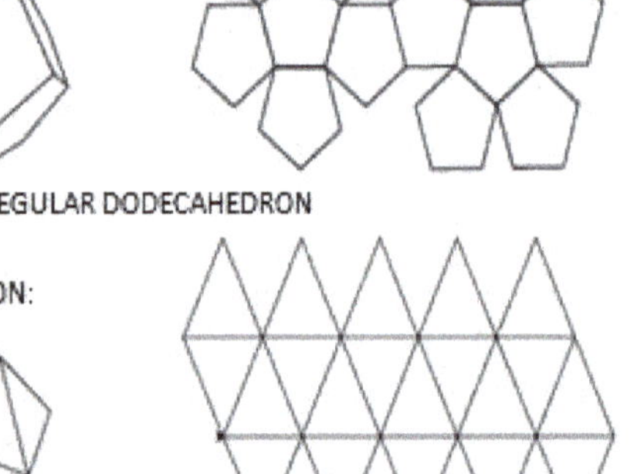

FIGURE 10 – EXPRESSION OF PLANETS BY
MEANS OF NOTES IN THE ANCIENT EGYPT

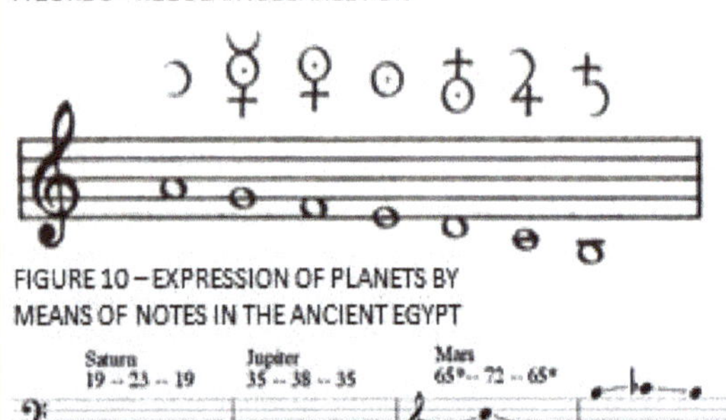

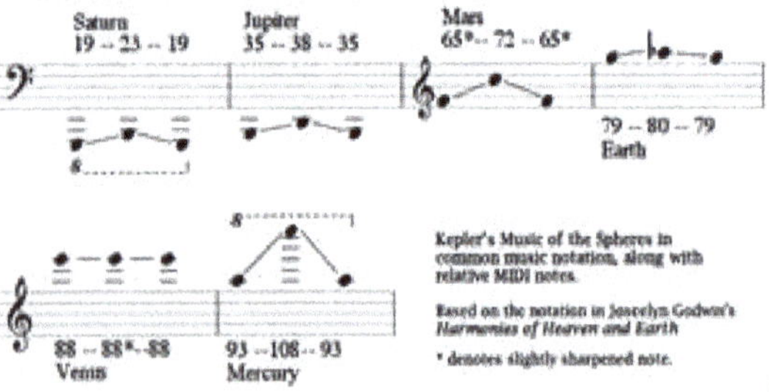

FIGURE 11 – MUSIC OF THE PLANETS ACCORDING TO KEPLER
(ANOTHER EXPRESSION OF "HARMONY OF THE WORLD" IN
- FIGURE 1)

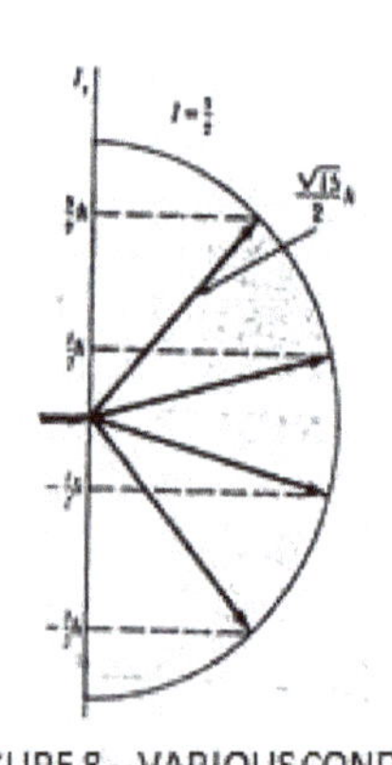

FIGURE 8 – VARIOUS CONDITIONS
OF SPIN (ANNEX Ic)

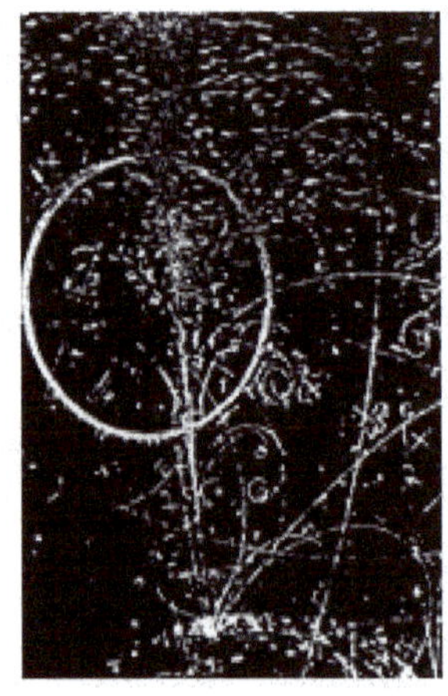

FIGURE 9 – A VIEW WHICH IS
OBSERVED AFTER COLLISION IN THE
PARTICLE COLLIDERS

TABLES IN THE ANNEXES OF THE BOOK

QUANTUMS OF UNIVERSES:

FACTORS CREATING THE ASSUMPTION WHICH WAS PREVIOUSLY KNOWN AS "PULSATING UNIVERSE THEORY" AND "OSCILLATION THEORY":

I- ADDITION OF "ELECTROMAGNETIC WAVE" AND "LIGHT QUANTUMS" WHICH HAVE BEEN PREVIOUSLY PROVED AND 3 DIMENSIONS OF UNIVERSE (X,Y,Z) IN THE TABLE; LOCATING THE RHOMBIC PRISMS (ANNEX IIb) INTO THE KINETICAL ENERGY PACKAGES, COLORING WITH GLUON COLORS, MAKING THEM THE KINETICAL ENERGY QUANTUMS OF UNIVERSES.

II- CHAIN-CONNECTION OF THE POTENTIAL ENERGY QUANTUMS OF UNIVERSES IN GRAY COLOR AS CAN BE SEEN IN THE TABLE.

III- WRITING MASS, ENERGY AND CRITICAL INTENSITY RATIOS OF OPEN, CLOSED AND FLAT UNIVERSES WHICH HAVE PREVIOUSLY BEEN KNOWN WITH A DENOMINATOR OF 3, ON THE TABLE IN THE CELLS THEY BELONG TO.

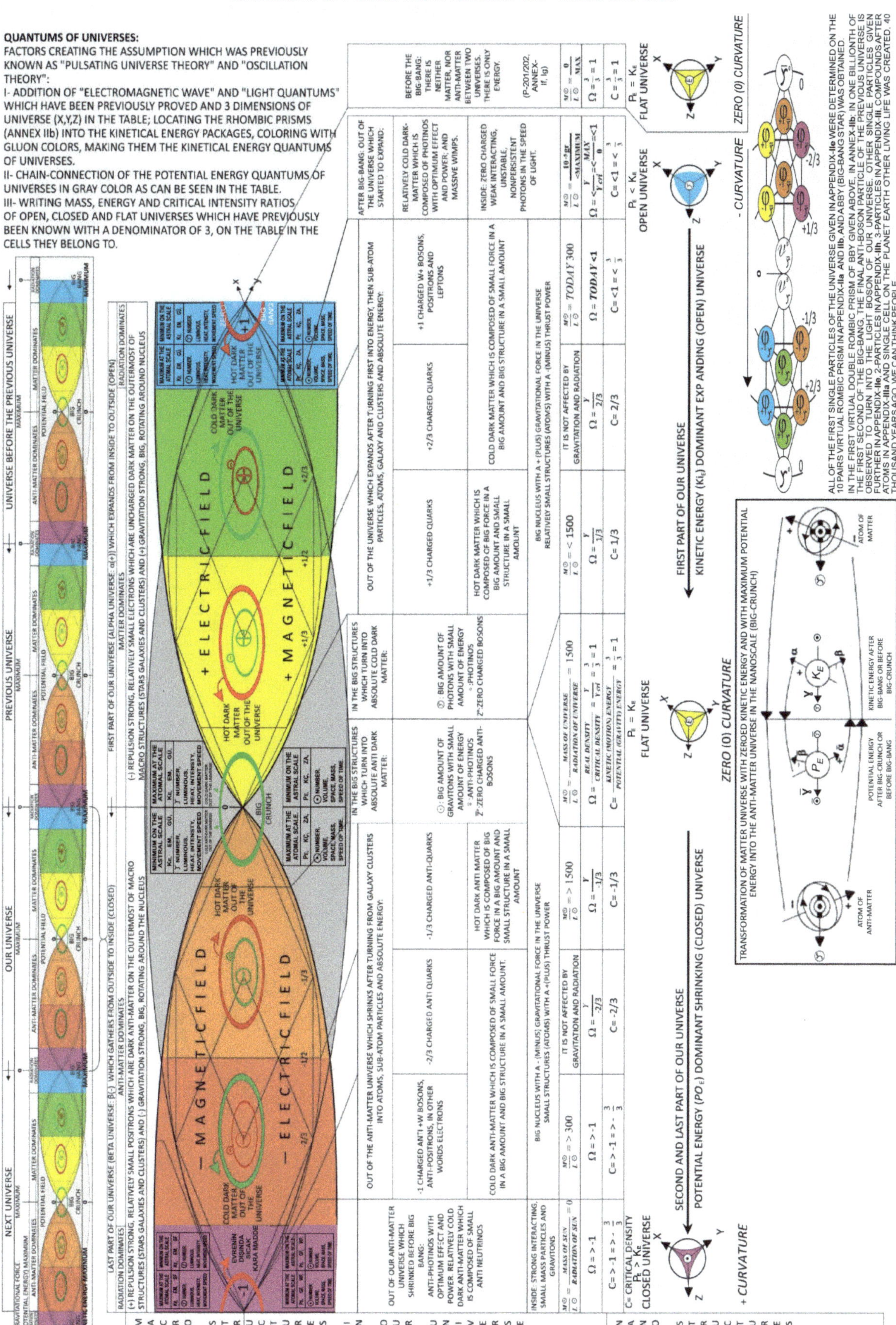

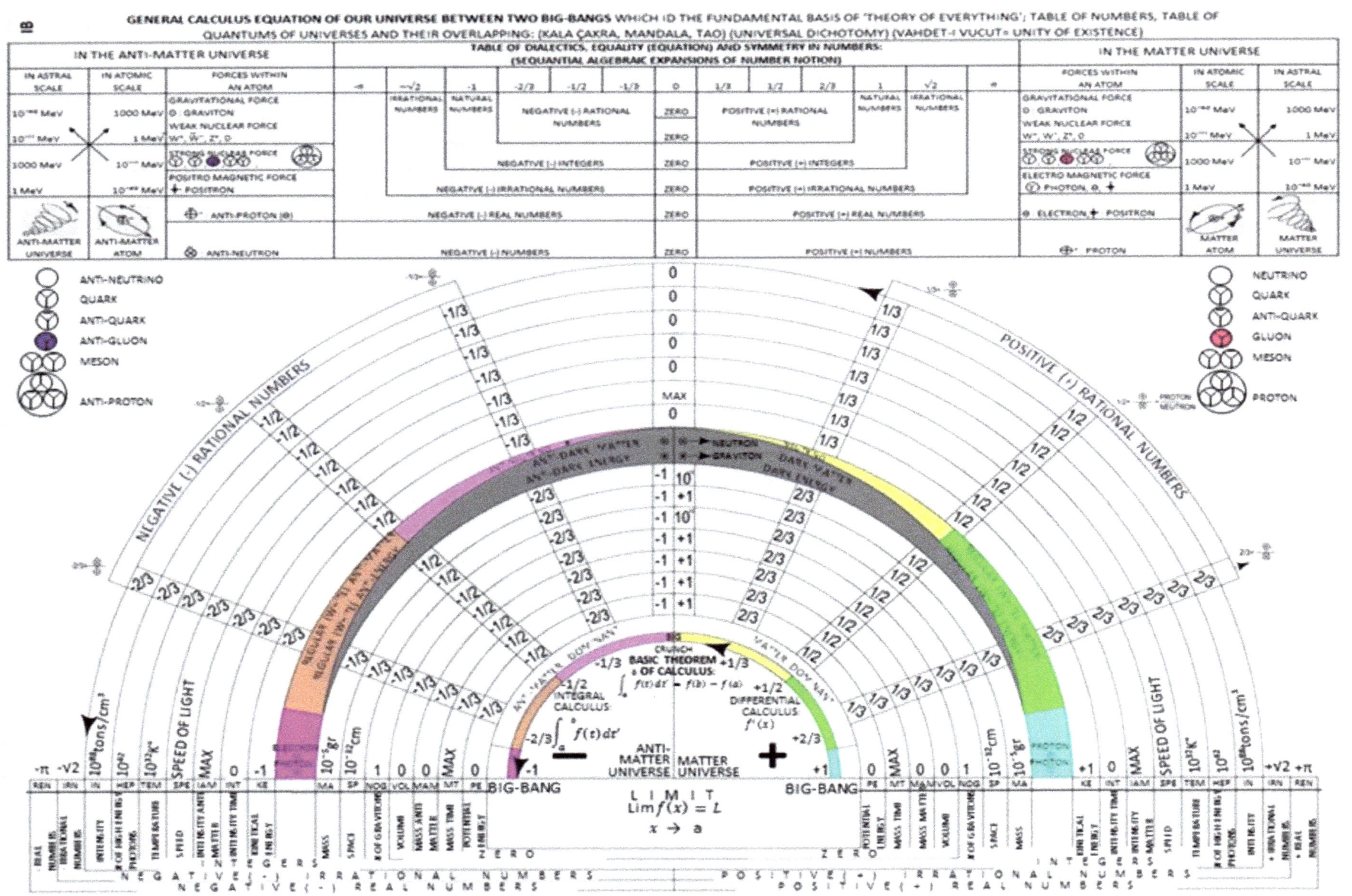
GENERAL CALCULUS EQUATION OF OUR UNIVERSE BETWEEN TWO BIG-BANGS WHICH ID THE FUNDAMENTAL BASIS OF 'THEORY OF EVERYTHING'; TABLE OF NUMBERS, TABLE OF
QUANTUMS OF UNIVERSES AND THEIR OVERLAPPING: (KALA ÇAKRA, MANDALA, TAO) (UNIVERSAL DICHOTOMY) (VAHDET-I VUCUT= UNITY OF EXISTENCE)
IN THE ANTI-MATTER UNIVERSE
TABLE OF DIALECTICS, EQUALITY (EQUATION) AND SYMMETRY IN NUMBERS:
(SEQUANTIAL ALGEBRAIC EXPANSIONS OF NUMBER NOTION)
IN THE MATTER UNIVERSE
IN ASTRAL SCALE
IN ATOMIC SCALE
FORCES WITHIN AN ATOM
GRAVITATIONAL FORCE
O · GRAVITON
WEAK NUCLEAR FORCE
STRONG NUCLEAR FORCE
POSITRO MAGNETIC FORCE
POSITRON
ANTI-PROTON (⊖)
ANTI-NEUTRON
ANTI-MATTER UNIVERSE
ANTI-MATTER ATOM
IRRATIONAL NUMBERS
NATURAL NUMBERS
NEGATIVE (-) RATIONAL NUMBERS
ZERO
POSITIVE (+) RATIONAL NUMBERS
NATURAL NUMBERS
IRRATIONAL NUMBERS
NEGATIVE (-) INTEGERS
ZERO
POSITIVE (+) INTEGERS
NEGATIVE (-) IRRATIONAL NUMBERS
ZERO
POSITIVE (+) IRRATIONAL NUMBERS
NEGATIVE (-) REAL NUMBERS
ZERO
POSITIVE (+) REAL NUMBERS
NEGATIVE (-) NUMBERS
ZERO
POSITIVE (+) NUMBERS
GRAVITATIONAL FORCE
O · GRAVITON
WEAK NUCLEAR FORCE
STRONG NUCLEAR FORCE
ELECTRO MAGNETIC FORCE
PHOTON, O, +
ELECTRON, + POSITRON
PROTON
MATTER ATOM
MATTER UNIVERSE
ANTI-NEUTRINO
QUARK
ANTI-QUARK
ANTI-GLUON
MESON
ANTI-PROTON
NEUTRINO
QUARK
ANTI-QUARK
GLUON
MESON
PROTON
NEGATIVE (-) RATIONAL NUMBERS
POSITIVE (+) RATIONAL NUMBERS
SPEED OF LIGHT
MAX
SPEED OF LIGHT
MAX
BIG-BANG
BIG-BANG
ANTI-MATTER UNIVERSE
MATTER UNIVERSE
BASIC THEOREM OF CALCULUS:
INTEGRAL CALCULUS
DIFFERENTIAL CALCULUS
f'(x)
CRUNCH
LIMIT
Lim f(x) = L
x → a
NEUTRON
GRAVITON
DARK MATTER
DARK ENERGY
ANTI-DARK MATTER
ANTI-DARK ENERGY
NEGATIVE (-) IRRATIONAL NUMBERS
NEGATIVE (-) REAL NUMBERS
POSITIVE (+) IRRATIONAL NUMBERS
POSITIVE (+) REAL NUMBERS
ZERO

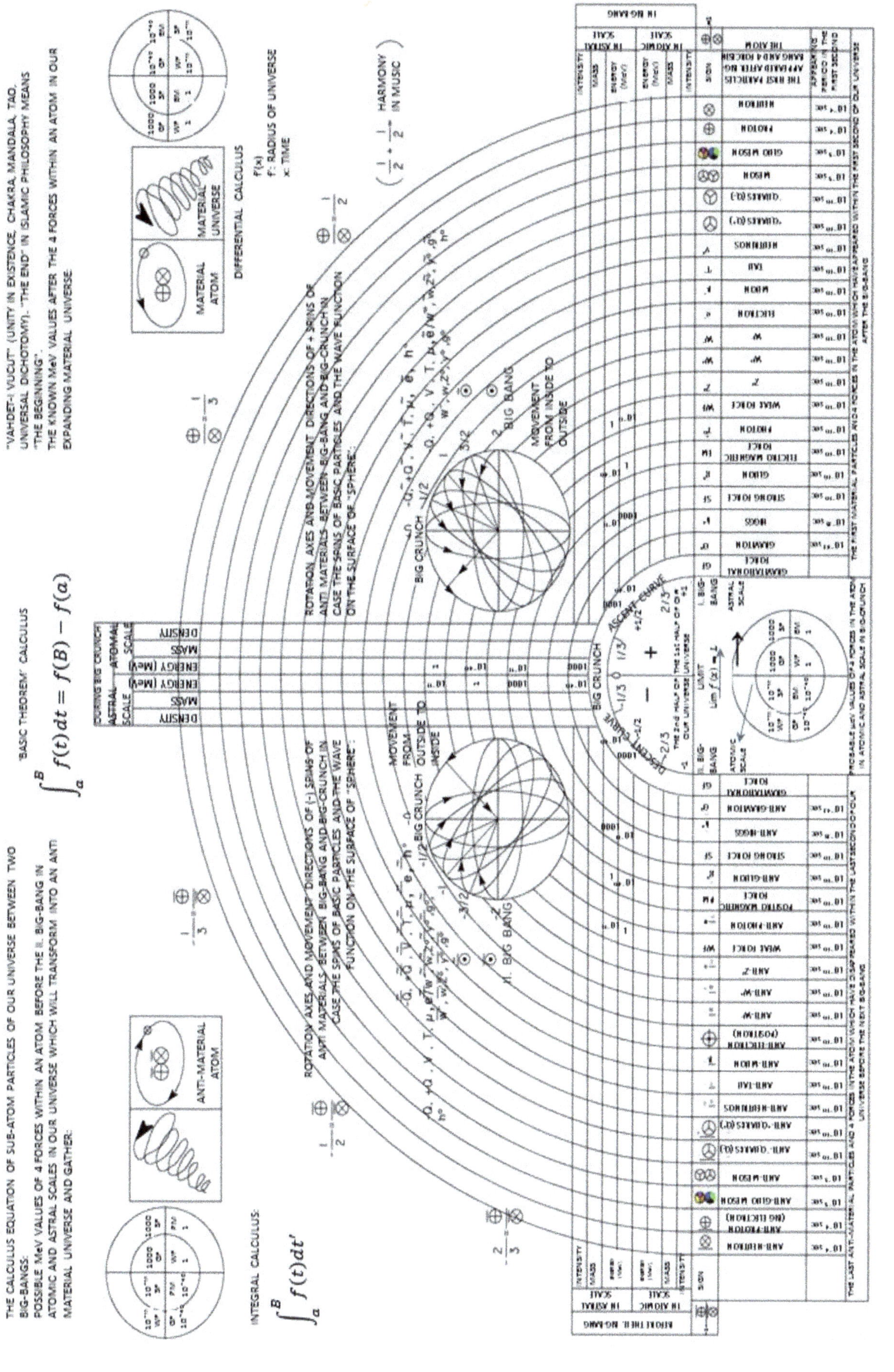

THE CALCULUS EQUATION OF SUB-ATOM PARTICLES OF OUR UNIVERSE BETWEEN TWO BIG-BANGS:

POSSIBLE MeV VALUES OF 4 FORCES WITHIN AN ATOM BEFORE THE II. BIG-BANG IN ATOMIC AND ASTRAL SCALES IN OUR UNIVERSE WHICH WILL TRANSFORM INTO AN ANTI MATERIAL UNIVERSE AND GATHER:

INTEGRAL CALCULUS:

$$\int_a^B f(t)\,dt'$$

"VAHDET-İ VUCUT" (UNITY IN EXISTENCE, CHAKRA, MANDALA, TAO, UNIVERSAL DICHOTOMY). "THE END" IN ISLAMIC PHILOSOPHY MEANS "THE BEGINNING".

THE KNOWN MeV VALUES AFTER THE 4 FORCES WITHIN AN ATOM IN OUR EXPANDING MATERIAL UNIVERSE.

DIFFERENTIAL CALCULUS

'BASIC THEOREM' CALCULUS

$$\int_a^B f(t)\,dt = f(B) - f(\alpha)$$

$$\left(\frac{1}{2} + \frac{1}{2} \right) \quad \text{HARMONY IN MUSIC}$$

f'(x)
f': RADIUS OF UNIVERSE
x: TIME

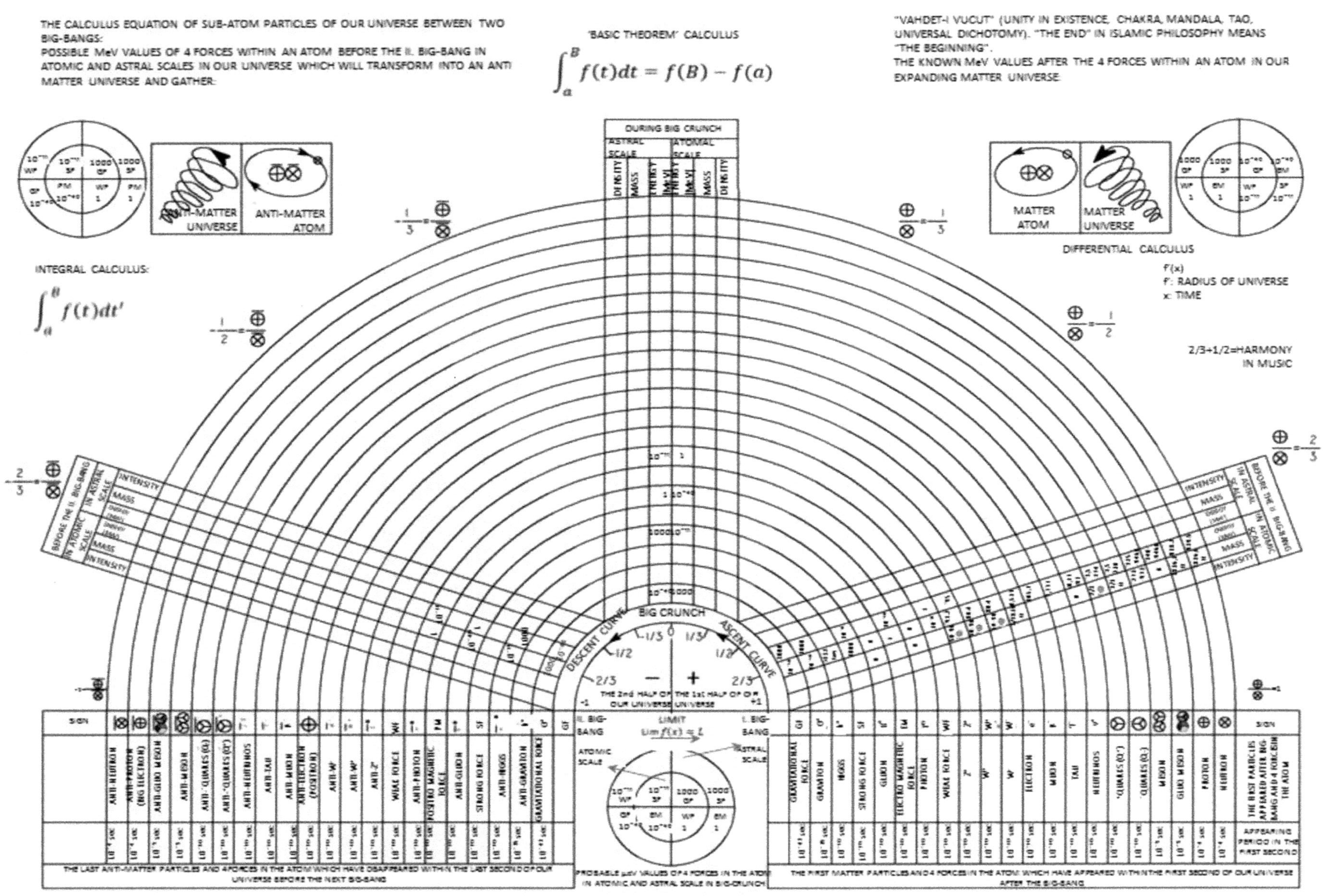
THE CALCULUS EQUATION OF SUB-ATOM PARTICLES OF OUR UNIVERSE BETWEEN TWO BIG-BANGS:
POSSIBLE MeV VALUES OF 4 FORCES WITHIN AN ATOM BEFORE THE II. BIG-BANG IN ATOMIC AND ASTRAL SCALES IN OUR UNIVERSE WHICH WILL TRANSFORM INTO AN ANTI MATTER UNIVERSE AND GATHER:
'BASIC THEOREM' CALCULUS
"VAHDET-I VUCUT" (UNITY IN EXISTENCE, CHAKRA, MANDALA, TAO, UNIVERSAL DICHOTOMY). "THE END" IN ISLAMIC PHILOSOPHY MEANS "THE BEGINNING".
THE KNOWN MeV VALUES AFTER THE 4 FORCES WITHIN AN ATOM IN OUR EXPANDING MATTER UNIVERSE:
ANTI-MATTER UNIVERSE
ANTI-MATTER ATOM
MATTER ATOM
MATTER UNIVERSE
DIFFERENTIAL CALCULUS
f'(x)
f: RADIUS OF UNIVERSE
x: TIME
INTEGRAL CALCULUS:
2/3+1/2=HARMONY IN MUSIC
DURING BIG CRUNCH
ASTRAL SCALE
ATOMAL SCALE
DENSITY
MASS
ENERGY
MeV
MeV
MASS
DENSITY
BEFORE THE II. BIG-BANG
IN ASTRAL SCALE
IN ATOMIC SCALE
INTENSITY
MASS
ENERGY
MeV
MASS
INTENSITY
DESCENT CURVE
ASCENT CURVE
BIG CRUNCH
THE 2nd HALF OF THE 1st HALF OF OUR UNIVERSE
II. BIG-BANG
I. BIG-BANG
LIMIT
lim f(x) = L
ATOMIC SCALE
ASTRAL SCALE
SIGN
ANTI-NEUTRON
ANTI-PROTON (BIG ELECTRON)
ANTI-GLUO MESON
ANTI-MESON
ANTI-"QUARKS (Q)"
ANTI-"QUARKS (Q')"
ANTI-NEUTRINOS
ANTI-TAU
ANTI-MUON
ANTI-ELECTRON (POSITRON)
ANTI-W'
ANTI-W
ANTI-Z
WEAK FORCE
ANTI-PHOTON
POSITIVE MAGNETIC FORCE
ANTI-GLUON
ANTI-GLUON FORCE
STRONG FORCE
ANTI-HIGGS
ANTI-GRAVITON
GRAVITATIONAL FORCE
THE LAST ANTI-MATTER PARTICLES AND 4 FORCES IN THE ATOM WHICH HAVE DISAPPEARED WITHIN THE LAST SECOND OF OUR UNIVERSE BEFORE THE NEXT BIG-BANG
PROBABLE μeV VALUES OF 4 FORCES IN THE ATOM IN ATOMIC AND ASTRAL SCALE IN BIG-CRUNCH
SIGN
GRAVITATIONAL FORCE
GRAVITON
HIGGS
STRONG FORCE
GLUON
ELECTRO MAGNETIC FORCE
PHOTON
WEAK FORCE
Z
W'
W
ELECTRON
MUON
TAU
NEUTRINOS
"QUARKS (Q')"
"QUARKS (Q)"
MESON
GLUO MESON
PROTON
NEUTRON
THE FIRST PARTICLES APPEARED AFTER BIG BANG AND 4 FORCES IN THE ATOM
APPEARING PERIOD IN THE FIRST SECOND
THE FIRST MATTER PARTICLES AND 4 FORCES IN THE ATOM WHICH HAVE APPEARED WITHIN THE FIRST SECOND OF OUR UNIVERSE AFTER THE BIG-BANG

HISTORY OF OUR UNIVERSE

PHASE: RADIATION DOMINANT

SIZE (m)	WAVE LENGTH (m)	TEMP. K°	MASS	TIME AFTER BIG-BANG
10^{-35}	10^{-35}	10^{23}	$(10^{19}\,\oplus)$ 10^{19} GeV $(10^{-5}$ gr$)$	AFTER 10^{-45} SECONDS
~20 cm		10^{10}	$10^{15}\,\oplus$	AFTER 10^{-35} SECONDS
10^{-30} cm			~200 GeV	AFTER 10^{-11} SECONDS
(W) 10^{-35} cm				
~10^{20}	~10^{20}	10^{13}	<200 GeV	AFTER 10^{-10} SECONDS
			<300 GeV	AFTER 10^{-5} SECONDS
		10^{10}	1 MeV	10^{-4} SECONDS
		10^{10}	< 1 MeV	AFTER 1 SECOND
				AFTER 1 MINUTE
			~100 KeV	AFTER 3 MINUTES
				AFTER 1 HOUR
				AFTER 1 MONTH
		2,73°		AFTER 1 YEAR
				100.000 YEARS

PHASE: MATERIAL DOMINANT

SIZE (m)	WAVE LENGTH (m)	TEMP. K°	MASS	TIME AFTER BIG-BANG
10^{-9}	10^{-9}	3000	<1 eV	AFTER 300.000 YEARS
				AFTER 1 MILLION YEARS
				1 BILLION YEARS
	10^{17}			2 BILLION YEARS
	10^{7}			3 BILLION YEARS
				4-5 BILLION YEARS
				6,7 BILLION YEARS
	10^{12}			9-10 BILLION YEARS
	10^{12}			10,7 BILLION YEARS
10^{20}	10^{20} **Ic**			TODAY

BEFORE BIG-BANG: WHILE THE STRONG NUCLEUS WHICH THE ANTI-BARYONS AND ANTI-MESONS OF THE PREVIOUS UNIVERSE REPRESENT GETS WEAKER, MAGNETIC FORCE WHICH THE ANTI-ELECTRONS (+) REPRESENT GETS STRONGER AND THOSE THREE FORCES UNITE.

PLANCK PERIOD AND BIG BANG: DESPITE THE ZEROED VOLUME, SIZE, TIME (SINGULARITY, WORMHOLES) MASS (GRAVITON NUMBER) AND GRAVITY POTENTIAL ENERGY; 4 FORCES UNITED WHEN MAXIMUM; HEAT, RADIATION, DENSITY TIMES VELOCITY, MOVEMENT SPEED, KINETICAL MOVEMENT ENERGY (-1) EXISTED; AND BIG-BANG HAPPENED BY MEANS OF CASIMIR EFFECT. KINETIC ENERGY TRANSFORMED FROM THE -1 CHARGED HOT RADIATION OF ANTI-MATERIAL TO +1 CHARGED HOT RADIATION OF OUR MATERIAL UNIVERSE. THEN, FOR EVERY DESTROYED RADIATION BELONGING TO THE KINETIC ENERGY, A MASS STARTED TO BE CREATED BELONGING TO THE POTENTIAL ENERGY. GENERATED MATERIALS WILL GO ON ACCUMULATING UNTIL THE EXPANDING OF OUR UNIVERSE COMES TO AN END AS ZERO CHARGED DARK-MATTER IN THE OUTERMOST OF THE UNIVERSE.

BIG INTEGRATION PERIOD: WIMPS (UNSTABLE WEAKLY INTERACTING MASSIVE PARTICLE) ARE THE APPROACHING OF WEAK AND STRONG NUCLEAR FORCES WHICH ARE OPENED TO DIFFERENT UNIVERSES (PREVIOUS UNIVERSE). THEY ARE IN SUSY (SUPER SYMMETRY) STATE. IN THIS STAGE, THERE ARE PHOTONS AS MUCH AS BARYONS (ENTROPi=0) ON THE UNIVERSE. IN THE SUSY PERIOD, THERE IS A COLD DARK MATERIAL IN THE UNIVERSE IN WHICH WEAK AND STRONG NUCLEAR FORCES AND ELECTROMAGNETIC FORCE ARE COMBINED.

LATTICES: 3D NODES WHICH ARE ASSUMED TO BE SPHERICAL ARE THE LIVING RESIDUALS OF BIG-COMBINATION PERIOD. (IT CAN BE THE BIG-BANG STAR IN ANNEX-IIa, B) IN THIS PHASE, LATTICES WERE DESTROYED AND GOLD STONE BOSONS STARTED TO BE EMMITTED (WEAK INTERACTIVE MASSLESS BOSONS) THE APPEARING FLUCTUATIONS, LATER FORMED GALAXIES AND THEIR CLUSTERS. TODAY'S GOLD STONE BOSON SEA WAS LEFT BEHIND.

STRINGS: WIDTH: 10^{-35} CM, LENGTH: INFINITE, 1D, HIGH ENERGY STRINGS HAVE POTENTIAL ENERGY DENSITY WHERE THERE ARE NO LATTICE AND STRINGS. "HIGGS AREAS" ARE FORMED.

SEPARATION PHASE OF 4 FORCES WITHIN THE ATOM: ELECTRON, POSITRON, QUARK, ANTI-QUARK COMBINATION (PROBABLY 160 PCS. SINGLE PARTICLES BELONGING TO BIG-BANG STAR)

TWO-PARTICLE MESONS AND ANTI-MESONS WERE CREATED.

THREE-PARTICLE BARYONS AND ANTI-BARYONS WERE CREATED. (NUCLEONS)

ELECTRONS CAUGHT (0) AND POSITRONS CAUGHT (+).

PARTICLE CREATION STOPPED: THERE IS 1 HELIUM NUCLEUS PER 10 PROTONS ON THE UNIVERSE.

NUCLEAR FUSION PHASE

THERE ARE 75% HYDROGEN, 25% HELIUM AND ELECTRONS ON THE UNIVERSE.

FLUCTUATIONS STARTED TO GET HIGHER.

THERMALIZATION PHASE: "COSMIC MICROWAVE RADIATION SIMILAR TO X RAYS RADIATIONS (DARK-MATTER RADIATIONS) HYDROGEN PLASMA, PHOTONS COLD AREAS. COLD POINTS, SPACES IN THE MATERIAL DISTRIBUTION, AS HOT POINTS.

HEAT AND DENSITY FLUCTUATIONS PHASE STARTED. RADIATION DOMINANCE PHASE ENDED. MATERIAL DOMINANCE STARTED. BARYON ORIGIN MACHOS (COAGULATION) WHICH HAVE DENSIFIED BEFORE THE GALAXIES ARE MASSIVE TIGHT AURA OBJECTS. (ASTRO PHYSICAL DARK MATTER). FLUCTUATIONS WHICH COULDN'T HAVE GROWN BIGGER SINCE THEY WERE PREVIOUSLY ADIABATIC (INCLUDING BOTH MATERIAL AND RADIATION) STARTED TO GROW.

JOINING PERIOD: ELECTRONS AND HELIUM NUCLEI JOIN TO FORM THE FIRST ATOMS. "LAST SCATTERING PERIOD" STARTED WITH THE DECREASE IN DENSITY AND TEMPERATURE. PHOTONS WERE RELEASED. MATERIALS TURNED FROM PLASMA INTO ATOMS. HYDROGEN ATOMS STARTED TO FORM. IN THIS PERIOD, UNIVERSE WAS 1000 TIMES SMALLER THAN IT IS TODAY. WHILE THE FIRST SMALL STRUCTURES OF 1 MILLION TIMES THE MASS OF SUN (MO) ARE FORMED, UNIVERSE BECAME RADIATION PERMEABLE, AND ENLIGHTENED AS HYDROGEN GAS CLUSTER.

FULL HOMOGEN DENSITY FLUCTUATIONS STARTED TO GO HIGHER.

PRIMITIVE GALAXIES: GRAVITATION OF THE GAS ON UNIVERSE FORMS THE GALAXIES.

HEAVY ELEMENTS STARTED TO BE FORMED IN GALAXIES AND MILKYWAY STARS.

QUARKS WERE FORMED IN THE CENTER OF GALAXIES AND MILKYWAY AND AURA WAS FORMED AROUND THEM.

DISCS WERE FORMED AROUND THE CENTER OF GALAXIES AND MILKYWAY.

GAS CLUSTERS (NEBULA) FORMED BY PURE HYDROGEN GAS IN THE UNIVERSE.

DWARF GALAXIES, STARS, SOLAR SYSTEM AND WORLD WERE CREATED.

THE FIRST PLANT IN SOUTH-EAST AFRICA. SINGLE-CELLED ALG WAS FORMED FROM COAL (FIRST CELL)

YEARS AGO= YA	
565 MILLION YA	THE FIRST MOVING MARINE ANIMAL
300 MILLION YA	THE FIRST LAND PLANTS
260-180 MILLION YA	THE ONCE ADJACENT CONTINENTS, AMERICA AND EUROPE ARE SEPARATED FROM EACH OTHER
225-75 MILLION YA	LIFETIME OF DINOSAURS ON THE WORLD
100 MILLION YA	THE FIRST INSECTS
80 MILLION YA	PALM TREES ON THE SOUTH-POLE. 30% OF ATMOSPHERE IS OXYGEN
60 MILLION YA	THE LAND GEOGRAPHY IS CREATED WHICH IS SIMILAR TO TODAY'S GEOGRAPHY.
55-30 MILLION YA	ANCESTOR OF ANCESTORS ON THE WORLD "MAYAHIPIS"
30 MILLION YA	HUMAN-LIKE CHIMP IN AL-FAYYUM OASIS IN CAIRO-EGYPT
10 MILLION YA	A HUMAN SKELETON IN THESSALONIKI.
8 MILLION YA	A SKULL IN CHINA.
7 MILLION YA	RIFT VALLEY AND LAKE TANGANYIKA (4000 M DEEP) WAS FORMED FROM THE CENTRAL SOUTH OF AFRICA TO SYRIA AND HATAY-ANATOLIA.
5 MILLION YA	AUSTRALOPITHECUS IN THE CENTRAL AFRICA. BRAIN VOLUME 500 CC.
3,2 MILLION YA	MOUNTAINS STARTED TO BE FORMED. HUMAN SKULL (CONSUL) FROM CENTRAL ASIA.
3 MILLION YA	A HUMAN SKELETON IN SOUTH-EAST AFRICA: LUCY (AUSTRALOPITHECUS). I. HUMANKIND "RAMAPITHECUS"
2 MILLION YA	"HOMO-HABILIS" HUMANKIND IN CENTRAL SOUTH AFRICA
1,5 MILLION YA	"HOMO-ERECTUS" IN CENTRAL SOUTH-AFRICA. BRAIN VOLUME 900 CC II. HUMANKIND
1 MILLION YA	THE OLDEST ICE AGE STARTED. NILUADES IS COVERED BY LAKES
600.000 YA	THE OLDEST ICE AGE IS OVER.
600.000-540.000 YA	I. ICE AGE
600.000-150.000 YA	HEIDELBERG MAN IN EUROPE
540.000-480.000 YA	HOT PERIOD. THE FIRST WILD HORSE
500.000 YA	"JAVA MAN" (PITHECANTHROPUS ERECTUS) IN JAVA ISLAND
480.000-430.000 YA	II.ICE AGE
400.000-240.000 YA	HOT PERIOD
370.000-240.000 YA	II. INTERMEDIATE ICE AGE
350.000 YA	LEARNING TO LIGHT A FIRE IN CENTRAL ASIA
250.000 YA	HUMAN FOOT PRINTS IN MANISA-ANATOLIA
240.000-180.000 YA	III. ICE AGE
180.000-120.000 YA	HOT PERIOD
150.000 YA	3RD HUMANKIND "NEANDERTHALS" IN ASIA, EUROPE AND AFRICA
120.000-12.000 YA	IV. ICE AGE
60.000-12.000 YA	THE LAST INTERMEDIATE ICE AGE. NORTHERN HEMISPHERE IS COVERED WITH ICE CAPS. SOUTHERN HEMISPHERE IS VERY HOT. TODAY'S HUMANKIND IN THE CENTRAL ASIA. "HOMO-SAPHIENS". THE FIRST THINKING AND TALKING HUMANKIND. IV. HUMANKIND
16.000 YA	IMMIGRATION FROM ASIA TO AMERICA VIA BERING STRAIT
15.000 YA	IMMIGRATION FROM ASIA TO THE MEDITERRANION REGION. THE OLDEST VILLAGE IN THRACE.
14.000 YA	IIIRD OLDEST VILLAGE "CATAL HOYUK" IN KONYA-ANATOLIA AND "GOBEKLI TEPE" (10000 B.C.) IN URFA-ANATOLIA
12.000 YA	ICE CAPS ARE MELTING IN THE NORTHERN HEMISPHERE. SEAS OF THE WORLD ARE GETTING 100 M HIGHER. TODAY'S SEA GEOGRAPHY IS BEING CREATED. (FLOOD. NOAH'S ARK). THE RECENT HOT PERIOD IS STARTING.
9.000 YA	THE 3RD OLDEST VILLAGE "CAYONU TEPESI" IN DIYARBAKIR-ANATOLIA AND "JERICHO VILLAGE" IN ISRAEL.
8.000 YA	INDO-EUROPEAN RACE IS COMING FROM CENTRAL ASIA TO THE MIDDLE EAST AND EUROPE.
TODAY	TODAY DENSITY OF UNIVERSE (13.7 BILLION YEARS OLD) IS ~10^{-29} gr/cm³. OXYGEN RATIO IN THE WORLD'S ATMOSPHERE IS %21. POLES WILL BE REPLACED ON THE WORLD 50 THOUSAND- 1 MILLION YEARS LATER. MOON WILL MOVE AWAY FROM THE WORLD, CONTINENTS WILL BE REPLACED, SUITABLE CLIMATE WILL BE DESTROYED 50 MILLION YEARS LATER.

WHEN THE UNIVERSE COMES TO ~45 BILLION YEARS, EXPANDING OF UNIVERSE WILL STOP. UNIVERSE WILL START TO BE GATHERED AGAIN (BIG CRUCH)

WHEN THE UNIVERSE COMES TO ~90 BILLION YEARS, THE NEXT UNIVERSE WILL START WITH A BIG-BANG.

LIMITS OF THE UNIVERSE

FREQUENCIES IN HERTZ → | 1 10^1 10^2 10^3 10^4 10^5 10^6 10^7 10^8 10^9 10^{10} 10^{11} 10^{12} 10^{13} 10^{14} 10^{15} 10^{16} 10^{17} 10^{18} 10^{19} 10^{20} 10^{21} 10^{22} 10^{23} 10^{24} 10^{25} 10^{26}

TABLE OF UNIVERSE AND ELECTROMAGNETIC WAVELENGTHS IN METERS

10^{22} 10^{21} 10^{20} 10^{19} 10^{18} 10^{17} 10^{16} 10^{15} 10^{14} 10^{13} 10^{12} 10^{11} 10^{10} 10^9 10^8 10^7 10^6 10^5 10^4 10^3 10^2 10^1 1 10^{-1} 10^{-2} 10^{-3} 10^{-4} 10^{-5} 10^{-6} 10^{-7} 10^{-8} 10^{-9} 10^{-10} 10^{-11} 10^{-12} 10^{-13} 10^{-14} 10^{-15} 10^{-16} 10^{-17} 10^{-18} 10^{-19} 10^{-20} 10^{-21} 10^{-22}

— THE QUARTED OF THE ATOMAL UNIVERSE (RABIAS) —

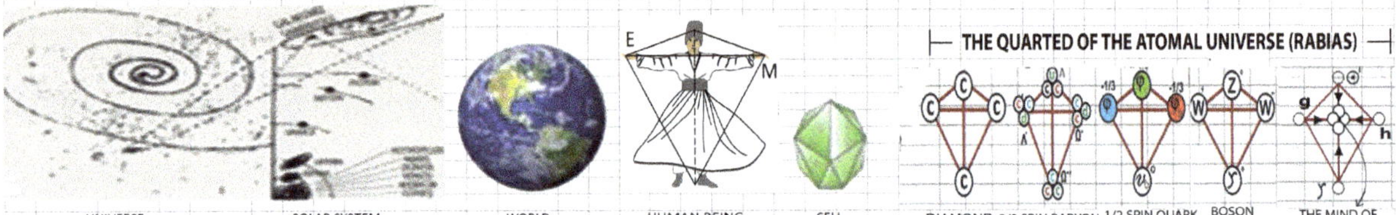

| UNIVERSE | SOLAR SYSTEM | WORLD | HUMAN BEING (E=M x C^2) | CELL | DIAMOND | 3/2 SPIN BARYON PYRAMID | 1/2 SPIN QUARK PYRAMID | BOSON PYRAMID | THE MIND OF THE UNIVERSES |

← FARTHERMOST GALAXIES AND QUASARS — 94 500 000 000 000 000 000 000 m — (10 000 000 000 LIGHT YEARS)
← DISTANCE OF THE NEAREST GALAXY "ANDROMEDA" — 7 087 500 000 000 000 000 m — (750 000 LIGHT YEARS)
←RADIUS OF OUR MILKY WAY GALAXY — 462 598 755 000 000 000 m — (15 000 PARSEC)
←MAXIMUM BOUNDARY OF COMETS — 15 000 000 000 000 000 m
←DISTANCE OF THE NEAREST STAR — 40 635 000 000 000 m — (4,3 LIGHT YEARS)
←DISTANCE OF THE OUTERMOST PLANET "PLUTO" FROM THE SUN — 5 910 000 000 000 m
←DISTANCE OF EARTH FROM THE SUN — 149 565 800 000 m
←RADIUS OF THE EARTH — 6 373 000 m
← HEIGHT OF PLANTS, ANIMALS AND HUMAN BEINGS
← LENGTH OF PROTISTS
←LENGTH OF ANIMAL AND VEGETABLE CELLS
←LENGTH OF BACTERIA AND NERVE CELLS
←AVERAGE WIDTH OF CELLS
←ELECTRON ORBITS OF ATOMS
←SIZE OF MOLECULES
←DIAMETER OF NUCLEUS OF URANIUM ATOM
←AVERAGE RADIUS OF ATOMS
←RADIUS OF ATOMIC NUCLEUS

TRIPLE PARTICLES OF ATOMIC NUCLEUS – MASS OF HYPERON	~ 2181-2583 ⊖	← ~ HYPERON
TRIPLE PARTICLES OF ATOMIC NUCLEUS – MASS OF NEUTRON	1838,6 ⊖	← ~ NEUTRON
TRIPLE PARTICLES OF ATOMIC NUCLEUS – MASS OF PROTON	1836,1 ⊖	← ~ PROTON
DUAL PARTICLES OF ATOMIC NUCLEUS – MASS OF K-MESON	968 ⊖	← ~ K-MESON
DUAL PARTICLES OF ATOMIC NUCLEUS – MASS OF +Pi-MESON	273 ⊖	← ~ Pi-MESON +
DUAL PARTICLES OF ATOMIC NUCLEUS – MASS OF -Pi-MESON	273 ⊖	← ~ Pi-MESON −
DUAL PARTICLES OF ATOMIC NUCLEUS – MASS OF Mu-MESON	207 ⊖	← ~ Mu-MESON
SINGLE PARTICLE OF ATOMIC NUCLEUS – MASS OF NEUTRINOS	NONE	
SINGLE PARTICLE OF ATOMIC NUCLEUS – MASS OF + QUARKS	+2/3 ⊕	← ~ +QUARKS
SINGLE PARTICLE OF ATOMIC NUCLEUS – MASS OF - QUARKS	-1/3 ⊖	← ~ - QUARKS
DUAL PARTICLES OF ATOMIC NUCLEUS – MASS OF GLUONS	NONE	
SINGLE PARTICLE OF ATOMIC NUCLEUS – MASS OF ELECTRON	1/1836 ⊕	← ~ ELECTRON
SINGLE PARTICLE OF ATOMIC NUCLEUS – MASS OF PHOTON	NONE	PHOTON →
SINGLE PARTICLE OF ATOMIC NUCLEUS – MASS OF +W PARTICLE	~ 50-90 ⊕	← ~ W PARTICLE
SINGLE PARTICLE OF ATOMIC NUCLEUS – MASS OF -W PARTICLE	~ 50-90 ⊕	← ~ W PARTICLE
SINGLE PARTICLE OF ATOMIC NUCLEUS – MASS OF °Z PARTICLE	~ 50-91,2 ⊕	
SINGLE PARTICLE OF ATOMIC NUCLEUS – MASS OF GRAVITON	NONE	GRAVITON →

Electron orbits box:
VII. ORBIT 24,5 μ
VI. ORBIT 18.0 μ
V. ORBIT 12.5 μ
IV. ORBIT 8.0 μ
III. ORBIT 4.5 μ
II. ORBIT 2.0 μ
I. ORBIT 0.5 μ

← RADIO WAVES (300 000 Hz)
← RADIATION OF HYDROGEN ATOM (1 420 000 Hz)
← WAVES ABSORBED IN IONOSPHERE →←WAVES→←→ ←WAVES ABSORBED IN ATMOSPHERE →
←γ BEAM W → ←GAMMA-RAY WAVES→
←X-RAY WAVE(BLACK)→
← (VHF) →←(UHF)→
←ULTRA SOUND→← AUDIBLE SOUND →← INFRA SOUND→ ← INFRARED BEAM → ← ULTRAVIOLET →
WAVE WAVE (7 NOTES) WAVE WAVE BEAM WAVE
(BETWEEN 20-20 000 Hz /min)
← COMMUNICATION WAVES →
←ELECTRICAL WAVES→
←MEDIUM WAVE →←RADAR WAVE→
(FM)
← TV → ←MICRO WAVE→
WAVE

ABSORBED IN ATMOSPHERE / OPTICAL WINDOW

VISIBLE LIGHT WAVE	WAVELENGTH	HEAT RADIATION WAVE	WAVELENGTH
RED	0,8 (0,75) μ	RED	0,75 μ
		MIDDLE RED	0,65 μ
ORANGE	0,63 (0,65) μ	ORANGE	0,60 μ
YELLOW	0,59 (0,55) μ	YELLOW	0,55 μ
GREEN	0,5 (0,5) μ	GREEN	0,51 μ
BLUE	0,45 (0,45) μ	BLUE	0,47 μ
		DARK BLUE	0,44 μ
PURPLE	0,4 (0,42) μ	PURPLE	0,42 μ

TABLE OF *SPEED* IN THE UNIVERSE IN m/sec

← SPEED OF PULSAR AND BLACK HOLES IN THE SPACE	1 060 000 000 m/sec	(10^9 m/sec)
← SPEED OF ELECTROMAGNETIC WAVE (PHOTON) NEUTRINO	299 793 000 m/sec	(10^8 m/sec)
← SPEED OF ELECTRONS	147 000 000 m/sec	(50 000 000 m/sec)
←SPEED OF OUR GALAXY MILKY-WAY IN THE SPACE	700 000 m/sec	(600 000 m/sec)
←SPEED OF SOLAR SYSTEM IN OUR GALAXY	216 000 m/sec	(400 000 m/sec)
←SPEED OF ROTATION OF THE WORLD AROUND THE SUN	29 760 m/sec	
←SPEED OF ROTATION OF THE WORLD AROUND ITS OWN AXIS	465 m/sec	(1 674 km/h)

TABLE OF *TIME* IN THE UNIVERSE IN seconds

BIG EXPLOSION (BIG-BANG)	473 040 000 000 000 000	→ (13,7 BILLION YEARS AGO)
FORMATION OF SOLAR SYSTEM	145 065 600 000 000 000	→
THE FIRST LIVING-BEINGS ON THE EARTH		→ (3 BILLION YEARS AGO)
THE FIRST ANIMALS ON THE EARTH		→ (400 MILLION YEARS AGO)
THE FIRST HUMAN-LIKE CHIMPANZEE		→ (30 MILLION YEARS AGO)
THE FIRST MAP OF THE WORLD SIMILAR TO THE CURRENT MAPS		→ (60 MILLION YEARS AGO)

← → THE SHORTEST CREATURE LIFE SECONDS
← THE SHORTEST LEPTON PARTICLE LIFE — $2,2 \times 10^{-6}$ sec
82 800 sec (25h 56m 4sec) — ← ROTATION OF THE WORLD AROUND ITS OWN AXIS — ← THE SHORTEST MESON PARTICLE LIFE — $2,2 \times 10^{-8}$ sec
31 536 000 sec (365 days, 5h 48m 46sec) — ← ROTATION OF THE WORLD AROUND THE SUN — ← THE SHORTEST BARYON PARTICLE LIFE — $>10^{-11}$ sec
1 892 160 000 sec (60 years) — ← AVERAGE HUMAN LIFE
HUNDRED YEARS — ← THE LONGEST PLANT-ANIMAL LIFE
← THE END OF SOLAR SYSTEM (4-5 BILLION YEARS LATER)
← THE END OF UNIVERSE (~86-100 BILLION YEARS LATER)

TABLE OF *TEMPERATURE* IN THE UNIVERSE IN degrees centigrade (° C)

← TEMPERATURE IN THE FIRST SECOND OF BIG-EXPLOSION (BIG-BANG) WHICH HAS CREATED THE UNIVERSE	100 000 000 000 °C (10^{11} °C)
← TEMPERATURE IN CENTER OF STARS	15 000 000 000 °C (10^{10} °C)
← TEMPERATURE IN THE UNDERGROUND (60 KM AND BELOW)	2↔3 000 °C (10^3 °C)
←TEMPERATURE LIMIT FOR LIFE→ (10^2 °C - 10^{-2} °C)	[BETWEEN -200↔600 °C]

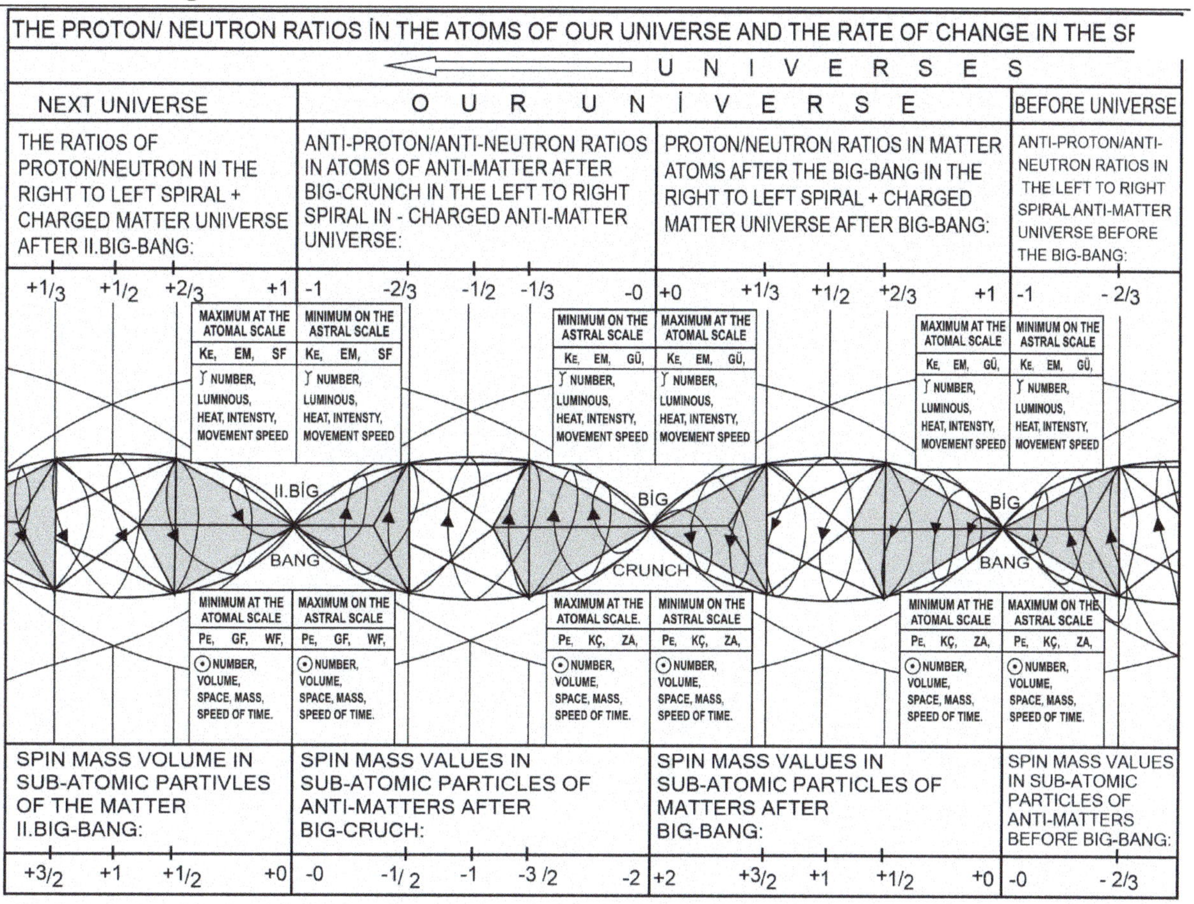

THE PROTON/ NEUTRON RATIOS İN THE ATOMS OF OUR UNIVERSE AND THE RATE OF CHANGE IN THE SF

⟵ U N I V E R S E S

NEXT UNIVERSE	O U R U N İ V E R S E	BEFORE UNIVERSE
THE RATIOS OF PROTON/NEUTRON IN THE RIGHT TO LEFT SPIRAL + CHARGED MATTER UNIVERSE AFTER II.BIG-BANG:	ANTI-PROTON/ANTI-NEUTRON RATIOS IN ATOMS OF ANTI-MATTER AFTER BIG-CRUNCH IN THE LEFT TO RIGHT SPIRAL IN - CHARGED ANTI-MATTER UNIVERSE:	PROTON/NEUTRON RATIOS IN MATTER ATOMS AFTER THE BIG-BANG IN THE RIGHT TO LEFT SPIRAL + CHARGED MATTER UNIVERSE AFTER BIG-BANG:	ANTI-PROTON/ANTI-NEUTRON RATIOS IN THE LEFT TO RIGHT SPIRAL ANTI-MATTER UNIVERSE BEFORE THE BIG-BANG:

Proton/neutron ratios:
$+1/3$ $+1/2$ $+2/3$ $+1$ | -1 $-2/3$ $-1/2$ $-1/3$ -0 | $+0$ $+1/3$ $+1/2$ $+2/3$ $+1$ | -1 $-2/3$

Boxes above the central axis (left to right):

- MAXIMUM AT THE ATOMAL SCALE — K_E, EM, SF — ɣ NUMBER, LUMINOUS, HEAT, INTENSTY, MOVEMENT SPEED
- MINIMUM ON THE ASTRAL SCALE — K_E, EM, SF — ɣ NUMBER, LUMINOUS, HEAT, INTENSTY, MOVEMENT SPEED
- MINIMUM ON THE ASTRAL SCALE — K_E, EM, GÜ — ɣ NUMBER, LUMINOUS, HEAT, INTENSTY, MOVEMENT SPEED
- MAXIMUM AT THE ATOMAL SCALE — K_E, EM, GÜ — ɣ NUMBER, LUMINOUS, HEAT, INTENSTY, MOVEMENT SPEED
- MAXIMUM AT THE ATOMAL SCALE — K_E, EM, GÜ — ɣ NUMBER, LUMINOUS, HEAT, INTENSTY, MOVEMENT SPEED
- MINIMUM ON THE ASTRAL SCALE — K_E, EM, GÜ — ɣ NUMBER, LUMINOUS, HEAT, INTENSTY, MOVEMENT SPEED

Central axis: II.BİG BANG — BIG CRUNCH — BİG BANG

Boxes below the central axis (left to right):

- MINIMUM AT THE ATOMAL SCALE — P_E, GF, WF — ⊙ NUMBER, VOLUME, SPACE, MASS, SPEED OF TIME.
- MAXIMUM ON THE ASTRAL SCALE — P_E, GF, WF — ⊙ NUMBER, VOLUME, SPACE, MASS, SPEED OF TIME.
- MAXIMUM AT THE ATOMAL SCALE. — P_E, KÇ, ZA — ⊙ NUMBER, VOLUME, SPACE, MASS, SPEED OF TIME.
- MINIMUM ON THE ASTRAL SCALE — P_E, KÇ, ZA — ⊙ NUMBER, VOLUME, SPACE, MASS, SPEED OF TIME.
- MINIMUM AT THE ATOMAL SCALE — P_E, KÇ, ZA — ⊙ NUMBER, VOLUME, SPACE, MASS, SPEED OF TIME.
- MAXIMUM ON THE ASTRAL SCALE — P_E, KÇ, ZA — ⊙ NUMBER, VOLUME, SPACE, MASS, SPEED OF TIME.

SPIN MASS VOLUME IN SUB-ATOMIC PARTIVLES OF THE MATTER II.BIG-BANG:	SPIN MASS VALUES IN SUB-ATOMIC PARTICLES OF ANTI-MATTERS AFTER BIG-CRUCH:	SPIN MASS VALUES IN SUB-ATOMIC PARTICLES OF MATTERS AFTER BIG-BANG:	SPIN MASS VALUES IN SUB-ATOMIC PARTICLES OF ANTI-MATTERS BEFORE BIG-BANG:

Spin mass values:
$+3/2$ $+1$ $+1/2$ $+0$ | -0 $-1/2$ -1 $-3/2$ -2 | $+2$ $+3/2$ $+1$ $+1/2$ $+0$ | -0 $-2/3$

Particle / force tables

PARTICLES	IN II.BIG-BANG — ON THE ASTRAL SCALE			AT THE ATOMAL SCALE		
g, ⊙, ⊙Ꙭ, Ꙭ	SF	1000	MeV	SF	10^{-11}	MeV
+, ɣ, ⊖	EM	1	MeV	EM	10^{-40}	MeV
W^{+}, W^{-}, Z°, ○	WF	10^{-11}	MeV	WF	1	MeV
⊙	GF	10^{-40}	MeV	GF	1000	MeV

IN BIG-CRUNCH — ON THE ASTRAL SCALE			AT THE ATOMAL SCALE		
SF	1000	MeV	SF	10^{-11}	MeV
EM	1	MeV	EM	10^{-40}	MeV
WF	10^{-11}	MeV	WF	1	MeV
GF	10^{-40}	MeV	GF	1000	MeV

IN BIG-BANG — ON THE ASTRAL SCALE			AT THE ATOMAL SCALE		
SF	10^{-11}	MeV	SF	1000	MeV
EM	10^{-40}	MeV	EM	1	MeV
WF	1	MeV	WF	10^{-11}	MeV
GF	1000	MeV	GF	10^{-40}	MeV

SF: STRONG NUCLEAR FORCE / **EM**: ELEKTRO MANYETIC FORCE **WF**: WEAK NUCLEAR FORCE / **GF**: GRAVİTY FORCE

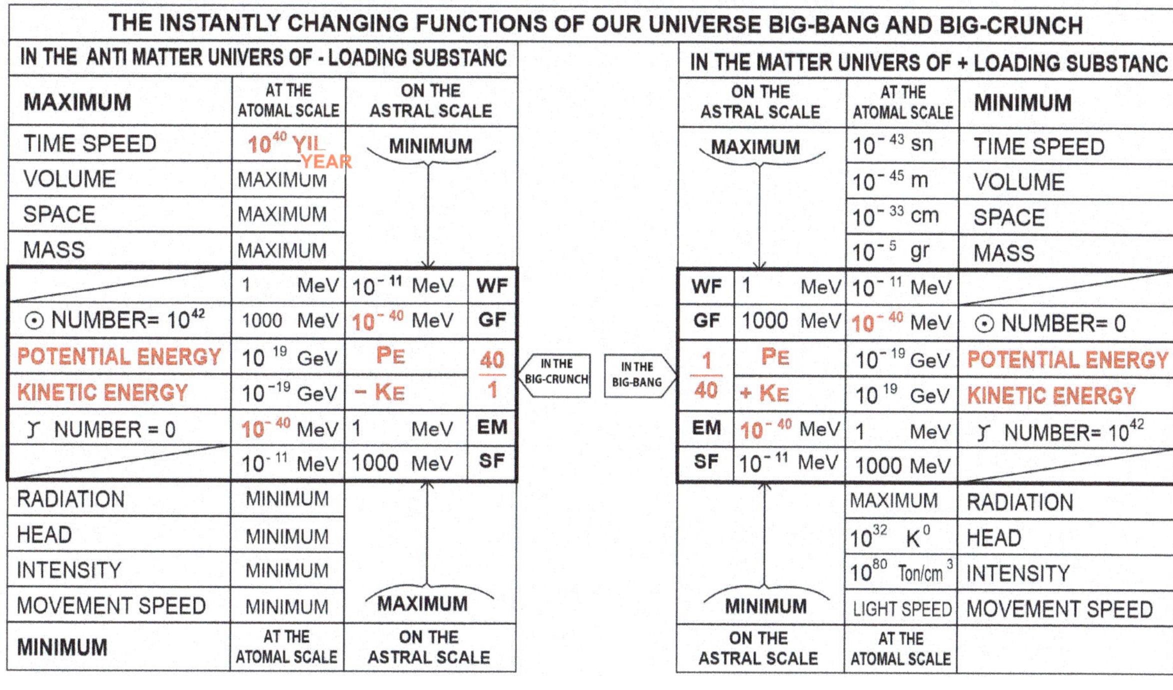

THE INSTANTLY CHANGING FUNCTIONS OF OUR UNIVERSE BIG-BANG AND BIG-CRUNCH

IN THE ANTI MATTER UNIVERS OF - LOADING SUBSTANC

MAXIMUM	AT THE ATOMAL SCALE	ON THE ASTRAL SCALE	
TIME SPEED	10^{40} YIL (YEAR)	MINIMUM	
VOLUME	MAXIMUM		
SPACE	MAXIMUM		
MASS	MAXIMUM		
	1 MeV	10^{-11} MeV	WF
⊙ NUMBER= 10^{42}	1000 MeV	10^{-40} MeV	GF
POTENTIAL ENERGY	10^{19} GeV	P_E	$\frac{40}{1}$
KINETIC ENERGY	10^{-19} GeV	$-K_E$	
ɣ NUMBER = 0	10^{-40} MeV	1 MeV	EM
	10^{-11} MeV	1000 MeV	SF
RADIATION	MINIMUM		
HEAD	MINIMUM		
INTENSITY	MINIMUM		
MOVEMENT SPEED	MINIMUM	MAXIMUM	
MINIMUM	AT THE ATOMAL SCALE	ON THE ASTRAL SCALE	

IN THE MATTER UNIVERS OF + LOADING SUBSTANC

	ON THE ASTRAL SCALE	AT THE ATOMAL SCALE	MINIMUM
	MAXIMUM	10^{-43} sn	TIME SPEED
		10^{-45} m	VOLUME
		10^{-33} cm	SPACE
		10^{-5} gr	MASS
WF	1 MeV	10^{-11} MeV	
GF	1000 MeV	10^{-40} MeV	⊙ NUMBER= 0
$\frac{1}{40}$	P_E	10^{-19} GeV	POTENTIAL ENERGY
	$+K_E$	10^{19} GeV	KINETIC ENERGY
EM	10^{-40} MeV	1 MeV	ɣ NUMBER= 10^{42}
SF	10^{-11} MeV	1000 MeV	
		MAXIMUM	RADIATION
		10^{32} K^{0}	HEAD
		10^{80} Ton/cm^{3}	INTENSITY
	MINIMUM	LIGHT SPEED	MOVEMENT SPEED
	ON THE ASTRAL SCALE	AT THE ATOMAL SCALE	

(Between the two halves: IN THE BIG-CRUNCH / IN THE BIG-BANG)

ANTI-MATERIAL PARTICLES

	ANTI GRAVITON SUPERCLUSTER
	ANTI HIGGS SUPERCLUSTER
	ANTI GLUON SUPERCLUSTER
	ANTI PHOTON SUPERCLUSTER
	ANTI Z° SUPERCLUSTER
	ANTI W⁺ SUPERCLUSTER
	ANTI W⁻ SUPERCLUSTER
	ANTI e SUPERCLUSTER
	ANTI μ SUPERCLUSTER
	ANTI τ SUPERCLUSTER
	ANTI GRAVITON SUPERCLUSTER NEUTRINO
	ANTI HIGGS SUPERCLUSTER NEUTRINO
	ANTI GLUON SUPERCLUSTER NEUTRINO
	ANTI PHOTON SUPERCLUSTER NEUTRINO
	ANTI Z° SUPERCLUSTER NEUTRINO
	ANTI W⁺ SUPERCLUSTER NEUTRINO
	ANTI W⁻ SUPERCLUSTER NEUTRINO
	ANTI ELECTRON SUPERCLUSTER NEUTRINO
	ANTI MUON SUPERCLUSTER NEUTRINO
	ANTI TAU SUPERCLUSTER NEUTRINO
	ANTI -CHARGED ⊙° QUARK SUPERCLUSTER
	ANTI -CHARGED h° QUARK SUPERCLUSTER
	ANTI -CHARGED g° QUARK SUPERCLUSTER
	ANTI -CHARGED γ° QUARK SUPERCLUSTER
	ANTI -CHARGED Z° QUARK SUPERCLUSTER
	ANTI -CHARGED W⁺ QUARK SUPERCLUSTER
	ANTI -CHARGED W⁻ QUARK SUPERCLUSTER
	ANTI -CHARGED e QUARK SUPERCLUSTER
	ANTI -CHARGED μ QUARK SUPERCLUSTER
	ANTI -CHARGED τ QUARK SUPERCLUSTER

I.st BIG-BANG STAR (1.BBS) (SUPERCLUSTER STAR): IT IS AN ICOSAHEDRON WHICH IS FORMED VIRTUALLY IN 10 AXES USING THE SUPERCLUSTER OF THOSE PARTICLES THAT ARE CREATED BEFORE THE SINGLE MATERIAL AND ANTI-MATERIAL PARTICLES AFTER THE BIG-BANG WHICH THE SCIENTISTS OF OUR AGE HAS DEFINED AS THE "STANDARD MODEL".

20 REGULAR TETRAHEDRONS ON THE OUTHER SURFACE OF THIS STAR ARE NOT SHOWN JUST FOR AN EASY UNDERSTANDING; 10 AXES ARE DEFINED WHICH ARE COMPOSED OF 20 REGULAR TETRAHEDRONS IN THE CENTER OF STAR; AND WHICH PASS FROM THE CENTERS OF THE TRIANGLES OF ICOSAHEDRON.

10 REGULAR TETRAHEDRONS WHICH ARE FORMED BY ANTI-MATERIAL PARTICLES ON THE UPPER PART OF THE CENTER OF STAR ARE SHOWN IN GRAY COLOR.

MATERIAL PARTICLES

	GRAVITON SUPERCLUSTER
	HIGGS SUPERCLUSTER
	GLUON SUPERCLUSTER
	PHOTON SUPERCLUSTER
	Z° SUPERCLUSTER
	W⁺ SUPERCLUSTER
	W⁻ SUPERCLUSTER
	e SUPERCLUSTER
	μ SUPERCLUSTER
	τ SUPERCLUSTER
	GRAVITON SUPERCLUSTER NEUTRINO
	HIGGS SUPERCLUSTER NEUTRINO
	GLUON SUPERCLUSTER NEUTRINO
	PHOTON SUPERCLUSTER NEUTRINO
	Z° SUPERCLUSTER NEUTRINO
	W⁺ SUPERCLUSTER NEUTRINO
	W⁻ SUPERCLUSTER NEUTRINO
	ELECTRON SUPERCLUSTER NEUTRINO
	MUON SUPERCLUSTER NEUTRINO
	TAU SUPERCLUSTER NEUTRINO
	- CHARGED ⊙° QUARK SUPERCLUSTER
	- CHARGED h° QUARK SUPERCLUSTER
	- CHARGED g° QUARK SUPERCLUSTER
	- CHARGED γ° QUARK SUPERCLUSTER
	- CHARGED Z° QUARK SUPERCLUSTER
	- CHARGED W⁺ QUARK SUPERCLUSTER
	- CHARGED W⁻ QUARK SUPERCLUSTER
	- CHARGED e QUARK SUPERCLUSTER
	- CHARGED μ QUARK SUPERCLUSTER
	- CHARGED τ QUARK SUPERCLUSTER

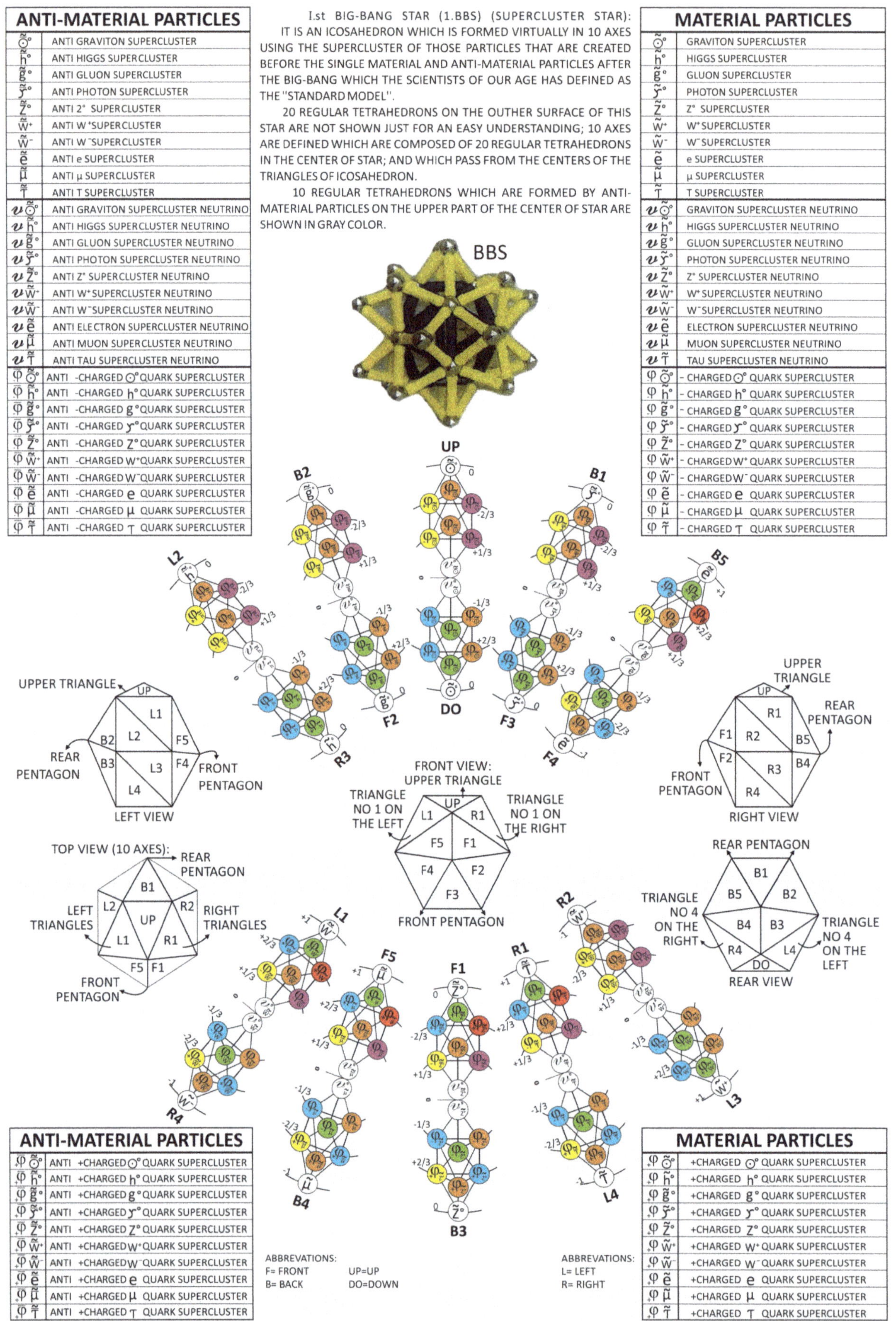

ANTI-MATERIAL PARTICLES

	ANTI +CHARGED ⊙° QUARK SUPERCLUSTER
	ANTI +CHARGED h° QUARK SUPERCLUSTER
	ANTI +CHARGED g° QUARK SUPERCLUSTER
	ANTI +CHARGED γ° QUARK SUPERCLUSTER
	ANTI +CHARGED Z° QUARK SUPERCLUSTER
	ANTI +CHARGED W⁺ QUARK SUPERCLUSTER
	ANTI +CHARGED W⁻ QUARK SUPERCLUSTER
	ANTI +CHARGED e QUARK SUPERCLUSTER
	ANTI +CHARGED μ QUARK SUPERCLUSTER
	ANTI +CHARGED τ QUARK SUPERCLUSTER

ABBREVATIONS:
F= FRONT   UP=UP
B= BACK   DO=DOWN

ABBREVATIONS:
L= LEFT
R= RIGHT

MATERIAL PARTICLES

	+CHARGED ⊙° QUARK SUPERCLUSTER
	+CHARGED h° QUARK SUPERCLUSTER
	+CHARGED g° QUARK SUPERCLUSTER
	+CHARGED γ° QUARK SUPERCLUSTER
	+CHARGED Z° QUARK SUPERCLUSTER
	+CHARGED W⁺ QUARK SUPERCLUSTER
	+CHARGED W⁻ QUARK SUPERCLUSTER
	+CHARGED e QUARK SUPERCLUSTER
	+CHARGED μ QUARK SUPERCLUSTER
	+CHARGED τ QUARK SUPERCLUSTER

ANTI-MATERIAL PARTICLES

Ō°	ANTI GRAVITON
h̄°	ANTI HIGGS
ḡ°	ANTI GLUON
ȳ°	ANTI PHOTON
z̄°	ANTI 2°
w̄⁺	ANTI W⁺
w̄⁻	ANTI W⁻
ē	ANTI ELECTRON
μ̄	ANTI MUON
τ̄	ANTI TAU
υŌ°	ANTI GRAVITON NEUTRINO
υh̄°	ANTI HIGGS NEUTRINO
υḡ°	ANTI GLUON NEUTRINO
υȳ°	ANTI PHOTON NEUTRINO
υz̄°	ANTI Z° NEUTRINO
υw̄⁺	ANTI W⁺ NEUTRINO
υw̄⁻	ANTI W⁻NEUTRINO
υē	ANTI ELECTRON NEUTRINO
υμ̄	ANTI MUON NEUTRINO
υτ̄	ANTI TAU NEUTRINO
φŌ°	ANTI -CHARGED Ō° QUARK
φh̄°	ANTI -CHARGED h̄° QUARK
φḡ°	ANTI -CHARGED ḡ° QUARK
φȳ°	ANTI -CHARGED ȳ° QUARK
φz̄°	ANTI -CHARGED z̄° QUARK
φw̄⁺	ANTI -CHARGED W⁺QUARK
φw̄⁻	ANTI -CHARGED W⁻ QUARK
φē	ANTI -CHARGED e QUARK
φμ̄	ANTI -CHARGED μ QUARK
φτ̄	ANTI -CHARGED τ QUARK

MATERIAL PARTICLES

O°	GRAVITON
h°	HIGGS
g°	GLUON
y°	PHOTON
Z°	Z°
W⁺	W⁺
W⁻	W⁻
e	ELECTRON (θ)
μ	MUON
τ	TAU
υO°	GRAVITON NEUTRINO
υh°	HIGGS NEUTRINO
υg°	GLUON NEUTRINO
υy°	PHOTON NEUTRINO
υZ°	Z° NEUTRINO
υW⁺	W⁺ NEUTRINO
υW⁻	W⁻NEUTRINO
υe	ELECTRON NEUTRINO
υμ	MUON NEUTRINO
υτ	TAU NEUTRINO
φO°	- CHARGED O° QUARK
φh°	- CHARGED h° QUARK
φg°	- CHARGED g° QUARK
φy°	- CHARGED y° QUARK
φZ°	- CHARGED Z° QUARK
φW⁺	- CHARGED W⁺ QUARK
φW⁻	- CHARGED W⁻QUARK
φe	- CHARGED e QUARK
φμ	- CHARGED μ QUARK
φτ	- CHARGED τ QUARK

II.BIG-BANG STAR (2.BBS) (STANDARS PARTICLES STAR):

THE FIRST PARTICLES WHICH ARE FORMED AFTER BIG-BANG WERE THE SUPERCLUSTER OF STANDARD PARTICLES TO BE FORMED LATER.

BELOW, ICOSAHEDRON STAR IS EXPLAINED SHICH WAS FORMED VIRTUALLY IN 10 AXES FROM THE STANDARD PARTICLES.

AS IN THE CRYSTAL ATOMS, PARTICLES ARE LOCATED ON THE CORNERS OF 10 PAIRS OF RHOMBIC PRISM.

INSTEAD OF USONG THE IMAGE OF STAR, FOR AN EASY UNDERSTANDING, WE CAN SEE THE FRONT, REAR, RIGHT AND LEFT WIEVS OF ICOSAHEDRON WHICH IS FORMED BY EACH NEUTRINO WITH 3 QUARKS IN THE CENTER OF THE STAR. SO, 10 AXES ARE DEFINED WHICH PASS THROUGH THE CENTERS OF EQUILATERAL TRIANGLES, PARALLEL TO EACH OTHER.

THE REGULAR TETRAHEDRONS FORMED BY THE ANTI-MATERIAL PARTICLES ON THE UPPER SIDE OF ICOSAHEDRON AND DOUBLR RHOMBIC PRISM ARE SHOW IN GRAY.

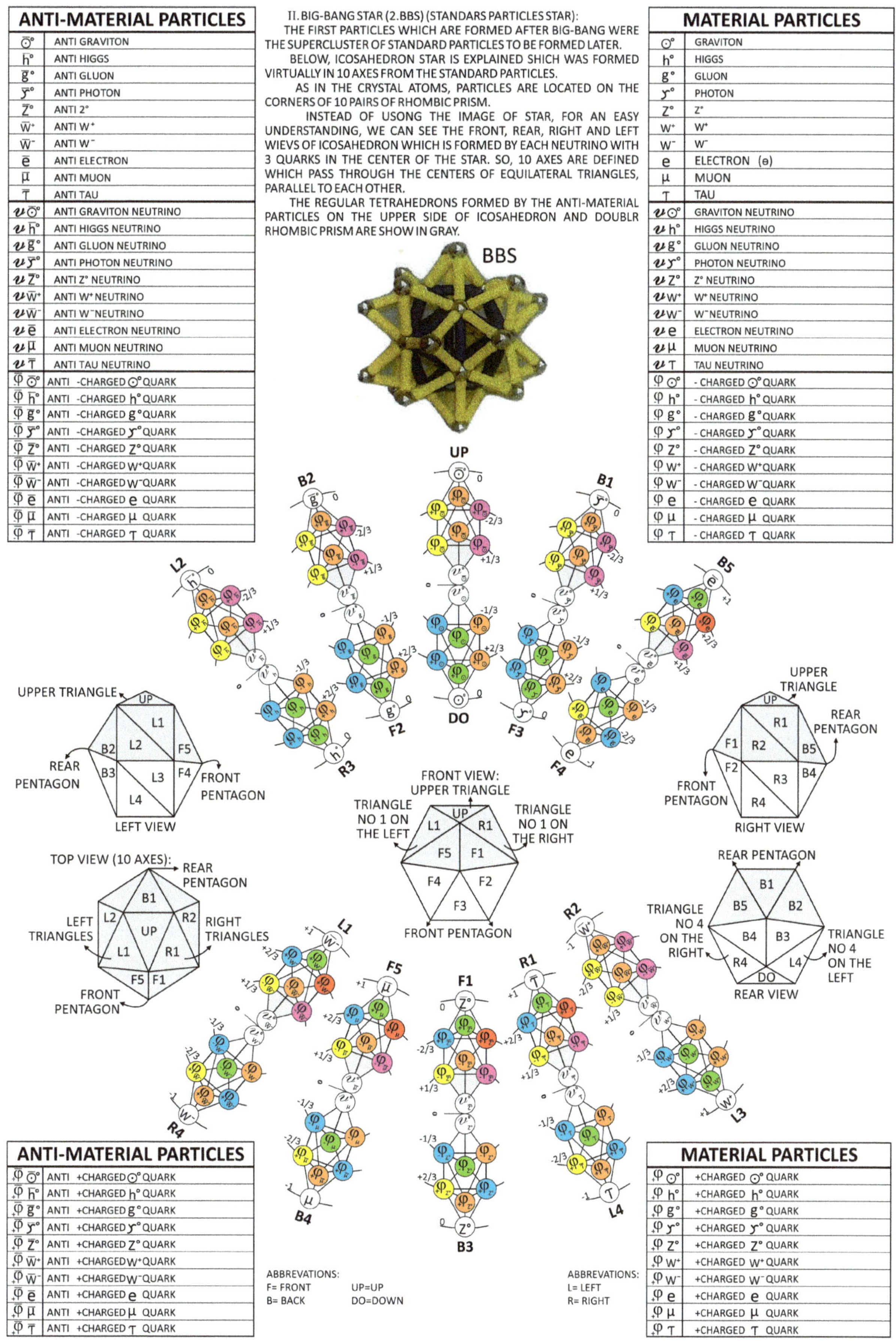

ABBREVATIONS:
F= FRONT UP=UP
B= BACK DO=DOWN

ABBREVATIONS:
L= LEFT
R= RIGHT

ANTI-MATERIAL PARTICLES

φŌ°	ANTI +CHARGED Ō° QUARK
φh̄°	ANTI +CHARGED h° QUARK
φḡ°	ANTI +CHARGED g° QUARK
φȳ°	ANTI +CHARGED y° QUARK
φz̄°	ANTI +CHARGED Z° QUARK
φw̄⁺	ANTI +CHARGED W⁺QUARK
φw̄⁻	ANTI +CHARGEDW⁻ QUARK
φē	ANTI +CHARGED e QUARK
φμ̄	ANTI +CHARGED μ QUARK
φτ̄	ANTI +CHARGED τ QUARK

MATERIAL PARTICLES

φO°	+CHARGED O° QUARK
φh°	+CHARGED h° QUARK
φg°	+CHARGED g° QUARK
φy°	+CHARGED y° QUARK
φZ°	+CHARGED Z° QUARK
φW⁺	+CHARGED W⁺ QUARK
φW⁻	+CHARGED W⁻QUARK
φe	+CHARGED e QUARK
φμ	+CHARGED μ QUARK
φτ	+CHARGED τ QUARK

BIG <u>CUBE COMPLEX</u> (SUPER CUBE) WITH EIGHT CUBES
WHICH IS COMPOSED OF A PART OF QUARK-LEPTON
FAMILY (8x8=64 PARTICLES) :

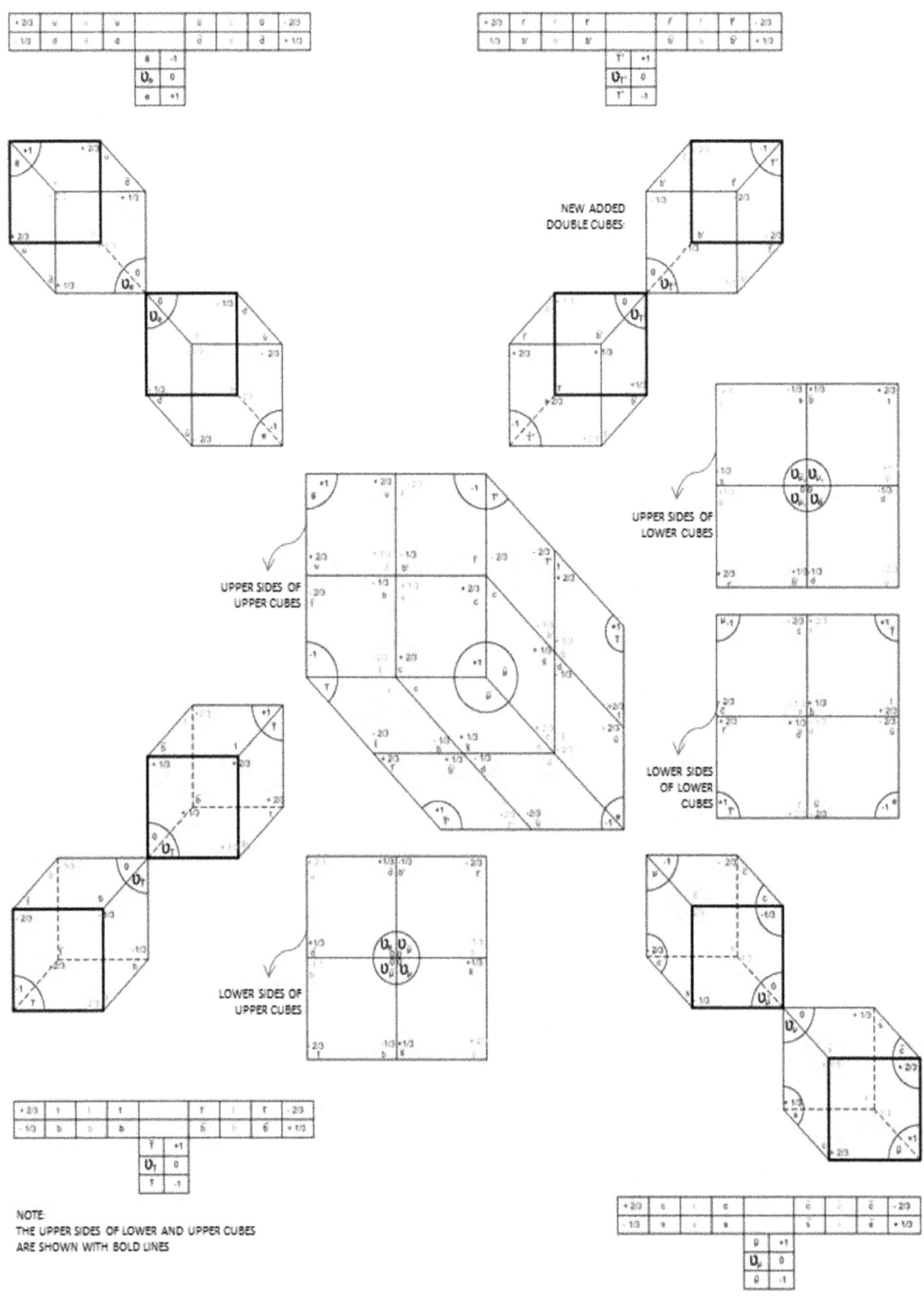

NOTE:
THE UPPER SIDES OF LOWER AND UPPER CUBES
ARE SHOWN WITH BOLD LINES

BIG HEXA-STAR PRISM MODEL WITH TWELVE
QUADRANGULAR PRISMS WHICH IS COMPOSED OF
THE COMPLETE QUARK-LEPTON FAMILY :16x6=96
PARTICLES
(96s TOPOGRAPHY)
(STAR OF DAVID PRISM)

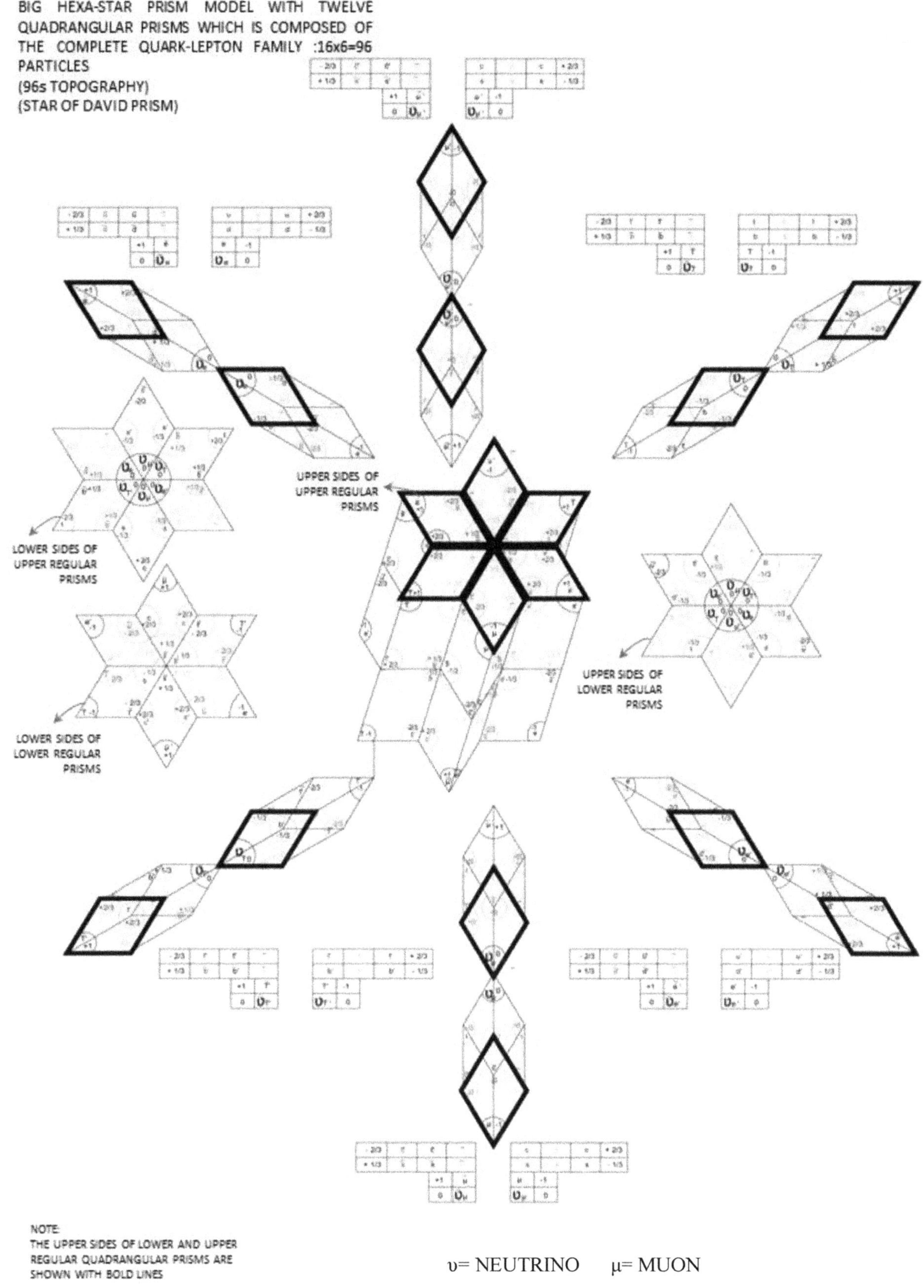

υ= NEUTRINO   μ= MUON

υ= ANTI-NEUTRINO   μ= ANTI-MUON

160 TYPES OF SINGLE PARTICLES OF ATOM

GROUP	SUPERPARTNERS SYMBOL (VIRTUAL)	(REAL)	SUPERPARTNER NAME	CHG	SPIN	MATTERS SYMBOL (VIRTUAL)	(REAL)	MATTER NAME	CHG	SPIN	MASS	ENERGY	LIFE(sec)	NO	ANTI-MATTERS SYMBOL (VIRTUAL)	(REAL)	ANTI-MATTER NAME	CHG	SPIN	ANTI SUPERPARTNERS SYMBOL (VIRTUAL)	(REAL)	ANTI SUPERPARTNER NAME	CHG	SPIN	IMPACT DISTANCE (m)	ATOMIC FORCE FEEL	ATOMIC FORCE TRANSMITTED
BOSONS	θ̄°	G̃°	GRAVITON SUPERPARTNER	0	3/2	θ°	G°	GRAVITON	0	2	0	<10⁻⁴⁰ MeV	STABLE	1	θ̄°	Ḡ°	ANTI GRAVITON	0	2	θ̄°	G̃°	ANTI θ° SUPERPARTNER	0	-3/2	INFINITE	GF	GF
	h̄°	H̃°	HIGGS SUPERPARTNER (HIGGSINO)	0	1/2	h°	H°	HIGGS	0	0	60000/1000000 MeV	>112 GeV	?	2	h̄°	H̄°	ANTI HIGGS	0	0	h̄°	H̃°	ANTI h° SUPERPARTNER	0	-1/2	INFINITE	WF,GF,EM	GF
	ḡ°	g̃°	GLUON SUPERPARTNER (GLUINO)	0	1/2	g°	g°	GLUON	0	1	0	0	STABLE	3	ḡ°	ḡ°	ANTI GLUON	0	1	ḡ°	g̃°	ANTI g° SUPERPARTNER	0	-1/2	10⁻¹⁵	SF	SF
	γ̄°	γ̃°	PHOTON SUPERPARTNER (PHOTINO)	0	1/2	γ°	γ°	PHOTON	0	1	0	0	STABLE	4	γ̄°	γ̄°	ANTI PHOTON	0	1	γ̄°	γ̃°	ANTI γ° SUPERPARTNER	0	-1/2	INFINITE	GF,EM	EM
	Z̄°	Z̃°	Z° SUPERPARTNER (ZINO)	0	1/2	Z°	Z°	Z°	0	1	~91,2 MeV	~91,2 MeV	?	5	Z̄°	Z̄°	ANTI Z°	0	1	Z̄°	Z̃°	ANTI Z° SUPERPARTNER	0	-1/2	10⁻¹⁸	WF,EM	WF
	W̄⁺	W̃⁺	W⁺ SUPERPARTNER (WINO⁺)	1	1/2	W⁺	W⁺	W⁺	1	1	~50-90 ⊕	80,4 MeV	?	6	W̄⁺	W̄⁺	ANTI W⁺	-1	1	W̄⁺	W̃⁺	ANTI W⁺ SUPERPARTNER	-1	-1/2	10⁻¹⁸	WF,EM	WF
	W̄⁻	W̃⁻	W⁻ SUPERPARTNER (WINO⁻)	-1	1/2	W⁻	W⁻	W⁻	-1	1	~50-90 ⊕	80,4 MeV	?	7	W̄⁻	W̄⁻	ANTI W⁻	1	1	W̄⁻	W̃⁻	ANTI W⁻ SUPERPARTNER	1	-1/2	10⁻¹⁸	WF,EM	WF
LEPTONS	ē	ẽ	ELECTRON SUPERPARTNER (SELECTRON)	-1	0	e	e	ELECTRON	-1	1/2	1/1836 ⊕	0,511 MeV	STABLE	8	ē	ē	ANTI ELECTRON	1	1/2	ē	ẽ	ANTI e SUPERPARTNER	1	0	INFINITE	WF,EM	
	μ̄	μ̃	MUON SUPERPARTNER (SMUON)	-1	0	μ	μ	MUON	-1	1/2		105,7 MeV	2,2x10⁻⁶	9	μ̄	μ̄	ANTI MUON	1	1/2	μ̄	μ̃	ANTI μ SUPERPARTNER	1	0	INFINITE	WF,EM	
	τ̄	τ̃	TAU SUPERPARTNER (STAU)	-1	0	τ	τ	TAU	-1	1/2		1,777 MeV	2,2x10⁻⁶	10	τ̄	τ̄	ANTI TAU	1	1/2	τ̄	τ̃	ANTI τ SUPERPARTNER	1	0	INFINITE	WF,EM	
LEPTONS	ϑθ̄°		θ° NEUTRINO SUPERPARTNER (SNEUTRINO)	0	0	ϑθ°		θ° NEUTRINO	0	1/2		? MeV	STABLE	1	ϑθ̄°		ANTI θ° NEUTRINO	0	1/2	ϑθ̄°		ANTI θ° NEUTRINO SUPERPARTNER	0	0	INFINITE	WF	
	ϑh̄°		h° NEUTRINO SUPERPARTNER (SNEUTRINO)	0	0	ϑh°		h° NEUTRINO	0	1/2		? MeV	STABLE	2	ϑh̄°		ANTI h° NEUTRINO	0	1/2	ϑh̄°		ANTI h° NEUTRINO SUPERPARTNER	0	0	INFINITE	WF	
	ϑḡ°		g° NEUTRINO SUPERPARTNER (SNEUTRINO)	0	0	ϑg°		g° NEUTRINO	0	1/2		? MeV	STABLE	3	ϑḡ°		ANTI g° NEUTRINO	0	1/2	ϑḡ°		ANTI g° NEUTRINO SUPERPARTNER	0	0	INFINITE	WF	
	ϑγ̄°		γ° NEUTRINO SUPERPARTNER (SNEUTRINO)	0	0	ϑγ°		γ° NEUTRINO	0	1/2	10-100 ⊕	0 MeV	STABLE	4	ϑγ̄°		ANTI γ° NEUTRINO	0	1/2	ϑγ̄°		ANTI γ° NEUTRINO SUPERPARTNER	0	0	INFINITE	WF	
	ϑZ̄°		Z° NEUTRINO SUPERPARTNER (SNEUTRINO)	0	0	ϑZ°		Z° NEUTRINO	0	1/2		~91 MeV	STABLE	5	ϑZ̄°		ANTI Z° NEUTRINO	0	1/2	ϑZ̄°		ANTI Z° NEUTRINO SUPERPARTNER	0	0	INFINITE	WF	
	ϑW̄⁺		W⁺ NEUTRINO SUPERPARTNER (SNEUTRINO)	0	0	ϑW⁺		W⁺ NEUTRINO	0	1/2		~91 MeV	STABLE	6	ϑW̄⁺		ANTI W⁺ NEUTRINO	0	1/2	ϑW̄⁺		ANTI W⁺ NEUTRINO SUPERPARTNER	0	0	INFINITE	WF	
	ϑW̄⁻		W⁻ NEUTRINO SUPERPARTNER (SNEUTRINO)	0	0	ϑW⁻		W⁻ NEUTRINO	0	1/2		~91 MeV	STABLE	7	ϑW̄⁻		ANTI W⁻ NEUTRINO	0	1/2	ϑW̄⁻		ANTI W⁻ NEUTRINO SUPERPARTNER	0	0	INFINITE	WF	
	ϑē	ϑe'	e NEUTRINO SUPERPARTNER (SNEUTRINO)	0	0	ϑe	ϑe'	e NEUTRINO	0	1/2		? MeV	STABLE	8	ϑē	ϑe'	ANTI e NEUTRINO	0	1/2	ϑē	ϑe'	ANTI e NEUTRINO SUPERPARTNER	0	0	INFINITE	WF	
	ϑμ̄	ϑμ'	μ NEUTRINO SUPERPARTNER (SNEUTRINO)	0	0	ϑμ	ϑμ'	μ NEUTRINO	0	1/2		<0.17 MeV	STABLE	9	ϑμ̄	ϑμ'	ANTI μ NEUTRINO	0	1/2	ϑμ̄	ϑμ'	ANTI μ NEUTRINO SUPERPARTNER	0	0	INFINITE	WF	
	ϑτ̄	ϑτ'	τ NEUTRINO SUPERPARTNER (SNEUTRINO)	0	0	ϑτ	ϑτ'	τ NEUTRINO	0	1/2		<15.5 MeV	STABLE	10	ϑτ̄	ϑτ'	ANTI τ NEUTRINO	0	1/2	ϑτ̄	ϑτ'	ANTI τ NEUTRINO SUPERPARTNER	0	0	INFINITE	WF	
FERMIONS — CHARGED QUARKS (−)	-Qθ°		- CHARGED θ° QUARK SUPERPARTNER	-1/3	0	-Qθ°		- CHARGED θ° QUARK	-1/3	1/2	-1/3 ⊖		STABLE	1	-Q̄θ°		ANTI - CHARGED θ° QUARK	1/3	1/2	-Q̄θ°		ANTI - CHARGED θ° QUARK SUPERPARTNER	1/3	0	10⁻¹⁵	SF,WF,EM	
	-Qh°		- CHARGED h° QUARK SUPERPARTNER (SQUARK)	-1/3	0	-Qh°		- CHARGED h° QUARK	-1/3	1/2	-1/3 ⊖		?	2	-Q̄h°		ANTI - CHARGED h° QUARK	1/3	1/2	-Q̄h°		ANTI - CHARGED h° QUARK SUPERPARTNER	1/3	0	10⁻¹⁵	SF,WF,EM	
	-Qg°		- CHARGED g° QUARK SUPERPARTNER (SQUARK)	-1/3	0	-Qg°	b	- CHARGED g° QUARK	-1/3	1/2	-1/3 ⊖	4,13-4,37 MeV	STABLE	3	-Q̄g°	b̄	ANTI - CHARGED g° QUARK	1/3	1/2	-Q̄g°		ANTI - CHARGED g° QUARK SUPERPARTNER	1/3	0	10⁻¹⁵	SF,WF,EM	
	-Qγ°		- CHARGED γ° QUARK SUPERPARTNER (SQUARK)	-1/3	0	-Qγ°	d	- CHARGED γ° QUARK	-1/3	1/2	-1/3 ⊖	3.5-6.0 MeV	STABLE	4	-Q̄γ°	d̄	ANTI - CHARGED γ° QUARK	1/3	1/2	-Q̄γ°		ANTI - CHARGED γ° QUARK SUPERPARTNER	1/3	0	10⁻¹⁵	SF,WF,EM	
	-QZ°		- CHARGED Z° QUARK SUPERPARTNER (SQUARK)	-1/3	0	-QZ°		- CHARGED Z° QUARK	-1/3	1/2	-1/3 ⊖		?	5	-Q̄Z°		ANTI - CHARGED Z° QUARK	1/3	1/2	-Q̄Z°		ANTI - CHARGED Z° QUARK SUPERPARTNER	1/3	0	10⁻¹⁵	SF,WF,EM	
	-QW⁺		- CHARGED W⁺ QUARK SUPERPARTNER (SQUARK)	-1/3	0	-QW⁺		- CHARGED W⁺ QUARK	-1/3	1/2	-1/3 ⊖		?	6	-Q̄W⁺		ANTI - CHARGED W⁺ QUARK	1/3	1/2	-Q̄W⁺		ANTI - CHARGED W⁺ QUARK SUPERPARTNER	1/3	0	10⁻¹⁵	SF,WF,EM	
	-QW⁻		- CHARGED W⁻ QUARK SUPERPARTNER (SQUARK)	-1/3	0	-QW⁻	s	- CHARGED W⁻ QUARK	-1/3	1/2	-1/3 ⊖	70-130 MeV	?	7	-Q̄W⁻	s̄	ANTI - CHARGED W⁻ QUARK	1/3	1/2	-Q̄W⁻		ANTI - CHARGED W⁻ QUARK SUPERPARTNER	1/3	0	10⁻¹⁵	SF,WF,EM	
	-Qe		- CHARGED e QUARK SUPERPARTNER (SQUARK)	-1/3	0	-Qe		- CHARGED e QUARK	-1/3	1/2	-1/3 ⊖		STABLE	8	-Q̄e		ANTI - CHARGED e QUARK	1/3	1/2	-Q̄e		ANTI - CHARGED e QUARK SUPERPARTNER	1/3	0	10⁻¹⁵	SF,WF,EM	
	-Qμ		- CHARGED μ QUARK SUPERPARTNER (SQUARK)	-1/3	0	-Qμ		- CHARGED μ QUARK	-1/3	1/2	-1/3 ⊖		2,2x10⁻⁶	9	-Q̄μ		ANTI - CHARGED μ QUARK	1/3	1/2	-Q̄μ		ANTI - CHARGED μ QUARK SUPERPARTNER	1/3	0	10⁻¹⁵	SF,WF,EM	
	-QT		- CHARGED T QUARK SUPERPARTNER (SQUARK)	-1/3	0	-QT		- CHARGED T QUARK	-1/3	1/2	-1/3 ⊖		2,2x10⁻⁶	10	-Q̄T		ANTI - CHARGED T QUARK	1/3	1/2	-Q̄T		ANTI - CHARGED T QUARK SUPERPARTNER	1/3	0	10⁻¹⁵	SF,WF,EM	
FERMIONS — CHARGED QUARKS (+)	+Qθ°		+ CHARGED θ° QUARK SUPERPARTNER (SQUARK)	2/3	0	+Qθ°		+ CHARGED θ° QUARK	2/3	1/2	2/3 ⊕		STABLE	1	+Q̄θ°		ANTI + CHARGED θ° QUARK	-2/3	1/2	+Q̄θ°		ANTI + CHARGED θ° QUARK SUPERPARTNER	-2/3	0	10⁻¹⁵	SF,WF,EM	
	+Qh°		+ CHARGED h° QUARK SUPERPARTNER (SQUARK)	2/3	0	+Qh°		+ CHARGED h° QUARK	2/3	1/2	2/3 ⊕		?	2	+Q̄h°		ANTI + CHARGED h° QUARK	-2/3	1/2	+Q̄h°		ANTI + CHARGED h° QUARK SUPERPARTNER	-2/3	0	10⁻¹⁵	SF,WF,EM	
	+Qg°		+ CHARGED g° QUARK SUPERPARTNER (SQUARK)	2/3	0	+Qg°	t	+ CHARGED g° QUARK	2/3	1/2	2/3 ⊕	169,1-173,3 MeV	STABLE	3	+Q̄g°	t̄	ANTI + CHARGED g° QUARK	-2/3	1/2	+Q̄g°		ANTI + CHARGED g° QUARK SUPERPARTNER	-2/3	0	10⁻¹⁵	SF,WF,EM	
	+Qγ°		+ CHARGED γ° QUARK SUPERPARTNER (SQUARK)	2/3	0	+Qγ°	u	+ CHARGED γ° QUARK	2/3	1/2	2/3 ⊕	1,5-3,3 MeV	STABLE	4	+Q̄γ°	ū	ANTI + CHARGED γ° QUARK	-2/3	1/2	+Q̄γ°		ANTI + CHARGED γ° QUARK SUPERPARTNER	-2/3	0	10⁻¹⁵	SF,WF,EM	
	+QZ°		+ CHARGED Z° QUARK SUPERPARTNER (SQUARK)	2/3	0	+QZ°		+ CHARGED Z° QUARK	2/3	1/2	2/3 ⊕		?	5	+Q̄Z°		ANTI + CHARGED Z° QUARK	-2/3	1/2	+Q̄Z°		ANTI + CHARGED Z° QUARK SUPERPARTNER	-2/3	0	10⁻¹⁵	SF,WF,EM	
	+QW⁺		+ CHARGED W⁺ QUARK SUPERPARTNER (SQUARK)	2/3	0	+QW⁺	c	+ CHARGED W⁺ QUARK	2/3	1/2	2/3 ⊕	1,16-1,34 MeV	?	6	+Q̄W⁺	c̄	ANTI + CHARGED W⁺ QUARK	-2/3	1/2	+Q̄W⁺		ANTI + CHARGED W⁺ QUARK SUPERPARTNER	-2/3	0	10⁻¹⁵	SF,WF,EM	
	+QW⁻		+ CHARGED W⁻ QUARK SUPERPARTNER (SQUARK)	2/3	0	+QW⁻		+ CHARGED W⁻ QUARK	2/3	1/2	2/3 ⊕		?	7	+Q̄W⁻		ANTI + CHARGED W⁻ QUARK	-2/3	1/2	+Q̄W⁻		ANTI + CHARGED W⁻ QUARK SUPERPARTNER	-2/3	0	10⁻¹⁵	SF,WF,EM	
	+Qe		+ CHARGED e QUARK SUPERPARTNER (SQUARK)	2/3	0	+Qe		+ CHARGED e QUARK	2/3	1/2	2/3 ⊕		STABLE	8	+Q̄e		ANTI + CHARGED e QUARK	-2/3	1/2	+Q̄e		ANTI + CHARGED e QUARK SUPERPARTNER	-2/3	0	10⁻¹⁵	SF,WF,EM	
	+Qμ		+ CHARGED μ QUARK SUPERPARTNER (SQUARK)	2/3	0	+Qμ		+ CHARGED μ QUARK	2/3	1/2	2/3 ⊕		2,2x10⁻⁶	9	+Q̄μ		ANTI + CHARGED μ QUARK	-2/3	1/2	+Q̄μ		ANTI + CHARGED μ QUARK SUPERPARTNER	-2/3	0	10⁻¹⁵	SF,WF,EM	
	+QT		+ CHARGED T QUARK SUPERPARTNER (SQUARK)	2/3	0	+QT		+ CHARGED T QUARK	2/3	1/2	2/3 ⊕		2,2x10⁻⁶	10	+Q̄T		ANTI + CHARGED T QUARK	-2/3	1/2	+Q̄T		ANTI + CHARGED T QUARK SUPERPARTNER	-2/3	0	10⁻¹⁵	SF,WF,EM	
FERMI GLUONS	M̄	M̃	BLUE GLUON SUPERPARTNER	0	1/2	M	M	BLUE GLUON	0	1	0	0	STABLE	1	M̄	M̄	ANTI BLUE GLUON (YELLOW)	0	1	M̄	M̃	ANTI BLUE GLUON SUPERPARTNER	0	-1/2	10⁻¹⁵	SF	SF
	Ȳ	Ỹ	GREEN GLUON SUPERPARTNER	0	1/2	Y	Y	GREEN GLUON	0	1	0	0	STABLE	2	Ȳ	Ȳ	ANTI GREEN GLUON (ORANGE)	0	1	Ȳ	Ỹ	ANTI GREEN GLUON SUPERPARTNER	0	-1/2	10⁻¹⁵	SF	SF
	K̄	K̃	RED GLUON SUPERPARTNER	0	1/2	K	K	RED GLUON	0	1	0	0	STABLE	3	K̄	K̄	ANTI RED GLUON (PURPLE)	0	1	K̄	K̃	ANTI RED GLUON SUPERPARTNER	0	-1/2	10⁻¹⁵	SF	SF
MESO GLUONS (DOUBLE GLUONS)	M̄M	M̃M	MM GLUON SUPERPARTNER	0	1/2	MM	MM	MM GLUON	0	1	0	0	STABLE	1	M̄M	M̄M	ANTI MM GLUON	0	1	M̄M	M̃M	ANTI MM GLUON SUPERPARTNER	0	-1/2	10⁻¹⁵	SF	SF
	M̄Y	M̃Y	MY GLUON SUPERPARTNER	0	1/2	MY	MY	MY GLUON	0	1	0	0	STABLE	2	M̄Y	M̄Y	ANTI MY GLUON	0	1	M̄Y	M̃Y	ANTI MY GLUON SUPERPARTNER	0	-1/2	10⁻¹⁵	SF	SF
	M̄K	M̃K	MK GLUON SUPERPARTNER	0	1/2	MK	MK	MK GLUON	0	1	0	0	STABLE	3	M̄K	M̄K	ANTI MK GLUON	0	1	M̄K	M̃K	ANTI MK GLUON SUPERPARTNER	0	-1/2	10⁻¹⁵	SF	SF
	ȲM	ỸM	YM GLUON SUPERPARTNER	0	1/2	YM	YM	YM GLUON	0	1	0	0	STABLE	4	ȲM	ȲM	ANTI YM GLUON	0	1	ȲM	ỸM	ANTI YM GLUON SUPERPARTNER	0	-1/2	10⁻¹⁵	SF	SF
	ȲY	ỸY	YY GLUON SUPERPARTNER	0	1/2	YY	YY	YY GLUON	0	1	0	0	STABLE	5	ȲY	ȲY	ANTI YY GLUON	0	1	ȲY	ỸY	ANTI YY GLUON SUPERPARTNER	0	-1/2	10⁻¹⁵	SF	SF
	ȲK	ỸK	YK GLUON SUPERPARTNER	0	1/2	YK	YK	YK GLUON	0	1	0	0	STABLE	6	ȲK	ȲK	ANTI YK GLUON	0	1	ȲK	ỸK	ANTI YK GLUON SUPERPARTNER	0	-1/2	10⁻¹⁵	SF	SF
	K̄M	K̃M	KM GLUON SUPERPARTNER	0	1/2	KM	KM	KM GLUON	0	1	0	0	STABLE	7	K̄M	K̄M	ANTI KM GLUON	0	1	K̄M	K̃M	ANTI KM GLUON SUPERPARTNER	0	-1/2	10⁻¹⁵	SF	SF
	K̄Y	K̃Y	KY GLUON SUPERPARTNER	0	1/2	KY	KY	KY GLUON	0	1	0	0	STABLE	8	K̄Y	K̄Y	ANTI KY GLUON	0	1	K̄Y	K̃Y	ANTI KY GLUON SUPERPARTNER	0	-1/2	10⁻¹⁵	SF	SF
	K̄K	K̃K	KK GLUON SUPERPARTNER	0	1/2	KK	KK	KK GLUON	0	1	0	0	STABLE	9	K̄K	K̄K	ANTI KK GLUON	0	1	K̄K	K̃K	ANTI KK GLUON SUPERPARTNER	0	-1/2	10⁻¹⁵	SF	SF

120 KINDS OF FERMİ (SINGLE) GLUONS ACCORDING TO THE VIRTUAL BIG-BANG STAR (BBS) IN ANNEX - IIa AND IIb

DIRECTION	UP	B1	B5	R2	R1	F1	F5	L1	L2	B2	CHARGE
ANTI-GLUONS	θ̄ θ̄ θ̄	Ȳ Ȳ Ȳ	ē ē ē	W̄⁺ W̄⁺ W̄⁺	τ̄ τ̄ τ̄	Z̄° Z̄° Z̄°	μ̄ μ̄ μ̄	W̄⁻ W̄⁻ W̄⁻	h̄° h̄° h̄°	ḡ° ḡ° ḡ°	-2/3
ANTI-GLUONS	θ̄ θ̄ θ̄	Ȳ Ȳ Ȳ	ē ē ē	W̄⁺ W̄⁺ W̄⁺	τ̄ τ̄ τ̄	Z̄° Z̄° Z̄°	μ̄ μ̄ μ̄	W̄⁻ W̄⁻ W̄⁻	h̄° h̄° h̄°	ḡ° ḡ° ḡ°	1/3
GLUONS	θ θ θ	Y Y Y	e e Y	W⁺ W⁺ W⁺	T T T	Z° Z° Z°	μ μ μ	W⁻ W⁻ W⁻	h° h° h°	g° g° g°	-1/3
GLUONS	θ θ θ	Y Y Y	e e Y	W⁺ W⁺ W⁺	T T T	Z° Z° Z°	μ μ μ	W⁻ W⁻ W⁻	h° h° h°	g° g° g°	2/3
DIRECTION	DO	F3	F4	L3	L4	B3	B4	R4	R3	F2	

SAME REAL QUARKS AND THEIR ABBREVIATIONS SEEN ABOVE

	CHARGE
s̄ : STRANGE	0, 2/3, 1,-1
C : CHARMED	+2/3
B : BOTTOM	-1/3
t : TOP	+2/3
d : DOWN	-1/3
u : UP	+2/3

⊕ : PROTON   ⊖ : ELEKTRON

THE 4 FORCES OF THE ATOM SEEN ABOVE AND THEIR ABBREVIATIONS

SYMBOL	FORCE	IMPACT DURATION (sec)
GF	GRAVITATIONAL FORCE	0
SF	STRONG NUCLEAR FORCE	10⁻¹⁸ - 10⁻²²
EM	ELECTROMAGNETIC FORCE	10⁻²⁰ - 10⁻¹⁸
WF	WEAK NUCLEAR FORCE	10⁻¹⁰ - 10⁻⁸

PARTICLES WHICH HAVE BEEN EXTINCTED IN THE LAST SECOND OF THE PREVIOUS UNIVERSE AND WHICH HAVE BEEN CREATED IN THE FIRST SECOND AFTER THE BIG - BANG:

TABLE OF ANTI - BOSONS, ANTI - FERMIONS AND THEIR POSSIBLE INTERACTIONS IN ACCORDANCE WITH THE EXTINCTION ORDER BEFORE BIG - BANG: (BIG - BANG STAR II IN ANNEX - IIb)

Extinction Order	Number of Dimensions	Extinction Time in the Last Sec.	Which Anti Fermion Sends Which Force to Which A. Boson	Mass	Energy	Last Subatom Particles (A. Bosons and A. Fermions)	Sign	Effecting Forces	Life Second (sec)	Spin	Impact Range Meters (M)	Duration of Impact Second (sec)
1		10^{-4}	10^{-4} sec		VARIABLE		ANTI - HEDRONS			VARIABLE	3/2, 1/2	10^{-15} m	10^{-23} - 10^{-22} sec
2		10^{-5}	10^{-5} sec		VARIABLE		ANTI - MESONS			VARIABLE	0,1,2, 1/2	10^{-15} m	10^{-23} - 10^{-22} sec
3		10^{-10}	10^{-10} sec		1/3 e	~5 - 150 MeV	ANTI - QUARKS		10 PCS	VARIABLE	1/2	10^{-15} m	10^{-23} - 10^{-22} sec
4		10^{-10}	10^{-10} sec		2/3 ⊕	~5 - 150 MeV	ANTI + QUARKS		10 PCS	VARIABLE	1/2	10^{-15} m	10^{-23} - 10^{-22} sec
5		10^{-10}	10^{-10} sec		VARIABLE		ANTI - NEUTRINOS		10 PCS	INFINITE	1/2	INFINITE	10^{-10} - 10^{-8} sec
6		10^{-10}	10^{-10} sec			1,777 MeV	ANTI - TAU			3×10^{-13} sec	1/2	INFINITE	10^{-20} - 10^{-18} sec
7		10^{-10}	10^{-10} sec			105,7 MeV	ANTI - MUON			$2{,}197\times10^{-6}$ sec	1/2	INFINITE	10^{-20} - 10^{-18} sec
8		10^{-10}	10^{-10} sec		1/1836 ⊕	0,511 MeV	ANTI - ELECTRON			INFINITE	1/2	INFINITE	10^{-20} - 10^{-18} sec
9		10^{-10}	10^{-10} sec		~50 - 90 ⊕	80,4 MeV	ANTI - W⁻			3×10^{-25} sec	1	10^{-18} m	10^{-10} - 10^{-8} sec
10		10^{-10}	10^{-10} sec	WE	~50 - 90 ⊕	80,4 MeV	ANTI - W⁺			3×10^{-25} sec	1	10^{-18} m	10^{-10} - 10^{-8} sec
11		10^{-10}	10^{-10} sec			~91,2 MeV	ANTI - Z°			3×10^{-25} sec	1	10^{-18} m	10^{-10} - 10^{-8} sec
12		10^{-10}	10^{-10} sec	EM	0	0 MeV	ANTI - PHOTON			INFINITE	1	INFINITE	10^{-20} - 10^{-18} sec
13		10^{-10}	10^{-10} sec	ST	0	0 MeV	ANTI - GLUON			INFINITE	1	10^{-15} m	10^{-23} - 10^{-22} sec
14		4-9	10^{-35} sec			>112 GeV	ANTI - HIGGS			INFINITE	0	INFINITE	MOMENTARY
15		10	10^{-43} sec	GF	0	<10^{-40} MeV	ANTI - GRAVITON			INFINITE	2	INFINITE	MOMENTARY

TABLE OF SUPERPARTNERS OF ANTI - BOSONS, ANTI - FERMIONS (IF PRESENT) AND THEIR POSSIBLE INTERACTIONS IN ACCORDANCE WITH THE EXTINCTION ORDER BEFORE BIG - BANG: (BIG - BANG STAR I IN ANNEX - IIa)

Extinction Order	Number of Dimensions	Extinction Time in the Last Sec.	Which A. Super Partner Sends Which Force to Which A.S. Partner	Mass	Energy	Last Particles (Superpartners of A. Bosons and A. Fermions)	Sign	Effecting Forces	Life Second (sec)	Spin	Impact Range Meters (M)	Duration of Impact Second (sec)
1	4	10^{-10} sec		1/3 e	~5 - 150 MeV	ANTI Q⁻ SUPERPARTNERS		10 PCS	VARIABLE	1/2	10^{-15} m	10^{-23} - 10^{-22} sec
2	4	10^{-10} sec		2/3 ⊕	~5 - 150 MeV	ANTI Q⁺ SUPERPARTNERS		10 PCS	VARIABLE	1/2	10^{-15} m	10^{-23} - 10^{-22} sec
3	4	10^{-10} sec		VARIABLE		ANTI ν° SUPERPARTNERS		10 PCS	INFINITE	1/2	INFINITE	10^{-10} - 10^{-8} sec
4	4	10^{-10} sec			1,777 MeV	ANTI τ⁻ SUPERPARTNERS			3×10^{-13} sec	1/2	INFINITE	10^{-20} - 10^{-18} sec
5	4	10^{-10} sec			105,7 MeV	ANTI μ⁻ SUPERPARTNERS			$2{,}197\times10^{-6}$ sec	1/2	INFINITE	10^{-20} - 10^{-18} sec
6	4	10^{-10} sec		1/1836 ⊕	0,511 MeV	ANTI e⁻ SUPERPARTNERS			INFINITE	1/2	INFINITE	10^{-20} - 10^{-18} sec
7	4	10^{-10} sec		~50 - 90 ⊕	80,4 MeV	ANTI W⁻ SUPERPARTNERS			3×10^{-25} sec	1	10^{-18} m	10^{-10} - 10^{-8} sec
8	4	10^{-10} sec	WE	~50 - 90 ⊕	80,4 MeV	ANTI W⁺ SUPERPARTNERS			3×10^{-25} sec	1	10^{-18} m	10^{-10} - 10^{-8} sec
9	4	10^{-10} sec			~91,2 MeV	ANTI Z° SUPERPARTNERS			3×10^{-25} sec	1	10^{-18} m	10^{-10} - 10^{-8} sec
10	4	10^{-10} sec	EM	0	0 MeV	ANTI γ° SUPERPARTNERS			INFINITE	1	INFINITE	10^{-20} - 10^{-18} sec
11	4	10^{-10} sec	ST	0	0 MeV	ANTI g° SUPERPARTNERS			INFINITE	1	10^{-15} m	10^{-23} - 10^{-22} sec
12	4-9	10^{-35} sec			>112 GeV	ANTI h° SUPERPARTNERS			INFINITE	0	INFINITE	MOMENTARY
13	10	10^{-43} sec	GF	0	<10^{-40} MeV	ANTI Θ° SUPERPARTNERS			INFINITE	2	INFINITE	MOMENTARY

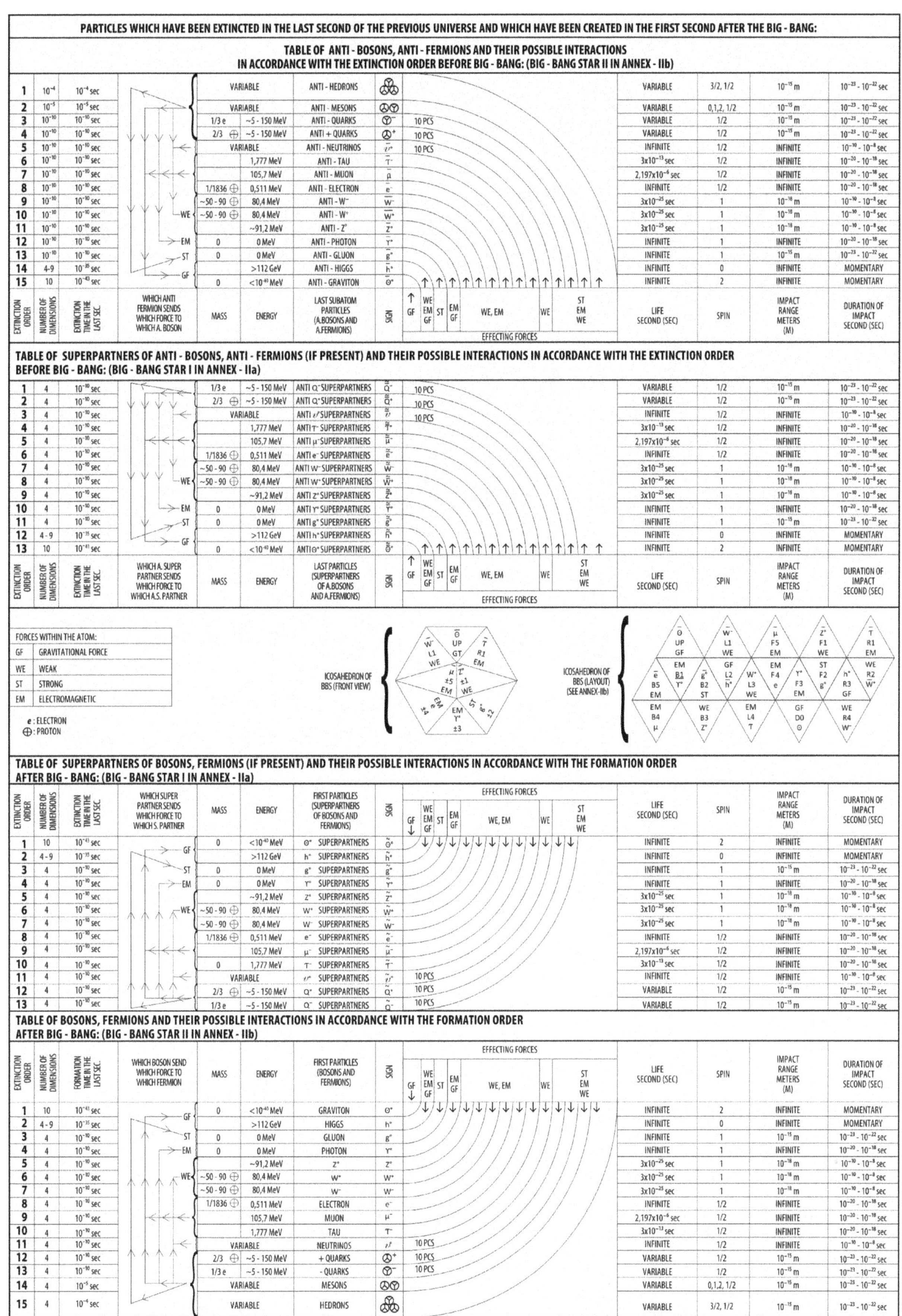

TABLE OF SUPERPARTNERS OF BOSONS, FERMIONS (IF PRESENT) AND THEIR POSSIBLE INTERACTIONS IN ACCORDANCE WITH THE FORMATION ORDER AFTER BIG - BANG: (BIG - BANG STAR I IN ANNEX - IIa)

Extinction Order	Number of Dimensions	Extinction Time in the Last Sec.	Which Super Partner Sends Which Force to Which S. Partner	Mass	Energy	First Particles (Superpartners of Bosons and Fermions)	Sign	Effecting Forces	Life Second (sec)	Spin	Impact Range Meters (M)	Duration of Impact Second (sec)
1	10	10^{-43} sec	GF	0	<10^{-40} MeV	Θ° SUPERPARTNERS			INFINITE	2	INFINITE	MOMENTARY
2	4 - 9	10^{-35} sec			>112 GeV	h° SUPERPARTNERS			INFINITE	0	INFINITE	MOMENTARY
3	4	10^{-10} sec	ST	0	0 MeV	g° SUPERPARTNERS			INFINITE	1	10^{-15} m	10^{-23} - 10^{-22} sec
4	4	10^{-10} sec	EM	0	0 MeV	γ° SUPERPARTNERS			INFINITE	1	INFINITE	10^{-20} - 10^{-18} sec
5	4	10^{-10} sec			~91,2 MeV	Z° SUPERPARTNERS			3×10^{-25} sec	1	10^{-18} m	10^{-10} - 10^{-8} sec
6	4	10^{-10} sec	WE	~50 - 90 ⊕	80,4 MeV	W⁺ SUPERPARTNERS			3×10^{-25} sec	1	10^{-18} m	10^{-10} - 10^{-8} sec
7	4	10^{-10} sec		~50 - 90 ⊕	80,4 MeV	W⁻ SUPERPARTNERS			3×10^{-25} sec	1	10^{-18} m	10^{-10} - 10^{-8} sec
8	4	10^{-10} sec		1/1836 ⊕	0,511 MeV	e⁻ SUPERPARTNERS			INFINITE	1/2	INFINITE	10^{-20} - 10^{-18} sec
9	4	10^{-10} sec			105,7 MeV	μ⁻ SUPERPARTNERS			$2{,}197\times10^{-6}$ sec	1/2	INFINITE	10^{-20} - 10^{-18} sec
10	4	10^{-10} sec		0	1,777 MeV	τ⁻ SUPERPARTNERS			3×10^{-13} sec	1/2	INFINITE	10^{-20} - 10^{-18} sec
11	4	10^{-10} sec		VARIABLE		ν° SUPERPARTNERS		10 PCS	INFINITE	1/2	INFINITE	10^{-10} - 10^{-8} sec
12	4	10^{-10} sec		2/3 ⊕	~5 - 150 MeV	Q⁺ SUPERPARTNERS		10 PCS	VARIABLE	1/2	10^{-15} m	10^{-23} - 10^{-22} sec
13	4	10^{-10} sec		1/3 e	~5 - 150 MeV	Q⁻ SUPERPARTNERS		10 PCS	VARIABLE	1/2	10^{-15} m	10^{-23} - 10^{-22} sec

TABLE OF BOSONS, FERMIONS AND THEIR POSSIBLE INTERACTIONS IN ACCORDANCE WITH THE FORMATION ORDER AFTER BIG - BANG: (BIG - BANG STAR II IN ANNEX - IIb)

Extinction Order	Number of Dimensions	Formation Time in the Last Sec.	Which Boson Send Which Force to Which Fermion	Mass	Energy	First Particles (Bosons and Fermions)	Sign	Effecting Forces	Life Second (sec)	Spin	Impact Range Meters (M)	Duration of Impact Second (sec)
1	10	10^{-43} sec	GF	0	<10^{-40} MeV	GRAVITON	Θ°		INFINITE	2	INFINITE	MOMENTARY
2	4 - 9	10^{-35} sec			>112 GeV	HIGGS	h°		INFINITE	0	INFINITE	MOMENTARY
3	4	10^{-10} sec	ST	0	0 MeV	GLUON	g°		INFINITE	1	10^{-15} m	10^{-23} - 10^{-22} sec
4	4	10^{-10} sec	EM	0	0 MeV	PHOTON	Y°		INFINITE	1	INFINITE	10^{-20} - 10^{-18} sec
5	4	10^{-10} sec			~91,2 MeV	Z°	Z°		3×10^{-25} sec	1	10^{-18} m	10^{-10} - 10^{-8} sec
6	4	10^{-10} sec	WE	~50 - 90 ⊕	80,4 MeV	W⁺	W⁺		3×10^{-25} sec	1	10^{-18} m	10^{-10} - 10^{-8} sec
7	4	10^{-10} sec		~50 - 90 ⊕	80,4 MeV	W⁻	W⁻		3×10^{-25} sec	1	10^{-18} m	10^{-10} - 10^{-8} sec
8	4	10^{-10} sec		1/1836 ⊕	0,511 MeV	ELECTRON	e⁻		INFINITE	1/2	INFINITE	10^{-20} - 10^{-18} sec
9	4	10^{-10} sec			105,7 MeV	MUON	μ⁻		$2{,}197\times10^{-6}$ sec	1/2	INFINITE	10^{-20} - 10^{-18} sec
10	4	10^{-10} sec			1,777 MeV	TAU	τ⁻		3×10^{-13} sec	1/2	INFINITE	10^{-20} - 10^{-18} sec
11	4	10^{-10} sec		VARIABLE		NEUTRINOS	ν°	10 PCS	INFINITE	1/2	INFINITE	10^{-10} - 10^{-8} sec
12	4	10^{-10} sec		2/3 ⊕	~5 - 150 MeV	+ QUARKS	⊕⁺	10 PCS	VARIABLE	1/2	10^{-15} m	10^{-23} - 10^{-22} sec
13	4	10^{-10} sec		1/3 e	~5 - 150 MeV	- QUARKS	∀⁻	10 PCS	VARIABLE	1/2	10^{-15} m	10^{-23} - 10^{-22} sec
14	4	10^{-5} sec		VARIABLE		MESONS			VARIABLE	0,1,2, 1/2	10^{-15} m	10^{-23} - 10^{-22} sec
15	4	10^{-4} sec		VARIABLE		HEDRONS			VARIABLE	3/2, 1/2	10^{-15} m	10^{-23} - 10^{-22} sec

DOUBLE PYTHAGORAS TRIANGULAR PYRAMIDS OF BIG - BANG:

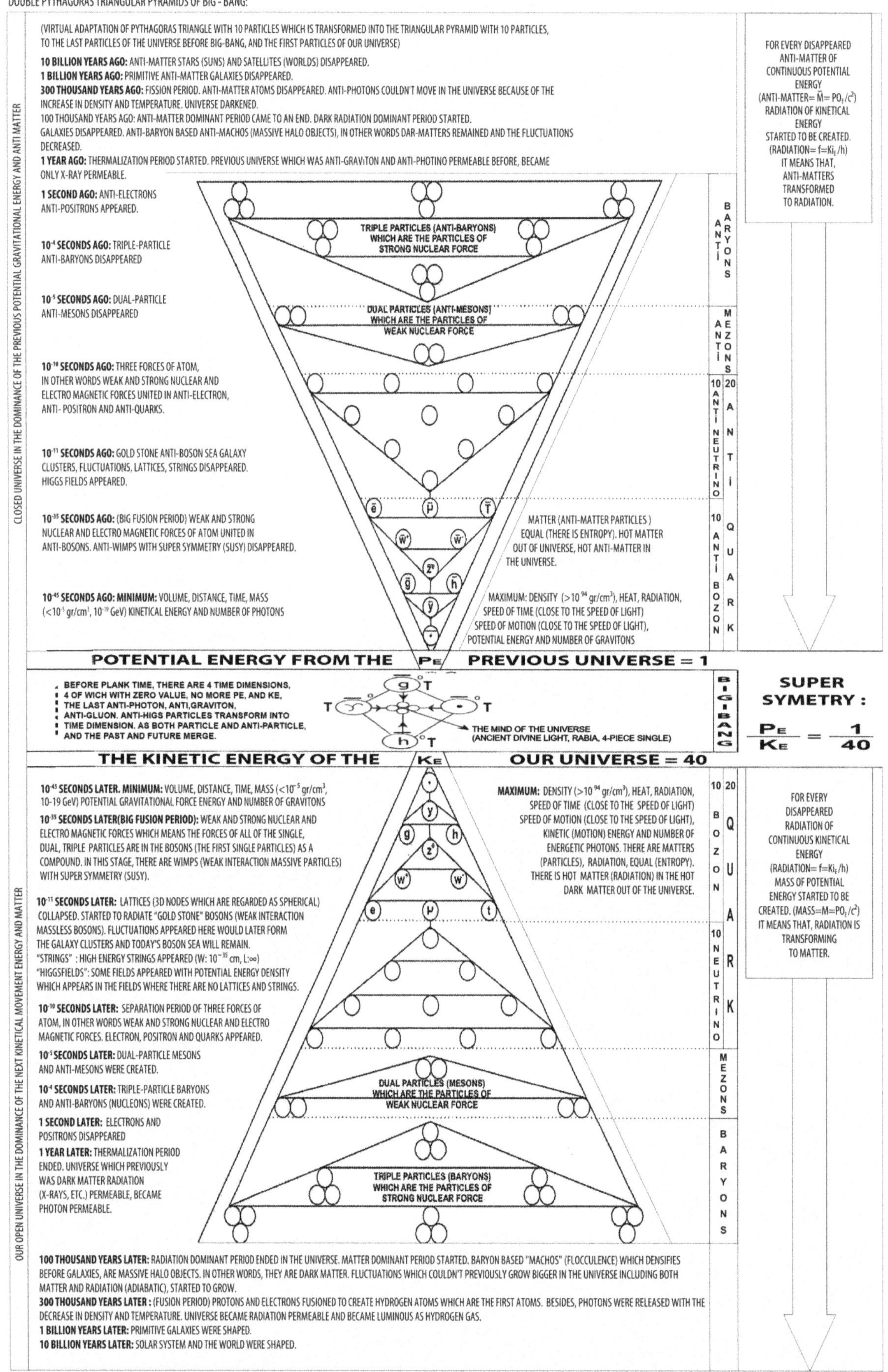

TABLE OF KNOWN DUAL BASIC PARTICLES (MESONS) OF ATOM

GROUP	ANTI-MATERIAL NAME	SYMBOL	CHARGE	LIFE/SEC	MATERIAL NAME	SYMBOL	CHARGE	SPIN	MASS	ENERGY	LIFE/SEC	RANGE	POWER
FERMIONS / QUARKS	ANTI d'	$\bar{d}'$	+1/3		d' QUARK	d'	-1/3	1/2	$-1/3\,\Theta$	360 MeV		10^{-15} MR	STRONG NUCLEUS POWER (DURATION OF EFFECT 10^{-23} 10^{-22} SECONDS)
	ANTI s'	$\bar{s}'$	+1/3		s' QUARK	s'	-1/3	1/2	$-1/3\,\Theta$	540 MeV		10^{-15} MR	
	ANTI b'	$\bar{b}'$	+1/3		b' QUARK	b'	-1/3	1/2	$-1/3\,\Theta$	5 MeV		10^{-15} MR	
	ANTI BLUE QUARK (YELLOW)	$\bar{M}$	0		BLUE QUARK	M	0	1/2		0 MeV		10^{-15} MR	
	ANTI GREEN QUARK (ORANGE)	$\bar{Y}$	0		GREEN QUARK	Y	0	1/2		0 MeV		10^{-15} MR	
	ANTI RED QUARK (PURPLE)	$\bar{K}$	0		RED QUARK	K	0	1/2		0 MeV		10^{-15} MR	
FERMIONS / GLUONS	ANTI -M$\bar{M}$ GLUON	M $\bar{M}$	0		M $\bar{M}$ GLUON	M $\bar{M}$	0	1	MASSLESS	0 MeV		10^{-15} MR	
	ANTI -M$\bar{Y}$ GLUON	M $\bar{Y}$	0		M $\bar{Y}$ GLUON	M $\bar{Y}$	0	1	MASSLESS	0 MeV		10^{-15} MR	
	ANTI -M$\bar{K}$ GLUON	M $\bar{K}$	0		M $\bar{K}$ GLUON	M $\bar{K}$	0	1	MASSLESS	0 MeV		10^{-15} MR	
	ANTI -Y$\bar{M}$ GLUON	Y $\bar{M}$	0		Y $\bar{M}$ GLUON	Y $\bar{M}$	0	1	MASSLESS	0 MeV		10^{-15} MR	
	ANTI -Y$\bar{Y}$ GLUON	Y $\bar{Y}$	0		Y $\bar{Y}$ GLUON	Y $\bar{Y}$	0	1	MASSLESS	0 MeV		10^{-15} MR	
	ANTI -Y$\bar{K}$ GLUON	Y $\bar{K}$	0		Y $\bar{K}$ GLUON	Y $\bar{K}$	0	1	MASSLESS	0 MeV		10^{-15} MR	
	ANTI -K$\bar{M}$ GLUON	K $\bar{M}$	0		K $\bar{M}$ GLUON	K $\bar{M}$	0	1	MASSLESS	0 MeV		10^{-15} MR	
	ANTI -K$\bar{Y}$ GLUON	K $\bar{Y}$	0		K $\bar{Y}$ GLUON	K $\bar{Y}$	0	1	MASSLESS	0 MeV		10^{-15} MR	
	ANTI -K$\bar{K}$ GLUON	K $\bar{K}$	0		K $\bar{K}$ GLUON	K $\bar{K}$	0	1	MASSLESS	0 MeV		10^{-15} MR	
BOSONIC HADRONS MESONS	ANTI Pi - MESON $\bar{\pi}$	d u	-1		Pi-MESON (Pion) π	u $\bar{d}$	1	0	$273\,\Theta$	140 MeV	$2{,}56\times10^{-8}$	10^{-15} MR	
	ANTI Pi - MESON $\bar{\pi}$	s s	-1		Pi-MESON (Pion) π	s s	1					10^{-15} MR	
	ANTI ETA -MESON $\bar{n}$	s s	-1		ETA-MESON (Ψ) n		1		547 MeV			10^{-15} MR	
	ANTI ETA'-MESON $\bar{n}'$	s s	-1		ETA'-MESON (Ψ) n'		1					10^{-15} MR	
	ANTI B-MESON $\bar{B}$	b u	-1		B-MESON B	u $\bar{b}$	1					10^{-15} MR	
	ANTI Bc-MESON $\bar{Bc}$	b c	-1		Bc-MESON Bc	c $\bar{b}$	1					10^{-15} MR	
	ANTI D-MESON $\bar{D}$	d c	-1		D-MESON D	c $\bar{d}$	1	0		1869 MeV		10^{-15} MR	
	ANTI D* (2010) MESON $\bar{D}^*$		-1		D* (2010)-MESON D*		1			2010 MeV		10^{-15} MR	
	ANTI Ds MESON $\bar{Ds}$	s c	-1		Ds-MESON Ds	c s	1			1968 MeV		10^{-15} MR	
	ANTI Ds* MESON $\bar{Ds}^*$		-1		Ds*-MESON Ds*		1			2112 MeV		10^{-15} MR	
	ANTI D2 (2460) MESON $\bar{D2}$		-1		D2* (2460)-MESON D*2		1			2459 MeV		10^{-15} MR	
	ANTI K MESON $\bar{K}$	s u	-1		K-MESON K	u s	1	0	$968\,\Theta$	494 MeV	$1{,}22\times10^{-8}$	10^{-15} MR	
	ANTI K* (1410) MESON $\bar{K}^*$	s u	-1		K* (1410)-MESON K*	u s	1			892 MeV		10^{-15} MR	
	ANTI K* (892) MESON $\bar{K}^*$	s u	-1		K* (892)-MESON K*	u s	1			892 MeV		10^{-15} MR	
	ANTI RHO MESON $\bar{R}$	d u	-1		RHO-MESON R	u $\bar{d}$	1	1		770 MeV		10^{-15} MR	
BOSONIC MESONS	ANTI Pi MESON $\bar{\pi}$		1		Pi MESON (Pion) π		-1	0	$273\,\Theta$	140 MeV	$2{,}56\times10^{-8}$	10^{-15} MR	STRONG NUCLEUS POWER (DURATION OF EFFECT 10^{-23} 10^{-22} SECONDS)
	ANTI ETA MESON $\bar{n}$		1		ETA MESON (Ψ) n		-1					10^{-15} MR	
	ANTI ETA' MESON $\bar{n}'$		1		ETA' MESON (Ψ) n'		-1					10^{-15} MR	
	ANTI B MESON $\bar{B}$	u b	1		B MESON B	b' u	-1					10^{-15} MR	
	ANTI Bc MESON $\bar{Bc}$	c b	1		Bc MESON Bc	b' c	-1					10^{-15} MR	
	ANTI D MESON $\bar{D}$	c d	1		D MESON D	d' c	-1			1869 MeV		10^{-15} MR	
	ANTI D* (2010) MESON $\bar{D}^*$		1		D* (2010) MESON D*		-1			2010 MeV		10^{-15} MR	
	ANTI Ds MESON $\bar{Ds}$	c s	1		Ds MESON Ds	s c	-1			1968 MeV		10^{-15} MR	
	ANTI Ds* MESON $\bar{Ds}^*$		1		Ds* MESON Ds*		-1			2112 MeV		10^{-15} MR	
	ANTI D2 (2460) MESON $\bar{D2}$		1		D2* (2460) MESON D2*		-1			2459 MeV		10^{-15} MR	
	ANTI K MESON $\bar{K}$	u s	1		K MESON (KAON) K	s u	-1			494 MeV	$1{,}22\times10^{-8}$	10^{-15} MR	
	ANTI K* (892) MESON $\bar{K}^*$		1		K* (892) MESON K*		-1			892 MeV		10^{-15} MR	
HADRONS	ANTI Pi MESON $\bar{\pi}$	d u	0		Pi MESON π	u $\bar{d}$	0	0	$264\,\Theta$	135 MeV	$2{,}2\times10^{-16}$	10^{-15} MR	
	ANTI ETA MESON $\bar{n}$	c c	0		ETA MESON (Ψ) nc	c c	0	0		2980 MeV		10^{-15} MR	
	ANTI B MESON $\bar{B}$	b d	0		B MESON B	d b	0			5279 MeV		10^{-15} MR	
	ANTI Bs MESON $\bar{Bs}$	b s	0		Bs MESON Bs	s b	0			5370 MeV		10^{-15} MR	
	ANTI D MESON $\bar{D}$	u c	0		D MESON D	c u	0			1864 MeV		10^{-15} MR	
	ANTI D* (2007) MESON $\bar{D}^*$	u c	0		D* (2007) MESON D*	c u	0			2007 MeV		10^{-15} MR	
	ANTI D2* (2460) MESON $\bar{D2}^*$	u c	0		D2* (2460) MESON D2*	c u	0			2459 MeV		10^{-15} MR	
	ANTI KL MESON $\bar{KL}$	s d	0		KL MESON KL	d' s	0	0	$973\,\Theta$	498 MeV	1×10^{-10}	10^{-15} MR	
	ANTI Ks MESON $\bar{Ks}$		0		Ks MESON Ks		0	0		498 MeV	6×10^{8}	10^{-15} MR	
	ANTI K* MESON $\bar{K}^*$		0		K* MESON K*		0			896 MeV		10^{-15} MR	
	ANTI Y (15) MESON $\bar{Y}$	b b	0		Y (15) MESON Y	b $\bar{b}$	0			9460 MeV		10^{-15} MR	

TABLE OF KNOWN TRIPLE BASIC PARTICLES (BARYONS) OF ATOM

GROUP	ANTI-MATERIAL NAME	ANTI SYMBOL	ANTI QUARKS	ANTI CHARGE	MATERIAL NAME	SYMBOL	QUARKS	CHARGE	SPIN	MASS	ENERGY	LIFE/SEC.	RANGE	POWER
NUCLEONS	ANTI PROTON (⊖)	P̄	ū ū d̄	-1	PROTON (⊕)	P	u u d	+1	+1/2	1836,1 Θ	938,2 MeV	STABLE	10^{-15} MR	STRONG NUCLEUS POWER (DURATION OF EFFECT 10^{-23} 10^{-22} SECONDS)
NUCLEONS	ANTI PROTON' (⊖')	P̄'	ū c̄ d̄	-1	PROTON' (⊕')	P'	u c d	+1			(1 GRUN)		10^{-15} MR	
NUCLEONS	ANTI NEUTRON (⊗)	N̄	ū d̄ d̄	0	NEUTRON (⊖)	N°	u d d	0	+1/2	1838,6 Θ	939,5 MeV	1,12X10³	10^{-15} MR	
NUCLEONS	ANTI NEUTRON' (⊖')	N̄'	s̄ c̄ d̄	0	NEUTRON' (⊕')	N'	s c d	0					10^{-15} MR	
HYPERONS	ANTI SIGMA	Σ̄	ū ū ū	-2	SIGMA	Σ⁺⁺	u u u	+2	3/2				10^{-15} MR	
HYPERONS	ANTI SIGMA	Σ̄	ū ū s̄	-1	SIGMA	Σ⁺	u u s	+1	3/2	2327 Θ	1189 MeV	0,9X10⁻¹⁰	10^{-15} MR	
HYPERONS	ANTI SIGMA /c	Σ̄c	ū ū c̄	-2	SIGMA /c	Σc⁺⁺	u u c	+2	3/2				10^{-15} MR	
HYPERONS	ANTI SIGMA /c	Σ̄c	ū d̄ c̄	-1	SIGMA /c	Σc⁺	u d c	+1	3/2				10^{-15} MR	
HYPERONS	ANTI XI	Ξ̄	ū ū c̄	-2	XI	Ξ⁺⁺	u u c	+2	3/2				10^{-15} MR	
HYPERONS	ANTI XI	Ξ̄	ū c̄ s̄	-1	XI	Ξ⁺	u c s	+1	3/2				10^{-15} MR	
HYPERONS	ANTI XI/c	Ξ̄	ū s̄ c̄	-1	XI/c	Ξc⁺	u s c	+1	3/2		2466 MeV		10^{-15} MR	
HYPERONS	ANTI OMEGA	Ω̄	c̄ c̄ c̄	-2	OMEGA (THE HEAVIEST HYPERON)	Ω⁺⁺	c c c	+2	3/2				10^{-15} MR	
HYPERONS	ANTI OMEGA	Ω̄	c̄ c̄ s̄	-1	OMEGA	Ω⁺	c c s	+1	3/2				10^{-15} MR	
HYPERONS	ANTI LAMDA	λ̄	c̄ c̄ ū	-2	LAMDA	λ⁺⁺	c c u	+2	3/2				10^{-15} MR	
HYPERONS	ANTI LAMDA	λ̄	c̄ c̄ d̄	-1	LAMDA	λ⁺	c c d	+1	3/2				10^{-15} MR	
HYPERONS	ANTI LAMDA/C	λ̄c	ū d̄ c̄	-1	LAMDA/C	λc⁺	u d c	+1	3/2		2285 MeV		10^{-15} MR	
HYPERONS	ANTI SIGMA	Σ̄	s̄ d̄ d̄	+1	SIGMA	Σ⁻	s d d	-1	3/2	2342 Θ	1197 MeV	1,74X10⁻¹⁰	10^{-15} MR	STRONG NUCLEUS POWER (DURATION OF EFFECT 10^{-23} 10^{-22} SECONDS)
HYPERONS	ANTI XI	Ξ̄	d̄ s̄ s̄	+1	XI	Ξ⁻	d s s	-1	3/2	2583 Θ	1321 MeV	1,8X10⁻¹⁰	10^{-15} MR	
HYPERONS	ANTI XI/b	Ξ̄	d̄ s̄ b̄	+1	XI/b	Ξb⁻	d s b	-1	3/2		5624 MeV		10^{-15} MR	
HYPERONS	ANTI OMEGA	Ω̄	s̄ s̄ s̄	+1	OMEGA	Ω⁻	s s s	-1	3/2		1672 MeV		10^{-15} MR	
HYPERONS	ANTI SIGMA	Σ̄	ū d̄ s̄	0	SIGMA	Σ⁰	u c s	0	3/2	2327 Θ	1193 MeV	>10⁻¹¹	10^{-15} MR	
HYPERONS	ANTI SIGMA/c	Σ̄	d̄ d̄ c̄	0	SIGMA/c	Σc⁰	d d c	0	3/2				10^{-15} MR	
HYPERONS	ANTI XI	Ξ̄	ū s̄ s̄	0	XI	Ξ⁰	u s s	0	3/2	2577 Θ	1315 MeV	1,9X10⁻¹⁰	10^{-15} MR	
HYPERONS	ANTI XI/b	Ξ̄	ū s̄ b̄	0	XI/b	Ξb⁰	u s b	0	3/2		5624 MeV		10^{-15} MR	
HYPERONS	ANTI XI/c	Ξ̄	d̄ s̄ c̄	0	XI/c	Ξc⁰	d s c	0	3/2		2472 MeV		10^{-15} MR	
HYPERONS	ANTI OMEGA/c	Ω̄	c̄ s̄ s̄	0	OMEGA/c	Ωc⁰	c s s	0	3/2		2698 MeV		10^{-15} MR	
HYPERONS	ANTI LAMDA	λ̄	ū d̄ s̄	0	LAMDA	λ⁰	u c s	0	3/2	2181 Θ	1115 MeV	3,03X10⁻¹⁰	10^{-15} MR	
HYPERONS	ANTI LAMDA/b	λ̄	ū d̄ b̄	0	LAMDA/b	λ⁰b	u d b	0	3/2		5624 MeV		10^{-15} MR	

Group (left margin): FERMIONIC HADRONS BARYONS

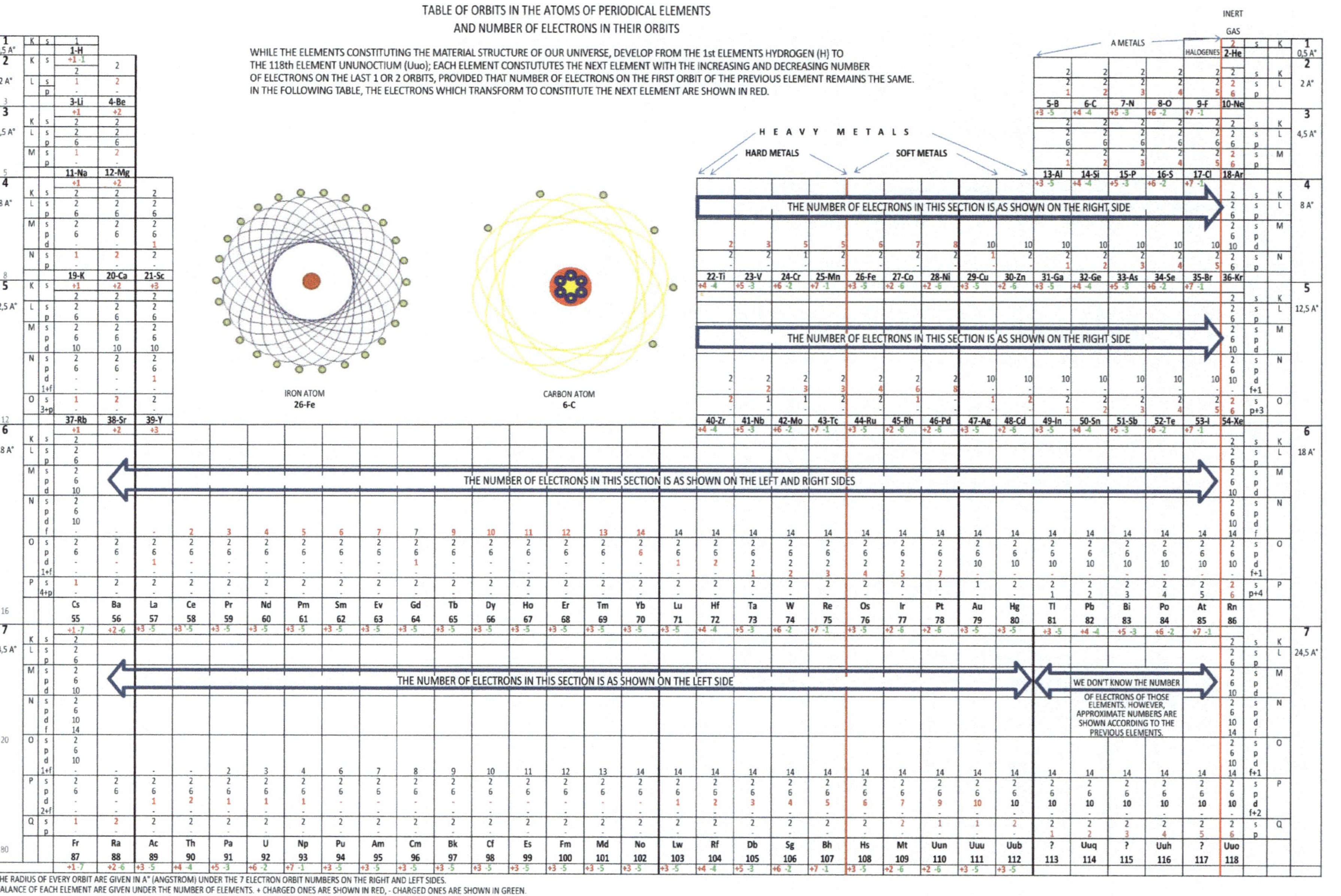
TABLE OF ORBITS IN THE ATOMS OF PERIODICAL ELEMENTS
AND NUMBER OF ELECTRONS IN THEIR ORBITS

WHILE THE ELEMENTS CONSTITUTING THE MATERIAL STRUCTURE OF OUR UNIVERSE, DEVELOP FROM THE 1st ELEMENTS HYDROGEN (H) TO THE 118th ELEMENT UNUNOCTIUM (Uuo); EACH ELEMENT CONSTUTUTES THE NEXT ELEMENT WITH THE INCREASING AND DECREASING NUMBER OF ELECTRONS ON THE LAST 1 OR 2 ORBITS, PROVIDED THAT NUMBER OF ELECTRONS ON THE FIRST ORBIT OF THE PREVIOUS ELEMENT REMAINS THE SAME. IN THE FOLLOWING TABLE, THE ELECTRONS WHICH TRANSFORM TO CONSTITUTE THE NEXT ELEMENT ARE SHOWN IN RED.

INERT GAS
A METALS
HALOGENES
HEAVY METALS
HARD METALS
SOFT METALS

THE NUMBER OF ELECTRONS IN THIS SECTION IS AS SHOWN ON THE RIGHT SIDE
THE NUMBER OF ELECTRONS IN THIS SECTION IS AS SHOWN ON THE RIGHT SIDE
THE NUMBER OF ELECTRONS IN THIS SECTION IS AS SHOWN ON THE LEFT AND RIGHT SIDES
THE NUMBER OF ELECTRONS IN THIS SECTION IS AS SHOWN ON THE LEFT SIDE

WE DON'T KNOW THE NUMBER OF ELECTRONS OF THOSE ELEMENTS. HOWEVER, APPROXIMATE NUMBERS ARE SHOWN ACCORDING TO THE PREVIOUS ELEMENTS.

IRON ATOM 26-Fe
CARBON ATOM 6-C

1) THE RADIUS OF EVERY ORBIT ARE GIVEN IN A° (ANGSTROM) UNDER THE 7 ELECTRON ORBIT NUMBERS ON THE RIGHT AND LEFT SIDES.
2) VALANCE OF EACH ELEMENT ARE GIVEN UNDER THE NUMBER OF ELEMENTS. + CHARGED ONES ARE SHOWN IN RED, - CHARGED ONES ARE SHOWN IN GREEN.

PERIODICAL ELEMENTS AND THEIR ISOTOPES IN ACCORDANCE WITH THE NUMBER OF PROTONS (⊕) AND NEUTRONS (⊗)

Column group 1

P (Protons)	N (Neutrons)	Z = P+N (Total)	Abbr.	Element Name
1	-	1	H	Hydrogen
1	1	2		Deuterium
1	2	3		Tritium
2	1	3	He	Helium
2	2	4		Helium 4
3	3	6	Li	Lithium
3	4	7		Lithium 7
4	5	9	Be	Beryllium
5	5	10	B	Boron
5	6	11		Boron 11
6	6	12	C	Carbon
6	7	13		Carbon 13
7	7	14	N	Nitrogen
7	8	15		Nitrogen 15
8	8	16	O	Oxygen
8	9	17		Oxygen 17
8	10	18		Oxygen 18
9	11	19	F	Fluorum
10	10	20	Ne	Neon
10	11	21		Neon 21
10	12	22		Neon 22
11	12	23	Na	Natrium
12	12	24	Mg	Magnesium
12	13	25		Magnesium 25
12	14	26		Magnesium 26
13	14	27	Al	Aluminium
14	14	28	Si	Silicium
14	15	29		Silicium 29
14	16	30		Silicium 30
15	16	31	P	Phosphorus
16	16	32	S	Sulphurium
16	17	33		Sulphurium 33
16	18	34		Sulphurium 34
16	19	35		Sulphurium 35
17	19	36	Cl	Chlorium
17	20	37		Chlorium 37
18	18	36	Ar	Argon
18	20	38		Argon 38
18	22	40		Argon 40
19	20	39	K	Kalium
19	21	40		Kalium 40
19	22	41		Kalium 41
20	20	40	Ca	Calcium
20	22	42		Calcium 42
20	23	43		Calcium 43
20	24	44		Calcium 44
20	26	46		Calcium 46
20	28	48		Calcium 48
21	24	45	Sc	Scandium
22	24	46	Ti	Titanium
22	25	47		Titanium 47
22	26	48		Titanium 48
22	27	49		Titanium 49
22	28	50		Titanium 50
23	28	51	V	Vanadium
24	26	50	Cr	Chromium
24	28	52		Chromium 52
24	29	53		Chromium 53
24	30	54		Chromium 54
25	30	55	Mn	Manganese
26	28	54	Fe	Iron
26	30	56		Iron 56
26	31	57		Iron 57
26	32	58		Iron 58
27	32	59	Co	Cobalt
28	30	58	Ni	Nickel
28	32	60		Nickel 60
28	33	61		Nickel 61
28	34	62		Nickel 62
28	36	64		Nickel 64
29	34	63	Cu	Copper
29	36	65		Copper 65
30	34	64	Zn	Zirconium
30	36	66		Zirconium 66
30	37	67		Zirconium 67
30	38	68		Zirconium 68
30	40	70		Zirconium 70

Column group 2

P (Protons)	N (Neutrons)	Z = P+N (Total)	Abbr.	Element Name
31	38	69	Ga	Gallium
31	40	71		Gallium 71
32	38	70	Ge	Germanium
32	40	72		Germanium 72
32	41	73		Germanium 73
32	42	74		Germanium 74
32	44	76		Germanium 76
33	42	75	As	Arsenic
34	40	74	Se	Selenium
34	42	76		Selenium 76
34	43	77		Selenium 77
34	44	78		Selenium 78
34	46	80		Selenium 80
34	48	82		Selenium 82
35	44	79	Br	Bromine
35	46	81		Bromine 81
36	42	78	Kr	Krypton
36	44	80		Krypton 80
36	46	82		Krypton 82
36	47	83		Krypton 83
36	48	84		Krypton 84
36	50	86		Krypton 86
37	48	85	Rb	Rubidium
37	50	87		Rubidium 87
38	46	84	Sr	Strontium
38	48	86		Strontium 86
38	49	87		Strontium 87
38	50	88		Strontium 88
39	50	89	Y	Yttrium
40	50	90	Zr	Zirconium
40	51	91		Zirconium 91
40	52	92		Zirconium 92
40	54	94		Zirconium 94
40	56	96		Zirconium 96
41	52	93	Nb	Niobium
42	50	92	Mo	Molybdenum
42	52	94		Molybdenum 94
42	53	95		Molybdenum 95
42	54	96		Molybdenum 96
42	55	97		Molybdenum 97
42	56	98		Molybdenum 98
42	58	100		Molybdenum 100
43			Tc	Technetium
44	52	96	Ru	Rubidium
44	54	98		Rubidium 98
44	55	99		Rubidium 99
44	56	100		Rubidium 100
44	57	101		Rubidium 101
44	58	102		Rubidium 102
44	60	104		Rubidium 104
45	58	103	Rh	Rhodium
46	56	102	Pd	Palladium
46	58	104		Palladium 104
46	59	105		Palladium 105
46	60	106		Palladium 106
46	62	108		Palladium 108
46	64	110		Palladium 110
47	60	107	Ag	Silver
47	62	109		Silver 109
48	58	106	Cd	Cadmium
48	60	108		Cadmium 108
48	62	110		Cadmium 110
48	63	111		Cadmium 111
48	64	112		Cadmium 112
48	65	113		Cadmium 113
48	66	114		Cadmium 114
48	68	116		Cadmium 116
49	64	113	In	Indium
49	66	115		Indium 115
50	64	112	Sn	Tin
50	64	114		Tin 114
50	65	115		Tin 115
50	66	116		Tin 116
50	67	117		Tin 117
50	68	118		Tin 118
50	69	119		Tin 119
50	70	120		Tin 120

Column group 3

P (Protons)	N (Neutrons)	Z = P+N (Total)	Abbr.	Element Name
50	72	122	Sn	Tin 122
50	74	124		Tin 124
51	70	121	Sb	Antimony
51	72	123		Antimony 123
52	70	122	Te	Tellurium
52	71	123		Tellurium 123
52	72	124		Tellurium 124
52	73	125		Tellurium 125
52	74	126		Tellurium 126
52	76	128		Tellurium 128
52	78	130		Tellurium 130
53	74	127	I	Iodine
54	70	124	Xe	Xenon
54	72	126		Xenon 126
54	74	128		Xenon 128
54	75	129		Xenon 129
54	76	130		Xenon 130
54	77	131		Xenon 131
54	78	132		Xenon 132
54	80	134		Xenon 134
54	82	136		Xenon 136
55	78	133	Cs	Caesium
56	74	130	Ba	Barium
56	76	132		Barium 132
56	78	134		Barium 134
56	79	135		Barium 135
56	80	136		Barium 136
56	81	137		Barium 137
56	82	138		Barium 138
57	82	139	La	Lanthanum
58	78	136	Ce	Cerium
58	80	138		Cerium 138
58	82	140		Cerium 140
58	84	142		Cerium 142
59	82	141	Pr	Praseodymium
60	82	142	Nd	Neodymium
60	83	143		Neodymium 143
60	84	144		Neodymium 144
60	85	145		Neodymium 145
60	86	146		Neodymium 146
60	88	148		Neodymium 148
60	90	150		Neodymium 150
61			Pm	Promethium
62	88	150	Sm	Samarium
62	85	147		Samarium 147
62	86	148		Samarium 148
62	87	149		Samarium 149
62	88	150		Samarium 150
62	90	152		Samarium 152
62	92	154		Samarium 154
63	88	151	Eu	Europium
63	90	153		Europium 153
64	88	152	Gd	Gadolinium
64	90	154		Gadolinium 154
64	91	155		Gadolinium 155
64	92	156		Gadolinium 156
64	93	157		Gadolinium 157
64	94	158		Gadolinium 158
64	96	160		Gadolinium 160
65	94	159	Tb	Terbium
66	92	158	Dy	Dysprosium
66	94	160		Dysprosium 160
66	95	161		Dysprosium 161
66	96	162		Dysprosium 162
66	97	163		Dysprosium 163
66	98	164		Dysprosium 164
67	98	165	Ho	Holmium
68	94	162	Er	Erbium
68	96	164		Erbium 164
68	98	166		Erbium 166
68	99	167		Erbium 167
68	100	168		Erbium 168
68	102	170		Erbium 170
69	100	169	Tm	Thulium
70	98	168	Yb	Ytterbium
70	100	170		Ytterbium 170
70	101	171		Ytterbium 171

Column group 4

P (Protons)	N (Neutrons)	Z = P+N (Total)	Abbr.	Element Name
70	102	172	Yb	Ytterbium 172
70	103	173		Ytterbium 173
70	104	174		Ytterbium 174
70	106	176		Ytterbium 176
71	104	175	Lu	Lutetium
71	105	176		Lutetium 176
72	102	174	Hf	Hafnium
72	104	176		Hafnium 176
72	105	177		Hafnium 177
72	106	178		Hafnium 178
72	107	179		Hafnium 179
72	108	180		Hafnium 180
73	108	181	Ta	Tantalum
74	106	180	W	Tungsten
74	108	182		Tungsten 182
74	109	183		Tungsten 183
74	110	184		Tungsten 184
74	112	186		Tungsten 186
75	110	185	Re	Rhenium
75	112	187		Rhenium 187
76	108	184	Os	Osmium
76	110	186		Osmium 186
76	111	187		Osmium 187
76	112	188		Osmium 188
76	113	189		Osmium 189
76	114	190		Osmium 190
76	116	192		Osmium 192
77	114	191	Ir	Iridium
77	116	193		Iridium 193
78	114	192	Pt	Platinum
78	116	194		Platinum 194
78	117	195		Platinum 195
78	118	196		Platinum 196
78	120	198		Platinum 198
79	118	197	Au	Gold
80	116	196	Hg	Mercury
80	118	198		Mercury 198
80	119	199		Mercury 199
80	120	200		Mercury 200
80	121	201		Mercury 201
80	122	202		Mercury 202
80	124	204		Mercury 204
81	122	203	Tl	Thallium
81	124	205		Thallium 205
82	122	204	Pb	Lead
82	124	206		Lead 206
82	125	207		Lead 207
82	126	208		Lead 208
83	126	209	Bi	Bismuth
84	126	210	Po	Polonium
84	127	211		AcC'
84	128	212		ThC'
84	130	214		RaC'
84	131	215		AcA
84	132	216		ThA
84	134	218		RaA
85	135	220	At	Astatine
86	136	222	Rn	Radon
86	134	220		Thoron
86	133	219		Action
87	137	224	Fr	Francium
88	138	226	Ra	Radium
88	136	224		Thx
88	135	223		AcX
89	138	227	Ac	Actinium
90	137	227		RdAc
90	138	228		RdTh
90	140	230		JO
90	142	232	Th	Thorium
91	140	231	Pa	Protactinium
92	142	234	U	Uranium (II)
92	143	235		Uranium (Ac)
92	146	238		Uranium (I)
93	144	237	Np	Neptunium
94	150	244	Pu	Plutonium
95	148	243	Am	Americium
96	151	247	Cm	Curium

Column group 5

P (Protons)	N (Neutrons)	Z = P+N (Total)	Abbr.	Element Name
97	150	247	Br	Berkelium
98	153	251	Cf	Californium
99	153	252	Es	Einsteinium
100	157	257	Fm	Fermium
101	157	258	Md	Mendelevium
102	157	259	No	Nobelium
103	159	262	Lw	Lawrencium
104	157	261	Rf	Rutherfordium
105	157	262	Db	Dubnium
106	157	263	Sg	Seaborgium
107	155	262	Bh	Bohrium
108	157	265	Hs	Hassium
109	157	266	Mt	Meitnerium
110	159	269	Uun	Ununnilium
111	161	272	Uuu	Ununnilium
112	165	277	Uub	Ununbium
113	?	?	?	?
114	171	285	Uug	Ununquadium
115	172	287	?	?
116	173	289	Uuh	Ununhexium
117	174	291	?	?
118	175	293	Uuo	Ununoktium
119	176	295	?	?
120	177	297	?	?
121	178	299	?	?
122	179	301	?	?
123	180	303	?	?

NOTE: 336 ELEMENTS IN TOTAL (INCLUDING ISOTOPES) IN 120 PROTONS ELEMENT

NOTE: PROTON/NEUTRON RATIO
RED COLORS: 1⊕/1⊗
GREEN COLORS: 2⊕/3⊗

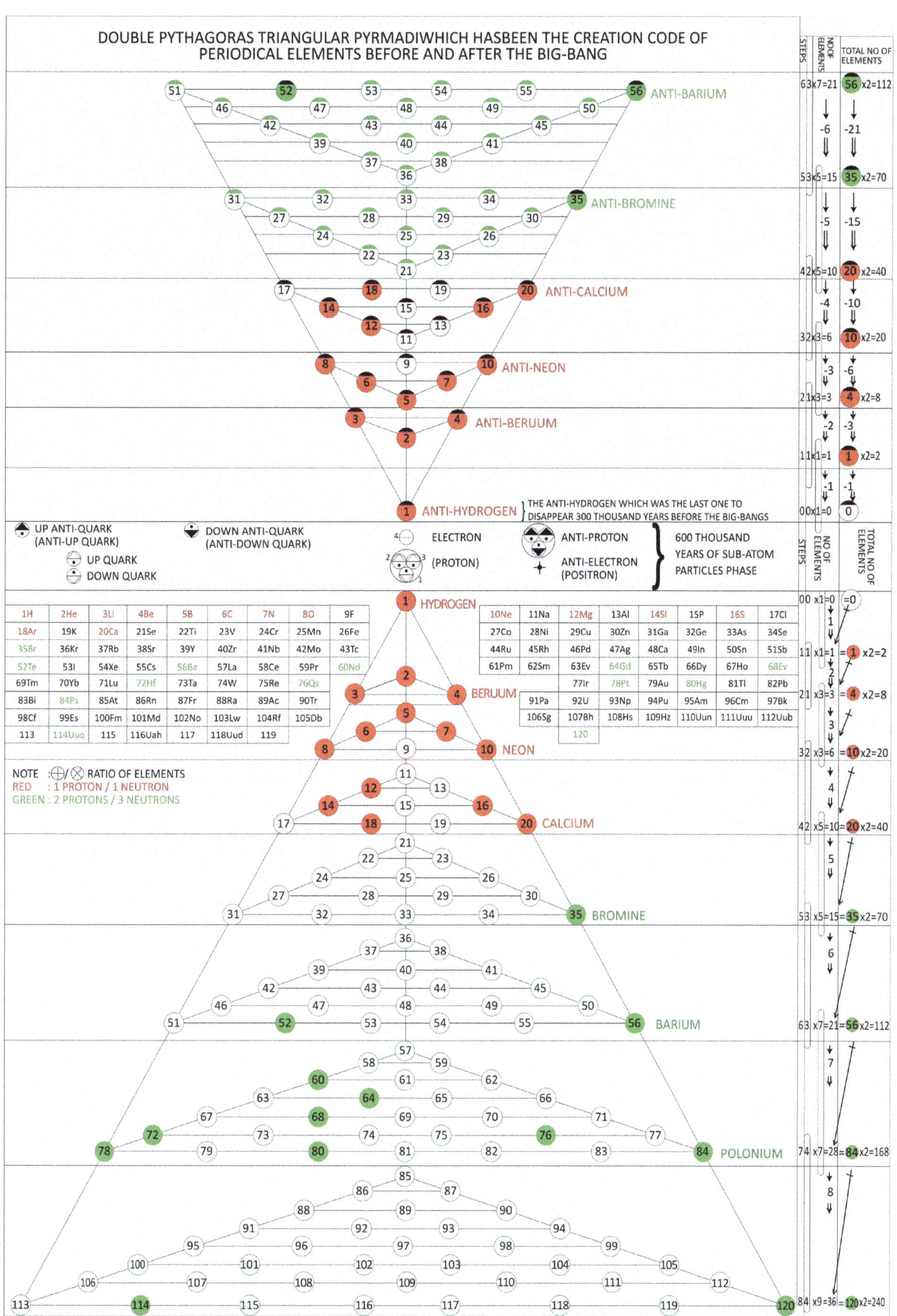
DOUBLE PYTHAGORAS TRIANGULAR PYRMADIWHICH HASBEEN THE CREATION CODE OF PERIODICAL ELEMENTS BEFORE AND AFTER THE BIG-BANG
STEPS
NO OF ELEMENTS
TOTAL NO OF ELEMENTS
ANTI-BARIUM
63x7=21
56 x2=112
-6
-21
ANTI-BROMINE
53x5=15
35 x2=70
-5
-15
ANTI-CALCIUM
42x5=10
20 x2=40
-4
-10
32x3=6
10 x2=20
ANTI-NEON
-3
-6
21x3=3
4 x2=8
ANTI-BERUUM
-2
-3
11x1=1
1 x2=2
ANTI-HYDROGEN
THE ANTI-HYDROGEN WHICH WAS THE LAST ONE TO DISAPPEAR 300 THOUSAND YEARS BEFORE THE BIG-BANGS
-1
-1
00x1=0
0
UP ANTI-QUARK (ANTI-UP QUARK)
DOWN ANTI-QUARK (ANTI-DOWN QUARK)
ELECTRON
ANTI-PROTON
600 THOUSAND
UP QUARK
(PROTON)
YEARS OF SUB-ATOM
DOWN QUARK
ANTI-ELECTRON (POSITRON)
PARTICLES PHASE
STEPS
NO OF ELEMENTS
TOTAL NO OF ELEMENTS
HYDROGEN
00 x1=0 =0
11 x1=1 = 1 x2=2
21 x3=3 = 4 x2=8
BERUUM
NEON
32 x3=6 = 10 x2=20
NOTE: RATIO OF ELEMENTS
RED : 1 PROTON / 1 NEUTRON
GREEN : 2 PROTONS / 3 NEUTRONS
CALCIUM
42 x5=10 = 20 x2=40
BROMINE
53 x5=15 = 35 x2=70
BARIUM
63 x7=21 = 56 x2=112
POLONIUM
74 x7=28 = 84 x2=168
84 x9=36 = 120 x2=240
215

DOUBLE PYTHAGORAS TRIANGULAR PYRAMIDS WHICH ARE THE CREATION CODES OF SUB-ATOM PARTICLES IN THE QUANTUMS OF UNIVERSES	STEP	ORDER OF APPEARANCE OR DISAPPEARANCE	NUMBER OF PARTICLES	TOTAL NUMBER AND VARIETY OF PARTICLES	OUR UNIVERSE IS BILLION YEARS OLD

+1   I.BIG BANG

AFTER THE BIG-BANG, WHILE FIRST THE SUB-ATOM PARTICLES AND LATER ELEMENT ATOMS WERE BEING FORMED IN AN INCREASING NUMBER, THE FIRST APPEARED SUB-ATOM PARTICLES STARTED TO BECOME ZERO CHARGED AND GATHERED OUT OF THE UNIVERSE AS DARK MATTER.

STEP	ORDER	NUMBER OF PARTICLES	VARIETY	TOTAL NUMBER AND VARIETY OF PARTICLES	AGE
1 / 2	I	7	BOSON		ZERO
3	II	3	LEPTON		
4	III	10	NEUTRINO		
5	V	15	MESON AND MESO-GLUON (36)	IV 120 QUARK VII 120 ELEMENT TYPES ARE BEING FORMED	
6		21			10^{10} (13,7)
7	VI	28	BARYON		
8		36 + / 120	IN TOTAL		10^{20}

AFTER 120 ELEMENTS, DIFFERENT ELEMENT FORMATION WILL STOP. BUT WHILE THE UNIVERSE CONTINUE TO EXPAND, PROTONS OF EXISTING 120 ELEMENTS WILL START TO BECOME NEUTRONS AND JOIN THE DARK-MATTER OUT OF OUR UNIVERSE.

AT LAST, WHEN ALL OF THE ELEMENTS IN THE UNIVERSE BURN IN THE STARS, DARK-MATTER OUT OF THE UNIVERSE WILL START TO COLLAPSE INSIDE WITH THE GRAVITATIONAL FORCE OF COLLECTED ZERO CHARGED PARTICLES, BY MEANS OF BIG-CRUNCH.

BIG-CRUNCH 0

STEP	ORDER	NUMBER OF PARTICLES	VARIETY	TOTAL NUMBER AND VARIETY OF PARTICLES	AGE
7	VI	28	BARYON		(27,4)
6	V	21	MESON AND MESO-GLUON (36)	IV 120 QUARK VII 120 ELEMENT TYPES WILL DISAPPEAR	10^{30}
5		15			(41,1)
4	III	10	NEUTRINO		
3	II	3	LEPTON		
2 / 1	I	7	BOSON		10^{40}

WHILE THE UNIVERSE START TO GATHER WITH BIG-CRUNCH; FIRST OF ALL, SUB-ATOM PARTICLES IN THE DARK-MATTER WILL BECOME ANTI-MATTERS, AND THEN THOSE PARTICLES WILL FORM THE ANTI-MATTER ELEMENTS AND START TO INCREASE.

STEP	ORDER	NUMBER OF PARTICLES	VARIETY	TOTAL NUMBER AND VARIETY OF PARTICLES	AGE
1 / 2	I	7	ANTI BOSON		(54,8)
3	II	3	ANTI LEPTON		
4	III	10	ANTI NEUTRINO		
5	V	15	ANTI MESON AND ANTI MESO-GLUON (36)	IV 120 ANTI-QUARK VII 120 ANTI-ELEMENT TYPES WILL BE FORMED	10^{50}
6		21			(68,5)
7	VI	28	ANTI BARYON		
8		36 + / 120	IN TOTAL		10^{60}

AFTER 120 ANTI-MATTER ELEMENTS ARE FORMED, DIFFERENT ANTI-MATTER ELEMENT FORMATION WILL STOP AND WHILE THE UNIVERSE CONTINUE TO GATHER, IT WILL BECOME A BLACK HOLE START TO SHATTER THE EXISTING ANTI-MATTER ELEMENTS IN AN ACCELERATING SPEED AND POWER. AFTER NO ELEMENTS ARE LEFT, ANTI-MATTER SUB-ATOM PARTICLES WILL START TO BE SHATTERED. SINCE THE LAST ANTI-MATTER PARTICLE (GRAVITON) EXCEEDED THE SPEED OF LIGHT, II. BIG BANG WILL HAPPEN WHEN GRAVITON IS TRANSFORMED TO ENERGY AND DISAPPEAR.

-1   II.BIG BANG

STEP	ORDER	NUMBER OF PARTICLES	VARIETY	TOTAL NUMBER AND VARIETY OF PARTICLES	AGE
7	VI	28	ANTI BARYON		(82,2)
6	V	21	ANTI MESON AND ANTI MESO-GLUON (36)	IV 120 ANTI-QUARK VII 120 ANTI-ELEMENT TYPES WILL DISAPPEAR	10^{70}
5		15			(95,9)
4	III	10	ANTI NEUTRINO		
3	II	3	ANTI LEPTON		
2 / 1	I	7	ANTI BOSON		10^{80} (109,6)

AN UTOPIC AND PERHAPS FANTASTIC TABLE OF CLASSIFICATION OF SCIENCES:

"MEANING" OR KNOWLEDGE ON THE RIGHT SIDE OF MEVLEVI DERVISH, AND THE MESSAGE ON THE LEFT SIDE HE GIVES AS A MATERIAL: PRODUCTION OF A PRODUCT OR SERVICE BY TREATING THE MATERIAL USING THE KNOWLEDGE A PERSON HAS GOT AS AN INDIVIDUAL. IN THE COMMUNITY MODEL ON THE RIGHT SIDE OF THE PAGE, 'GOVERNMENT' WAS REPLACED WITH 'PUBLIC'; THERE ARE 'THE ONES WHO DO' (PROFESSIONALS, WORKERS) ON THE LEFT SIDE, AND 'THE ONES WHO KNOW' (SCHOOLS, UNIVERSITIES AND STUDENTS) ON THE RIGHT SIDE. THE DUTY OF GOVERNMENT IS MAKING THE 'ONES WHO KNOW' INFORM "THE ONES WHO DO" ABOUT THE LATEST CONTEMPORARY IMPROVEMENTS IN THEIR PROFESSION. SOME OF THE PERSONS AND CORPORATIONS WHO HAVE CLASSIFIED THE SCIENCES BEFORE:

I- 7 DISCIPLINES WHICH THE PRIESTS HAVE STUDIED, IN THE TEMPLES OF EGYPT: 1- ASTRONOMY 2- MUSIC 3 - GEOMETRY 4- MATHEMATICS 5-RHETORIC 6-LOGIC 7-GRAMMAR

II- R. DESCARTES – FRANCE (1596 – 1650) HE PREPARED A GENERAL SCIENCE AND SIGN LANGUAGE (SCIENTIA GENERALIS)

III- JOHN LOCKE-ENGLAND (1632- 1704) HE CLASSIFIED ALL SCIENCES SEMIOTICALLY.

IV- G. W. LEIBNITZ -GERMAN PHILOSOPHER (1646-1716): HE APPLIED MATHEMATICS INTO PHILOSOPHY TO CONSTITUTE A UNIVERSAL LANGUAGE FOR SCIENCE, PHILOSOPHY AND SIGNS: "CHARACTERISTICA UNIVERSALIS".

V- AUGUSTE COMTE - FRENCH PHILOSOPHER (1798-1857) HE CLASSIFIED THE SCIENCES IN 7 GROUPS: 1- MATHEMATICS 2- ASTRONOMY 3- PHYSICS 4- CHEMISTRY 5- BIOLOGY 6- SOCIOLOGY 7- ETHICS. BESIDES, HE ASSERTED THAT THE COMMUNITY SHOULD BE ORGANIZED BY MEANS OF SCIENCE AND HE TRIED TO BUILD THE SCIENCE ON VARIOUS TRILOGIES. HE COLLECTED ALL SCIENCES IN PHILOSOPHY.

VI- MELOIL DEWEY: HE CLASSIFIED THE SCIENCES IN THE NAME OF "DECIMAL CLASSIFICATION" IN 1895, BRUSSELS. THIS CLASSIFICATION WAS APPROVED BY INTER-NATIONAL "BIBLIOGRAFISCHE INSTITUT". TODAY, BEING APPLIED THROUGHOUT THE LIBRARIES OF THE WORLD, IT WAS ALSO APPROVED BY INTERNATIONAL "AUSSCHUSS FÜR UNIVERSAL KLASSIFIKATION" IN NETHERLANDS.

VII- EXPORT PROMOTION CENTER (IGEME) IN ANKARA AND THE CLASSIFICATION IN THE EXPORT DIRECTORY, TURKEY

VIII- STANDARD AND INTERNATIONAL TRADE CLASSIFICATION (SITC) IN ACCORDANCE WITH BRUSSELS TARIFF NOMENCLATURE (BTN)

IX- ISTANBUL CHAMBER OF COMMERCE DIRECTORY; IT WAS CLASSIFIED IN ACCORDANCE WITH THE INTER - NATIONAL STANDARD INDUSTRIAL CLASSIFICATION OF ALL COMMERCIAL ACTIVITIES, WHICH UNITED NATIONS HAVE PREPARED. X- INTERNATIONAL PATENT CLASSIFICATION.

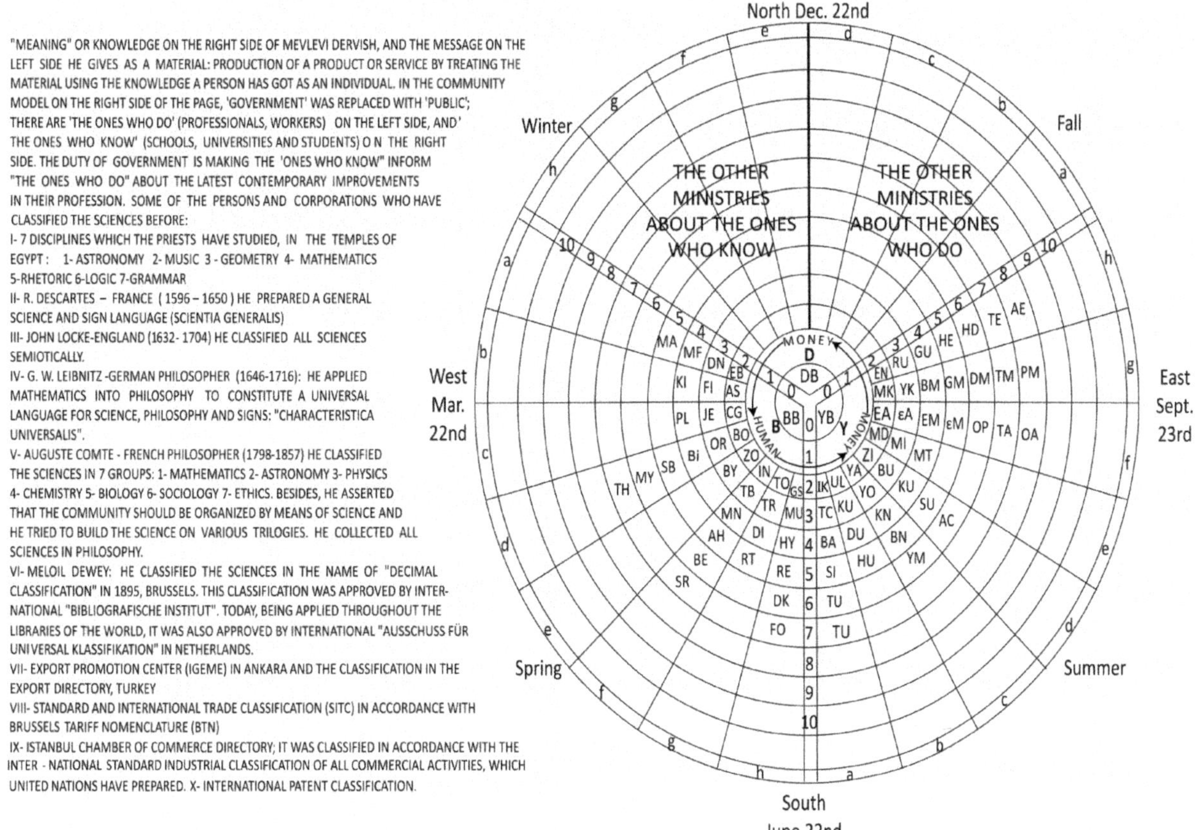

ABBREVIATION CODES OF SCIENCES WHICH CAN BE SEEN IN THE TABLE ABOVE:

MASTER OF THE KNOWERS (BB) (B0) — 0 / THE ONES WHO KNOW (B) (B1) — 1

	a	b	c	d	e	f	g	h
2	EB B2A	AS B2b	CG B2c	BO B2d	ZO B2d	IN B2f	TO B2g	GS B2h
3	DN B3a	Fi B3b	JE B3c	OR B3d	BY B3e	TB B3f	TR B3g	Mu B3h
4	MF B4a	Ki B4b	PL B4c	Bi B4d		MN B4f	Di B4g	HY B4h
5	MA B5a			SB B5d		AH B5f	RT B5g	RE B5h
6				MY B6d		BE B6f		DK B6h
7				TH B7d		SR B7f		FO B7h
8								

HEAD OF THE STATE-PRESIDENT (DB) (D0) — 0 / GOVERNMENT (D) (D1) — 1

	h	g	f	e	d	c	b	a
2	IT D2h	IH D2g	IS D2f	IG D2e	DG D2d	DS D2c	DH D2b	DT D2a
3								
4		THE OTHER						
5		MINISTRIES			THE OTHER			
6		ABOUT THE ONES WHO KNOW			MINISTRIES ABOUT THE ONES WHO DO			
7								
8								

MASTER OF THE ONES WHO DO (YB) (Y0) — 0 / THE ONES WHO DO (Y) (Y1) — 1

	a	b	c	d	e	f	g	h
2	IK Y2a	UL Y2b	YA Y2c	Zi Y2d	MD Y2e	EA Y2f	MK Y2g	EN Y2h
3	TC Y3a	KU Y3b	YO Y3c	Bu Y3d	Mi Y3e	EA Y3f	YM Y3g	Ru Y3h
4	BA Y4a	DU Y4b	KN Y4c	Ku Y4d	MT Y4e	EM Y4f	BM Y4g	Gu Y4h
5	Si Y5a	HU Y5b	BN Y5c	Su Y5d		EA Y5f	GM Y5g	HE Y5h
6	TU Y6a		YM Y6c	AC Y6d		OP Y6f	DM Y6g	HD Y6h
7	KE Y7a					TA Y7f	TM Y7g	TE Y7h
8						OA Y8f	PM Y8g	AE Y8h

NO	CODE	SCIENCE
	A	
1	AH	ETHIC
2	AS	ASTRONOMY
3	AE	ATOMIC ENERGY
4	AC	BEEKEEPING
	B	
5	B	THE ONES WHO KNOW (AND THE ONES WHO LEARN)
6	BB	MASTER OF THE KNOWERS
7	BA	BANKING
8	BM	PRINT-PRESS
9	BN	BUILDING CONSTRUCTION
10	BE	NUTRITION
11	BU	BOVINE ANIMAL
12	BO	BOTANIC
13	BY	BIOLOGY
14	BI	PLANT
	C	
15	CG	GEOGRAPHY
	D	
16	D	GOVERNMENT (DEPARTMENT OF FINANCE)
17	DB	HEAD OF THE STATE-PRESIDENT
18	DT	INTERNATIONAL TRADE
19	DH	INTERNATIONAL LAW
20	DS	FOREIGN POLICY
21	DG	FOREIGN SECURITY
22	DN	RELIGION
23	DI	LANGUAGE
24	DK	DECORATION
25	DU	MARINE TRANSPORTATION
26	DM	SEWING MACHINE
	E	
27	EB	COSMOLOGY
28	EN	ENERGY
29	EM	ELECTRICAL TOOLS
30	EA	ELECTRICAL APPLIANCE
31	EA	ELECTRONIC TOOLS
32	EM	ELECTRONIC APPLIANCE
	F	
33	FI	PHYSICS
34	FO	PHOTOGRAPHY
	G	
35	GÜ	SOLAR ENERGY
36	GS	FINE ARTS
37	GM	FOOD MACHINE
	H	
38	HY	SCULPTURE
39	HU	AIR TRANSPORTATION
40	HD	HYDROGEN ENERGY
41	HE	HYDROELECTRIC POWER
	I	
42	IN	HUMAN
43	IK	ECONOMY
44	IG	LOCAL SECURTY
45	IS	LOCAL POLICY
46	IH	LOCAL LAW
47	IT	LOCAL TRADE
	J	
48	JE	GEOLOGY
	K	
49	KI	CHEMISTRY
50	KE	CULTURE-ENTERTAINMENT
51	KU	SHEEPS, GOATS
52	KU	LAND TRANSPORT
53	KN	CITY-STRUCTURE
	M	
54	MN	LOGIC
55	MA	MATHEMATICS
56	MU	MUSIC
57	MY	FRUIT GROWING
58	MK	MACHINE
59	MT	METALLURGY
60	MA	MINE
61	MF	METAPHYSICS
62	MI	MINERALS
	O	
63	OP	OPTIC
64	OR	FOREST
	O	
65	OA	MEASURING INSTRUMENTS
	P	
66	PM	PACKAGING MACHINE
67	PL	PALEONTOLOGY
	R	
68	RU	WIND POWER
69	RE	PAINTING
70	RT	RHETORIC
	S	
71	SB	VEGETABLE GARDENING
72	SR	SPORT
73	SU	WATER PRODUCTS
74	SN	INDUSTRY
75	SI	INSURANCE
	T	
78	TH	CEREAL
79	TB	MEDICAL
80	TO	SOCIETY
81	TC	TRADE
82	TA	MEDICAL TOOLS
83	TU	TOURISM
84	TE	THERMAL ENERGY
	U	
85	UL	TRANSPORTATION
	Y	
86	Y	THE ONES WHO DO
87	YB	MASTER OF THE ONES WHO DO
88	YA	BUILDING
89	YK	BUILDING MACHINERY
90	YM	BUILDING MATERIALS
	Z	
91	ZO	ZOOLOGY
92	ZI	AGRICULTURE

TABLE OF HISTORY OF TODAY'S PEOPLE ON THE WORLD

RACES: UA: URAL ALTAIC / IE: INDO EUROPEAN / SA: SAMI / HM: HAMI / CA: CENTRAL ASIAN / M: MONGOL / T: TURKISH / SCA: SA+CA+IE
LOCATIONS: MS: MESOPOTAMIA / ASM: AS, AM / MEX: MEXICO / AME: AMERICA / ASI: ASIA / EU: EUROPE / AU: AUSTRALIA / AF: AFRICA / AM: ASIA MINOR / ASVF: ASI+EU+AF / FR: FRANCE / AN: ANATOLIA
RELIGIONS: CHF: CAVE, HUNTING, FETISHISM, MAGIC / MG: MOTHER GODDESS / FG: FATHER GODDESS / SH: SHAMANISM / PT: POLYTHEISTIC / MT: MONOTHEISTIC / EG: EARTH GODDESS / SO: SOUL / LE: LEGEND / OR: OR
Numbered columns (SG+EG+SOUL): 1 MOON, 2 SUN, 3 FIRE, 4 FIREPLACE, 5 MOUNTAIN, 6 SOIL, 7 WATER, 8 TREE, 9 WIND, 10 STORM, 11 AGRICULTURE, 12 PLENTIFULNESS, 13 LOVE-BEAUTY, 14 DESTINY, 15 HUNTING, 16 SHEPHERDS, 17 DISEASE, 18 WAR, 19 PEACE, 20 CITY, 21 WISDOM-MIND, 22 UNDERGROUND, 23 TIME AND DEATH, 24 ANGELS (MESSENGERS OF GOD), 25 DOOR OF THE HOUSE (HOLY), 26 WOLF MYTH, 27 BABY LEFT TO RIVER' MYTH, 28 FLOOD MYTH, 29 CREATION MYTH, 30 SEA

Text columns

DATE	DATE	PEOPLE-COMMUNITIES	RACES	LOCATION	RELIGION	SKY GOD (FG)	EARTH GODDESS (MG)	OTHER
B.C.		RAMAPITHECUS MAN						
B.C.		HOMO-ERECTUS MAN						
B.C.	150-50000	NEANDERTHAL MAN						
B.C.	50000	NEANDERTHAL MAN						
B.C.	40000+	HOMO-SAPIENS (US)		ASI	CHF			
B.C.	30000	IMMIGRATION IN EUROPE TO THE SOUTH		EU	CHF			
B.C.	28000	ERGANI-AN		ASM	CHF			
B.C.	16000	IMMIGRATION TO AMERICA VIA BERING SEA		ASM	CHF			
B.C.	15000	IMMIGRATION FROM ASIA TO EUROPE		ASI, EU	CHF			
B.C.	10000	ZAHO-NORTHERN IRAQ		MS				
B.C.	10000	GOBEKLITEPE-URFA-AN		AN	.			
B.C.	12000-6500	LA MANDELAINE-FR		EU.FR	CHF, MG			
B.C.	9000	ALPINS-AS	UA	CA	PT			
B.C.	7000	ALPINS-EU		EU				
B.C.	7000	ÇATALHÖYÜK (KONYA)		AN	MG			
B.C.	6000	ÇAYÖNÜ TEPESİ-DİYARBAKIR		AN	MG			
B.C.	← 6000	INDIAN-EUROPEANS	IE	ASI, EU	SH			
B.C.	5000	SYBARIS (HURRIS)		AN	PT			
B.C.	← 5000	INDUS CULT (DRAVIDIANS)		INDIA	PT		MOTHER GODDESS	
B.C.	5000	JERICHO		AM	MG			
B.C.	← 5000-4000	ANAV CULT		ASI				
B.C.	← 5000	MINUSINSK CULT	EU, M	ASI	MG, PT			
B.C.	4500-3000	SUMERIANS, ELAMITE (GUTIANS)	UA	IRAN	MG, PT	ANU (ANSAR)	KISAR (MAYADA)	
B.C.	4000-2200	SUMERIANS, ELAMITE (GUTIANS) (WRITING)	UA	MS	MG, PT			
B.C.	4000-525	EGYPTIONS	IE, M	AF	MG, PT	HORUS	OSIRIS	PHTHA
B.C.	4000-1200	CRETANS		AEGEAN	MG	BESE (BULL)	REHEA	
B.C.	4000-2750	ACCADIANS	SA	AEGEAN	PT	ANU		
B.C.	3500-2000	JAPANS	JAPAN	ASI		IDYA NAGI	IOZA NAMI	
B.C.	3000	ANCIENT PERCIANS	IE	IRAN	PT	AHURA (ASURA)	ANA HITA	
B.C.	3000-1200	TROY	-	AN	MG			
B.C.	3000	URAL-ALTAIC	UA	ASI	SH	BAYULGEN (KUDAY)		
B.C.	3000-1800	BABEL	SA	MS	PT	AN	ISHTAR	MARDUK GOD (FG)
B.C.	3000	HITTITES, LELEGIANS, CARIANS	IE, CA	AN, EU	PT			
B.C.	3000	PELASGIANS, LELEGIANS, CARIANS	IE, CA	GREECE	PT			
B.C.	3000	SAMI	SA	AN	PT			
B.C.	3000-1200	ITALICS	IE	IT, EU	PT			
B.C.	3000-2000	SYBARIS (HURRIS)	UA	AN	PT	TESHUP (STORM)	HEPAT (PLENTIFULNESS)	
B.C.	3000	SYBARIS (HURRIS)	UA	IRAN	PT			
B.C.	3000	PELASGIANS	UA	EU, AN	PT			
B.C.	3000	ELAMITE (GUTIANS)	UA	IRAN	PT	BAAL		
B.C.	3000-1200	PALESTINIANS	SA	AM	PT			
B.C.	3000	PARTS		IRAN	PT			
B.C.	3000	SELEUCID EMPIRE		IRAN	PT			
B.C.	3000	ARAMEANS	SA	AM	PT	TESHUP (HURRIS)	HEPAT (PLENTIFULNESS)	
B.C.	2600-640	ELAMITE (GUTIANS)	UA	AM	PT	HURBAN	PINENKUR (PLENTIFULNESS)	
B.C.	2000-612	ASSYRIANS	SCH	AM	PT	ADAT (THUNDER)	ISHTAR	ASUR GOD (FG)
B.C.	2000-IV.y.y.	PERSIANS	IE	IRAN	MG, PT	AHURA (HURMUS)	ANAHITA	ARINNA (MG)
B.C.	2000	GREEKS	IE	EU	MG, PT	ZEUS (URANUS)	DEMETER	
B.C.	2000-VIII.y.y.	PHOENICIANS		AM	MG, PT	BAAL (SUN GOD)	BAALAT (ASTARTE)	
B.C.	2000-600	INDIA (VEDISM)	IE	ASI	MG, PT	DIAUS PITAR	PRITIRA MATAR	ADITA (MG)
B.C.	2000	ARYANS (INDO-EUROPEAN)	IE	ASI	MG, PT	MITRA (DAY SKY)	INDRA (NIGHT SKY)	VARUNA (MG)
B.C.	2000	GERMENS	IE	EU	MG, PT	DONAR (THOR)		ODIN (WOTAN)
B.C.	2000	CELTS	IE	EU	MG, PT	TARANIS	MATRES	DRUDE STONES
B.C.	1800-1200	HITTITES	IE	AN	MG, PT	KUPAPA (HEPA)	ARINNA	TESHUP (STORM)
B.C.	1800-1200	ACHAEANS (INDO-GERMEN)	IE	EU	PT			
B.C.	1800-1580	HYKSOS (HURRI, SAMI)	SCH	AM				
B.C.	1700-522	SOGDIA (MASSAGETAE, SUGITS)	IE	ASI				
B.C.	1680-1160	KASSITES	IE	AM		SURIAS (SUN)		
B.C.	1600-1200	MITANNIS	CA	AN, MS		KUMARVE	HEPAT (PLENTIFULNESS)	TESHUP (STORM)
B.C.	1600-XIII.y.y.	JEWISH	SA	AM	MT	JEHOVAH	-	
B.C.	1200	SLAVS (DORS)	IE	EU, GREECE				
B.C.	1200-IV.y.y.	SCYTHIANS (SAKA, KIMER)	IE, UA	AM				
B.C.	1200-VIII.y.y.	URARTIANS	UA	AN		HALDI (WAR GOD)		
B.C.	1200-676	PHRYGIANS (VENUS)		AN	MG, PT	ATTIS	KYBELE	PHRYGIAN (VENUS MG)
B.C.	1200-546	LYDIANS		AN		LETO (WAR GOD)		SANTAS
B.C.	XII.-VIII.y.y.	AZTECS		AME	PT	KUKULKAN		
B.C.	1000-295	ETRUSCAN (TUSCANS)	AM	ITALY	PT	TINIA	TURAN (YUNO)	
B.C.	1000-245	HUNS	T	ASI, EU	SH	BAYULGEN	YERSU	
B.C.	X.y.y.	GERMENS	IE	EU	PT	DONAR (THOR)		
B.C.	1059-249	PROTO TURKS	T	CHINA	PT	SKY GOD		
B.C.	1000-VI.y.y.→	CHINESE	CHINA	CHINA	PT	YANG (MALE)	YING (FEMALE)	
B.C.	→	TIBETIANS	UA	AM				
B.C.	→	AUSTRALIANS		AUSTRALIA				
B.C.	→	INDIANS (SIOUX)	M, CA	AME				VAKAN
B.C.	X.y.y.-A.D.476	ROMANS	ETRUSKAN	EU	PT	JUPITER (SKY GOD)	YUNO (ISIS)	SATURNUS
B.C.	IX.y.y.-550 →	MEDS	IE	IRAN				
B.C.	VI.y.y. →	PERSIANS	IE	IRAN, AN				
B.C.		INKAS	CA	AME	PT			
B.C.	500 →	WALES (CELTS)	IE	EU	PT			
B.C.	IV.y.y.-A.D.106	NABATAEANS	SA	AM	PT			
B.C.	II.-VI.y.y. →	VIKINGS	IE	EU	PT			ODIN (WOTAN)
B.C.	I.-XIII.y.y. →	UIGHURS	T	ASI	SH			
B.C.	← →	ESKIMOS	M	AME	SH			
A.D	II.-IX.y.y. →	AVARS (ROURAN KHAGANATE)	T	AM	SH			
A.D	III.y.y. →	VISIGOTHS	IE	EU	PT			
A.D	III.y.y. →	OSTROGOTHS	IE	EU	PT			
A.D	III.y.y. →	ANGLO-SAXONS	IE	EU	PT			
A.D	III.y.y. →	FRANKS	IE	EU	PT			
A.D	III.y.y. →	VANDALS		EU	PT			
A.D	IX.y.y. →	HUNGARIAN		EU	PT			
A.D	960-1299 →	SELJUKIAN	T	AM, AN	ISLAM			
A.D	552-659 →	GOKTURKS	T	AM, AN	PT, SH	BAYULGEN	YERSU (IDUK)	UMAY
A.D	VIII.y.y. →	YAKUTS	T	AM	PT, SH	TOYON AGA	EARTH GODDESS (EG)	SECEN
A.D	X.-VI.y.y. →	OGHUZS	T	AM, AN	ISLAM			
A.D		KARLUKS	T	AM		KEVALIN		TUHSIN
A.D	1206-1226 →	MONGOLS	M	AM, AN	PT, SH	ULGEN		
A.D	1299-1920 →	OTTOMANS	T	AM, AN	ISLAM			

Numbered-concept columns (1–30) — rows bearing X marks

(column positions read from the scan grid; alignment approximate in the densest rows)

PEOPLE-COMMUNITIES	1	2	3	4	5	6	7	8	9	10	11	12	13	14	15	16	17	18	19	20	21	22	23	24	25	26	27	28	29	30
SUMERIANS, ELAMITE (GUTIANS)	X	X			X	X				X	X	X			X				X											
EGYPTIONS	X	X			X		X			X		X		X		X		X	X											
CRETANS															X															
JAPANS	X	X	X				X	X																		X				
ANCIENT PERCIANS	X	X		X			X			X						X		X	X	X										
URAL-ALTAIC	X	X	X	X	X		X				X			X	X				X			X		X			X			
BABEL	X	X	X				X				X		X						X											
SYBARIS (HURRIS) (3000-2000)	X	X			X						X				X															
ARAMEANS	X	X			X						X																			
ELAMITE (GUTIANS) (2600-640)	X	X										X								X										
ASSYRIANS	X	X											X					X			X									
PERSIANS (2000-IV.y.y.)	X	X			X					X											X									
GREEKS	X	X	X	X	X			X	X		X	X	X	X	X	X		X	X		X	X	X	X						X
PHOENICIANS	X	X														X														X
INDIA (VEDISM)	X	X	X			X		X	X					X						X		X	X						X	
ARYANS (INDO-EUROPEAN)		X			X			X																						
GERMENS (2000)	X				X													X	X										X	
CELTS	X	X	X											X	X			X		X	X									
HITTITES (1800-1200)	X	X	X	X				X	X		X	X	X	X					X											
MITANNIS												X		X		X														
PHRYGIANS (VENUS)		X			X						X								X											
AZTECS		X			X																									
ETRUSCAN (TUSCANS)	X	X	X			X					X				X					X										
HUNS	X	X			X			X	X			X																		
CHINESE						X	X	X		X		X																	X	
ROMANS	X	X	X	X			X				X		X		X		X		X		X		X	X	X					
INKAS	X	X																												
VIKINGS																					X			X						
GOKTURKS	X	X	X			X	X			X			X						X											X
YAKUTS	X	X	X	X				X		X		X	X	X	X	X		X	X	X			X					X		
MONGOLS	X	X	X				X				X				X			X					X				X			

NOTE: ALTHOUGH THERE ARE SOME MISSING INFORMATION, WE TRIED TO SHOW THE LOCATIONS, RACES AND RELIGIONS OF ALL OF THE COMMUNITIES AND CULTURES WHICH HAVE LIVED UP TO TODAY IN A DATE ORDER. WOULDN'T THE INTEGRATION OF ALL THOSE OLD AND NEW RELIGIONS, SECTS AND RACES BRING PEACE AND HAPPINESS TO OUR WORLD BY MEANS OF MAWLANA'S WORDS: "COME, IT DOESN'T MATTER WHO OR WHAT YOU ARE, COME"!

BIBLIOGRAPHY

[1]. Evrenin Kısa Tarihi, Joseph Silk, Tübitak Yay. 2000
[2]. Atomaltı parçacıklar, Steven Weinberg, Tübitak yayınları 2002
[3]. Kozmos evrenin ve yaşamın sırları, Carl Sagan, Altın kitaplar, 2009
[4]. Süper simetri, Gordon Kane, Tübitak Yayınları 2009
[5]. Kuantum fiziğinin ufkunda Dünya gerçekten var mı? Sven Ortoli-Jean Pierre Pharabod, Dharma Yayınları, 2010
[6]. Kuantum benlik- Danah Zohar- Doruk yayınları, 2007
[7]. Entropi dünyaya yeni bir bakış, Jeremy Rifkin& Ted Howard, İz yayınları 2010
[8]. Maddenin son yapı taşları, Gerardt Hofft, Tübitak yayınları 2003
[9]. X ışınlarından Kuarklara çağdaş fizikçiler ve buluşları, Emilio Segre, Sarmal yay. 1995
[10]. Kristalografi ders notları Prof. Işık Kumbasar, İtü maden fakültesi 1983.
[11]. Mineraloji, prof. Işık Kumbasar-Prof. Atilla Aykol, İtü Maden fakültesi 1993.
[12]. İzafiyet Teorisi, Albert Einstein, say Dağıtım ltd. Şti. 1993
[13]. Nanobilim ve Nanoteknoloji, Şakir Erkoç, ODTÜ yayınları 2007
[14]. Astronomi (lise III fen), Doç Dr. M. Dizer, MEB 1964
[15]. Astroloji, yıldızlar ve geleceğimiz, Sheila Geddes, Remzi kitabevi 1978
[16]. Modern Fizikokimya, prof. Alirıza berkem, İstanbul üni.
[17]. Fizikokimya, prof. Ali Rıza Berkem-prof. Sacide Baykut, İstanbul Ün. Müh. Fak. 1984
[18]. Modern fiziğe giriş, prof. Erol Gündüz, Ege Üni. Fen Fak. Fizik bl. 1988
[19]. Crystal stuructures, Ralph W. G. Wyckoff, Usa MackPrinting co. Easton. Pa. 1963
[20]. Kristallerin esneklik özellikleri, Doç. Dr. Mustafa Dikici, 19 Mayıs Üni. Eğitim Fak Samsun 1993
[21]. Lise Biyoloji 1-Sevgi Börü-Emine Öztürk- Şermin Cavak, MEB Basımevi İstanbul 2001
[22]. Lise Biyoloji 2 Ayten Sucu- Semra Bayor- Melahat Küpeli, MEB Basımevi 1999
[23]. Lise Biyoloji 3, Özer Bulut- Davut Sağdıç- Selim Korkmaz, MEB Basımevi 2000
[24]. Yaşam Enerjisi, Prof. Ahmet Maranki-Elmas Maranki, Mozaik Yayınları 2009
[25]. Atom, Ümit Şimşek, Yeniasya Yayınları 1979
[26]. Beynimiz sinirlerimiz, prof. Ayhan Sungar, Yeniasya Yayınları 1979
[27]. Big Bang Kâinatın doğuşu, Ümit Şimşek, Yeniasya Yayınları 1984
[28]. Yaşayan gezegen, prof. Yılmaz Muslu, Yeniasya Yayınları 1979
[29]. Atomdan Hücreye, Prof. Münip Yeğin Yeniasya yayınları 1980
[30]. Hayatın kökleri, Mahlon B. Hoagland, Tübitak yayınları 1995
[31]. Genel Matematik, Hüseyin Yıldırım, Ahmet Ergülen, Afyon Kocatepe üni. 2000
[32]. Doğada bilimde sanatta Altın Oran, Mehmet Suat Bergil, Arkeoloji ve sanat yay. 1993
[33]. Trigonometrinin tarihi gelişimi, Mehmet Ekmekçioğlu, MEB yayınları 1991
[34]. Matematik sanatı, Jerry P. King, Tübitak yayınları Ankara 1997
[35]. Uzak doğudan Maya ülkesine bir, iki, üç…, Georges Ifrah, Tübitak yay. Ankara 1996
[36]. Hind uygarlığının sayısal simgeler sözlüğü, Georges Ifrah, Tübitak yay. Ankara 1998
[37]. Sıfırın gücü, Georges Ifrah, Tübitak yayınları Ankara 1997
[38]. Sayıların gizemi, Anne Marie Schimmer, Kabalcı yayınları 2000
[39]. Bir sayı tut, Malcolme E. Lines, Tübitak yayınları Ankara 1998
[40]. Matematik tarihi, Marcel Böll, İletişim yayınları- Presses Üniversitaires de France 1991[41]. Büyü Gizem Bilim batı uygarlığında Okült- Dan Burton-David Grandy Varlık ya. 2005
[42]. Bir, iki, üç… sonsuz, George Gamow, MEB yayınları
[43]. Kim korkar matematikten- Nazif Tepedelenlioğlu, Sarmal yayınları, İstanbul 1995
[44]. Dr. Matrix ve gizemli sayılar, Martin Gardner, Güncel yayınları 1985
[45]. Matematik sözlüğü, Kamil İşcan, İTÜ İstanbul 1967
[46]. Matematiksel düşünce, Cemal Yıldırım, Remzi Kitabevi İstanbul 1988
[47]. Matematiğin aydınlık dünyası, Sinan Sertöz, Tübitak yayınları Ankara 1996
[48]. Pi coşkusu, David Blather, Tübitak yayınları Ankara 2003
[49]. Bilimler Bilimi Sibernetik, Toygar Akman, Milliyet yayınları 1977
[50]. Güneş Ay ve Dünya, Robin Heath, Dharma yayınları İstanbul 2003
[51]. Platon'un ve Arşimet'in çok yüzlü cisimleri, Daud Sutton, Dahrma yay. İstanbul, 2004
[52]. Kutsal Geometri, Miranda Lundy, Dharma yayınları İstanbul 2003

[53]. Maddenin özü elementler, matt twweedi dharma yay.

[54]. Yararlı Matematik ve Fizik formülleri, Matthew Watkins, Dharma yay. İstanbul 2004

[55]. Armonograf Müzikten Matematiğin görsel bir rehberi, Anthony Ashton, Dharma yay. İstanbul 2003

[56]. Tesadüfler kitabı, John Martineau, Dharma yayınları İstanbul 2003

[57]. Müzik neyi anlatır, Sidney Finkelstein, Kaynak yayınları İstanbul 2000

[58]. Çağdaş sanat kuramı, Paul Klee, Dost kitabevi Ankara

[59]. Evrensel eşit kuyruklu canlı, Nusret Kaya, Abis yayınları Ankara 2010

[60]. Evrenin dili, Doç. Dr. Nusret Kaya, Sistem yayınları İstanbul 2004

[61]. Felsefe sözlüğü, Orhan Hançerlioğlu, Remzi kitabevi yayınları İstanbul 1979

[62]. İnsan ve kültür, Bozkurt Güvenç, Remzi kitabevi yayınları İstanbul 1979

[63]. İnanç sözlüğü dinler, mezhepler, tarikatler, efsaneler, Orhan Hançerlioğlu, Remzi
      kitabevi yayınları İstanbul, 1975

[64]. Müslüman ilim öncüleri ansiklopedisi, Şaban Döğen, Yeni Asya yayınları İstanbul 1987

[65]. Mısır tarihi, Ord. Prof Yusuf Ziya Özer, Türk Tarih Kurumu yayınları Ankara 1987

[66]. Hindistan tarihi, Y. Hikmet Bayur, Türk Tarih kurumu yayınları Ankara 1946

[67]. Emir Timur ve mirası, Doç. Dr. Abdulvahap Kara- Yard. Doç. Dr. Ömer İşbilir, Mimar
      Sinan Üniversitesi İstanbul İstanbul 2005

[68]. Tasavvuf ve tarikatlar, Prof. H. Kâmil Yılmaz, Ensar Neşriyat İstanbul 2002

[69]. Tasavvuf ve tarikatlar, Selçuk Eraydın, Marifet yayınları İstanbul 1981

[70]. Sayıların gizemi ve Tasavvufun dinamikleri, Bayram Ali Çetinkaya, İnsan yayınları

[71]. Ekonomi Doktrinleri tarihi, Prof. Mahmut Koloğlu, Doğuş matbaacılık Ankara 1963

[72]. Eski Türk Dini, Prof. İ. Kafesoğlu, Kültür Bakanlığı yayınları Ankara 1980

[73]. Çakıl taşlarından Babil Kulesine, George İfrah, Tübitak yayınları 1996

[74]. Bir gölgenin peşinde, George İfrah, Tübitak yayınları 1995

[75]. İslamic spirituality Seyyed Hüssein Nasr, Scm Pres ltd.

[76]. Taschenbuch für chemiker und physiker, Dr. İng. Jean D'ans und Dr. Phil Ellen Lax
      Springer verlag, 1949

[77]. Grosser historisches Weltatlas, Bayerischer Schulbuch Verlag München 1964

[78]. Das bild der modernen Physik, H. Linder-Urania Verlag Leibzig, Jena, Berlin

[79]. Spektrum der Wissenchatt, Spektrum der Wissenschaft Verlag Geselschaft MbH&co Manheim

[80]. Unsere welt ein vernetztes system, Frederic Vester, Ernst Klett Stuttgart

[81]. Mensch und Kybernetik, Gilbert Obermair, Wilhelm Heyne Verlag München

[82]. Denken, lernen, vergessen, Frederic Vester, DtV Verlag München

[83]. Im amfang war der Wasserstoff, Hoffmann und Campe Verlag Hamburg, Hoimar V. Ditfurth

[84]. Der denk prozess, Edward de Bono, Rowohlt Taschenbuch Verlag GmbH Reinbeck bei Hamburg

[85]. Lexikotek 1)Spektrum der Naturwissenschaften, 2)Panaroma der Welt Geschichte ilh.
      Bartelmann Verlag Guterloh

[86]. Das will ich von Natür und Technik Wissen, Axel Rex, Südwest Verlag München

[87]. Yaratıcı tekâmül, Henry Bergson, Bilim ve kültür eserleri dizisi, Milli Eğitim Basımevi İstanbul 1986

[88]. Peygamberler Tarihi, Enver Bahman Sapolyo, Ön Asya Yayınları 1968

[89]. Dinler Tarihi, Felicien Challage, Varlık Yayınları

[90]. Dinler Tarihi araştırmaları, Dr. Hikmet Tanju, Ankara Üniversitesi İlâhiyat Fakültesi Yayınları

[91]. Büyük Dünya Tarihi, Jacques Pirenne

[92]. Büyük Filozoflar Ansiklopedisi, Cemil Sena, Nebioğlu Matbaası

[93]. Medeniyet Tarihi, Louis Frederic, Doğan kardeş Yayınları İstanbul

[94]. İlk üretimciliğe geçiş evresinde Anadolu ve Güneydoğu Avrupa, Prof. Dr. Ufuk Esin

[95]. Eski Mısır tarih ve Medeniyeti, Prof. Afet İnan

[96]. Der Mensch kam aus Sibirien, Alexey Pawlowitsch Oklodnikow, Verlag Fritz Molden

[97]. Zentral Asien, Aleksandr Belenickıy, Wilhelm Heyne Verlag München

[98]. Meine Welt in bildern, Econ Verlag Düsseldorf, Wien

[99]. Der grose kültür Fahrplan, Werner Stein, F. A. Herbig Verlag Buchhandlung Welsermühl Austria

[100]. Der Mensch auf seiner Erde, George Gerster, Atlantis Verlag Zürich

[101]. İnsan ve Kültür, Bozkurt Güvenç, Remzi Kitabevi 1979

[102]. Tarih atlası, Faik Reşit Unat, Kanaaat Yayınları

[103]. Başlangıçtan bugüne Dünya tarihi, Nezihe Aras, Prof. Dr. Şehabettin Tekirdağ,
    Gülseren Devrim, Kaynak kitablar Basın yayın ve Tic. A. Ş. İstanbul

[104]. Mettoden des ZF(Zukunft Forschung), Prof. Gerhard Bruckmann, Westdeutsche Verlag 1975-1976

[105]. Friden durch ein neues Weltrecht, Grenville Clark und Louis B. Lohn, Afred Metzner
    Verlag/Berlin, 1961

[106]. Flaggen Wappen Daten, Karl Heinz Hesner, Bertelmann Lexikon Verlag München, Wien

[107]. Yearbook of international Organizations, Eyvind S. Tem, 1968-1969

[108]. Am ende unsere zukunft, Rudiger Proske, Verlag Olde Hansen Hamburg

[109]. Verzeichnes Rechner Gestützter Umwelt modelle, Erick Schmidt Verlag Berlin

[110]. Das große Bilder Lexikon der flammen, Von F. A. Nowak Vorwort von K. H. Hanisch,
    Bertelman lexikonVerlag

[111]. Sanskrit Alphabet "Das Kosmische Ballet", Denis Postle, Verlag Macmillan London   Limited

[112]. Die idee einer Systematische Phlosophy, A. Diemer 1967

[113]. Strüktür und aufbau der Wissenschaftslische Teorien, W. Lein Fellner 1965

[114]. Einführung in die Wissenschaftstheorie, H. Seiffert 1971-1972

[115] Kozmik Bilim ışığında şifalı taşlar, Ahmet & Elmas Maranki, Hayat yay. İstanbul 2010

[116] Geleceğe notlar, Eskimeyen kırk makale, Serdar H. Yıldırım 2018.

[117] Türk Musikisi nazariyatı ve usulleri, kudüm velveleleri, İsmail Hakkı Özkan 1984

[118] "Costing Big Bang" (with M.Arik and E.Rizaoglu) Phys. Lett. 321B, 329 (1995).